T0122461

Lecture Notes on Data Engineering and Communications Technologies

Volume 16

The aim of the book series is to present cutting edge engineering approaches to data technologies and communications. It publishes latest advances on the engineering task of building and deploying distributed, scalable and reliable data infrastructures and communication systems.
The series has a prominent applied focus on data technologies and communications with aim to promote the bridging from fundamental research on data science and networking to data engineering and communications that lead to industry products, business knowledge and standardisation.

More information about this series at http://www.springer.com/series/15362

Durgesh Kumar Mishra · Xin-She Yang
Aynur Unal
Editors

Data Science and Big Data Analytics

ACM-WIR 2018

Editors
Durgesh Kumar Mishra
Department of Computer Science and Engineering
Sri Aurobindo Institute of Technology
Indore, Madhya Pradesh
India

Aynur Unal
Department of Mechanical Engineering
Indian Institute of Technology Guwahati
Guwahati, Assam
India

Xin-She Yang
School of Science and Technology
Middlesex University
London
UK

ISSN 2367-4512 ISSN 2367-4520 (electronic)
Lecture Notes on Data Engineering and Communications Technologies
ISBN 978-981-10-7640-4 ISBN 978-981-10-7641-1 (eBook)
https://doi.org/10.1007/978-981-10-7641-1

Library of Congress Control Number: 2018941227

Printed on acid-free paper

This Springer imprint is published by the registered company Springer Nature Singapore Pte Ltd.
The registered company address is: 152 Beach Road, #21-01/04 Gateway East, Singapore 189721, Singapore

Preface

The pervasive nature of digital technologies as witnessed in industry, services, and everyday life has given rise to an emergent, data-focused economy stemming from many aspects of human individual and commercial activity. The richness and vastness of these data are creating unprecedented research opportunities in some fields including urban studies, geography, economics, finance, and social science, as well as physics, biology and genetics, public health, and many others. In addition to big data-inspired research, businesses have seized on big data technologies to support and propel growing business intelligence needs. As businesses build out big data hardware and software infrastructure, it becomes increasingly important to anticipate technical and practical challenges and to identify best practices learned through experience. Big data analytics employ software tools from advanced analytics disciplines such as data mining, predictive analytics, and machine learning. At the same time, the processing and analysis of big data present methodological and technological challenges. The goal of this lecture series and symposium is to present both novel solutions to challenging technical issues and compelling big data use cases. This special issue contains many papers that provide deep research results to report the advances in big data analytics, infrastructure, and applications. The goal of this special issue is to crystallize the emerging big data technologies and efforts to focus on the most promising solutions in the industry. The papers provide clear proof that big data technologies are playing a more and more important and critical role in supporting various applications in the industry. It is also believed that the papers will further research openings in new best practices and directions in this emerging research discipline.

Indore, India
London, UK
Guwahati, India

Durgesh Kumar Mishra
Xin-She Yang
Aynur Unal

Contents

About the Editors

Dr. Durgesh Kumar Mishra is a professor (CSE) and director of the Microsoft Innovation Centre at Sri Aurobindo Institute of Technology, Indore, India, and visiting faculty at IIT Indore. He has 24 years of teaching and 12 years of research experience. He has published more than 90 papers in refereed international/national journals and conferences including IEEE, ACM and organized many conferences as General Chair and Editor. He is a senior member of IEEE, CSI, ACM, chairman of IEEE MP subsection, IEEE Computer Society Bombay Chapter. At present, he is the chairman of CSI Division IV Communication at the National Level and ACM Chapter Rajasthan and MP State.

Prof. Xin-She Yang is an associate professor of Simulation and Modelling at Middlesex University, London. His main interests are applied mathematics, algorithm development, computational intelligence, engineering optimization, mathematical modeling, optimization, and swarm intelligence. His research projects have been supported by the National Measurement Office, BIS, Southwest Development Agency (UK), Euro Met, EPSRC, NPL, and the National Science Foundation of China. He is EEE CIS Task Force Chair of the BIKM, Technical Committee of Computational Finance and Economics of IEEE Computational Intelligence Society; Advisor to the *International Journal of Bio-Inspired Computation*; Editorial Board Member of *Elsevier's Journal of Computational Science*; and Editor-in-Chief of the *International Journal of Mathematical Modelling and Numerical Optimisation*.

Dr. Aynur Unal is a strategic adviser and visiting full professor at the IIT Guwahati, India. She has created a product-focused engineering program using the cloud-based infrastructure. Her main interests include ecologically and socially responsible engineering, zero waste initiative, and sustainable green engineering. Her research focuses on both rural and urban sustainable development, renewable energy, solar towers and pumps. She has taught at Stanford University and worked in Silicon Valley to develop products for data mining from big data (Triada's Athena I & II), collaborative design and manufacturing, secure and private communication, and collaboration software platforms (Amteus, listed in LSE AIM).

Keynote Speakers

Analysis and Evaluation of Big Data Video Quality in Wireless Networks

Talks Abstract

With the development of the Internet technologies and online multimedia applications of the Internet, video becomes the main source of online-generated data. To measure the quality of received video, computer simulation is the research tool of choice for a majority of the wired and wireless network research community. These days, most of the research works on the network adopt computer simulation to verify novel ideas. The presentation introduces a framework and a tool-set for evaluating the quality of video transmitted over a simulated wireless network. Besides measuring the quality of service (QoS) parameters of the underlying network, such as loss rates, delays, and jitter, frameworks also support a video quality evaluation of the received video. On wired as well as wireless networks, the medium access control (MAC) plays an important role in the performance of video transmission. MAC handles allocating the resources to the different types of applications or wireless stations. Many types of researches are conducted for video quality measurement on network and application layer. The framework can be used for research and evaluating new techniques for MAC layer optimizations. This talk will present an overview of the framework and a tool-set for evaluating the quality of video transmitted over a simulated wireless network. Finally, the future research directions will be discussed.

Keywords Big data video, Networks and communications, QoS, Multimedia technologies

Dr. Dharam Singh Jat ACM Distinguished Speaker and Professor, Department of Computer Science, Namibia University of Science and Technology, Windhoek, Namibia.

Dharm Singh Jat received his degrees Master of Engineering and Ph.D. in Computer Science and Engineering from prestigious universities in India. He is a professor in the Department of Computer Science at Namibia University of Science and Technology. From 1990 to 2014, he was with College of Technology and Engineering, MPUAT, Udaipur, India, and has more than 27 years of academic experience. He was instrumental in the setting up of MPUAT Intranet. He has given several guest lectures/invited talks at various prestigious conferences such as 45th Annual National Convention of Computer Society of India (2010), 26th Indian Engineering Congress at Bangalore (2011), and various ACM conferences as ACM speaker. More than 8 Ph.D. and 30 Master students have been guided by him who are occupying leading positions including professor in academia and industry. His interests span the areas of multimedia communications, wireless technologies, mobile communication systems, roof computing, and video transmission over wired–wireless networks, software-defined networks, network security, Internet of things, and ICT Applications.

Publishing Ethics and Author Services

Talks Abstract

The importance of research publishing can be defined by a simple quote of Gerard Piel, which says "Without publication, science is dead." The first scientific journal was published in 1665, and we have traveled 350 years since then. In the last 20 years, science and reporting of science have undergone revolutionary changes. Computerization and the Internet have changed the traditional ways of reading and writing. Hence, it is very important for scientists and students of the sciences in all disciplines to understand the complete process of writing and publishing of a scientific paper in good journals. There is also a downside of digital publishing. The principal challenge for publishers is to handle ethical issues, and it is of utmost importance for the authors to understand the ethical practices involved in the process.

Keyword Author services for the publishing work, Computerization and science

Mr. Aninda Bose Senior Publishing Editor with Springer India Pvt. Ltd.

Aninda Bose is presently working as a senior publishing editor with Springer India Pvt. Ltd. He is part of the Global Acquisition Team at Springer and responsible for acquisition of scientific content across the globe. The disciplines he is responsible for are chemistry, material sciences, earth sciences, and computational intelligence. He has more than 23 years of industrial experience in marketing and different fields of publishing. He has completed Master's in organic chemistry from Delhi University and Master's in marketing research from Symbiosis Institute of Management Studies, Pune. He has published books for secondary level in chemistry and a member of American Chemical Society, USA. He has delivered more than 50 talks on Scientific Writing and Publishing Ethics in reputed universities, international conferences, and author workshops.

Applying Machine Learning Techniques for Big Data Analytics

Talks Abstract

Big data is gaining tremendous importance in today's digital world. Big data is a collection of data sets that are so voluminous and complex that traditional data processing application software seems inadequate to deal with them. This data is diverse in nature and grows exponentially. It includes both organized and unorganized formats. Some of the data collected from the transactions like credit/debit cards, e-commerce database, social networking sites, patient data from hospitals, and student's data from MOOCs/colleges could be stated as examples of big data. Handling this data type of data and more importantly analyzing this data to predict the near future is a tedious task for any analyst. Predictive analytics is a technique that uses both new and historical data to forecast future activity, behavior, and trends. It involves applying statistical analysis techniques, analytical queries, and predictive algorithms to the data sets to create models that show a numerical value, or score on the likelihood of a particular event happening. Predictive analytics deals with extracting the patterns from the data available so that it would be presented in a useful way to the concerned people, be it in businesses, health care, or any other field. As the domain changes, the properties of the data set also changes, and hence, different predictive models seem to be suitable for different domains.

In our present research, we have applied many machine learning techniques and tools. The talk would focus on the experimental results obtained in some of the domains stated as above.

Dr. Annapurna P. Patil Professor, CSE, M.S. Ramaiah Institute of Technology, Bangalore, India.

Dr. Annapurna P. Patil is working as a professor in the Department of Computer Science and Engineering at M.S. Ramaiah Institute of Technology, Bangalore, India. She has an experience of 23 years in academics. She is a senior IEEE member, LMCSI, LMISTE, member ACM, faculty advisor of IEEE, WiE, MSRIT. She has published several papers in various national and international conferences and in reputed journals and chaired sessions. Her teaching interest spans areas of mobile ad hoc networks, protocol engineering, artificial intelligence, and predictive models for data analytics, advanced algorithms, software engineering, distributed computing, bio-inspired computing, IoT, and cloud.

She has been involved in collaborative works with industries like CISCO, IBM, HPE, TransNeuron and Nihon Communications Ltd, Bangalore, in the areas of wireless networks, cognitive computing, IoT, and cloud.

Outlier Detection in Big Data

Talks Abstract

According to Hawkins (Hawkins 1980): "An outlier is an observation which deviates so much from the other observations as to arouse suspicions that it was generated by a different mechanism." Outlier detection extends traditional database monitoring with increased intelligence that helps security analysts understand risk based on relative change in behavior. Finding outliers may generate high value if they are found, the value in terms of cost savings, improved efficiency, compute time savings, fraud reduction, and failure prevention. According to IBM statistics, "Every day, we create 2.5 quintillion bytes of data—so much that 90% of the data in the world today has been created in the last two years alone. This data comes from everywhere: sensors used to gather climate information, posts to social media sites, digital pictures and videos, purchase transaction records, and cell phone GPS signals to name a few. This data is big data." These large-scale heterogeneous types of data appear problematic for traditional outlier detection methods to process. In this talk, we will discuss some of the most relevant outlier detection techniques for big data, research issues, and its applications.

Dr. Ramadoss Balakrishnan Professor, National Institute of Technology, Tiruchirappalli.

Dr. (Mr.) Ramadoss Balakrishnan received M.Tech. degree in Computer Science and Engineering from IIT Delhi in 1995 and earned Ph.D. degree in Applied Mathematics from IIT Bombay in 1983. Currently, he is working as a professor of Computer Applications at National Institute of Technology, Tiruchirappalli. His research interests include software testing methodologies, security and privacy in big data and cloud, data mining, predictive analytics in big data, and multimedia mining.

Under his guidance, eight candidates have successfully completed Ph.D. and five candidates are pursuing Ph.D. He has more than 35 refereed international journal publications and more than 40 papers presented in reputed international conferences.

He is a recipient of Best Teacher Award at National Institute of Technology, Tiruchirappalli, India, during 2006–2007. He is a member of IEEE; Life Member (LM) of ISTE, New Delhi; Life Member (LM), Computer Society of India. For more details, visit https://www.nitt.edu/home/academics/departments/faculty/Brama.pdf.

Identification and Prevention of Internet Addiction in Youth

Talks Abstract

With the advancement of the Internet technology and available resources, the accessibility and attractiveness of youth toward using Internet have been enhanced. It has an impact on individual's cognitive learning, physical and mental development for adolescents. Addiction to the Internet can negatively impact family relationships and adolescents' behavior. Early stage identification of Internet addiction and its risk factor is therefore a clinical significance for the prevention of the Internet addiction in youth. This presentation explores the impact of level of the Internet addiction and aggression as well as the prevalence of the different forms of aggression in youth. Addiction to the Internet causes behavioral and emotional problems. Teenagers and youth feel restless if they are restricted to use the Internet. For reducing behavioral problems, counseling seems to be a significant method.

Keywords Internet addiction, Psychological well-being, Social networks, Internet addiction prevention

Poonam Dhaka Department of Human Sciences, University of Namibia, Windhoek, Namibia.

Dr. Poonam Dhaka received her Ph.D. degree in Psychology from MLSU, India, and Master of Science in Home Science from College of Home Science, Udaipur, India. She has more than 20 years of academic and research experience, and presently she is a senior lecturer in clinical psychology with the Department of Human Sciences, University of Namibia. Her interests span the areas of mind power techniques, cyberpsychology and behavior, health psychology, mental health disorders, psychological assessment, gender identity, cognitive psychology, parent–child relationships. She has developed clinical psychological tools and mind power techniques for teenagers to increase intelligence, creativity, and learning. She has authored and co-authored over 37 research publications in peer-reviewed reputed journals, chapters, and conference proceedings. She is a member of Psychology Association of Namibia, Computer Society of India and served as a member of various committees in the national and international events.

Android: Security Issues

Talks Abstract

Android provides built-in security at the kernel level and framework level. Android Linux kernel is responsible for provisioning application sandboxing and enforcement of some Linux permissions. Sandboxing provides isolation where each application runs within its own user ID and virtual machine. At framework level, system-built and user-defined permissions make the access to software and hardware resources restricted. Android also provides app signing where installation of only signed apps is permissible. In spite of built-in security features, malware developers have succeeded in implementing various attacks on Android. One of the reasons is mistake made by the end user who normally grants permissions without any judgement. The other reasons are Android's architecture which permits app repackaging, dynamic code loading, reflection, use of native code, etc. Some intelligent malware writers utilize covert channels also to perform attacks. These attacks include stealing of private information, resource drainage, frauds, premium-rate SMSs.

Our work is focused on meliorating detection of malware that uses reflection as the means of hiding leaks. Reflection is a programming language feature that permits analysis and transformation of the behavior of classes used in programs in general and in apps in particular at the runtime. Reflection facilitates various features such as dynamic class loading, method invocation, and attribute usage at runtime. Unfortunately, malware authors leverage reflection to subvert the malware detection by static analyzers as reflection can hinder taint analysis used by static analyzers for the analysis of sensitive leaks. Even the latest and probably the best performing static analyzers are not able to detect information leaks in the malware via reflection. We proposed a system that combines dynamic analysis with code instrumentation for a more precise detection of leaks in malicious apps via reflection with code obfuscation. The evaluation of approach shows substantial improvement in detection of sensitive leaks via reflection.

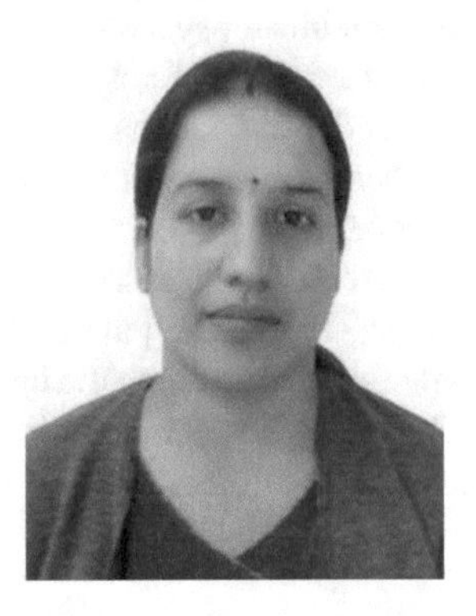

Dr. Meenakshi Tripathi Associate Professor, Department of Computer Science, MNIT, Jaipur, India.

Dr. Meenakshi Tripathi has more than 10+ years of experience. She has completed M.Tech. degree in 2005 at Computer Science and Engineering, Bansthali Vidyapith, Banasthali, and Ph.D. degree in 2014 at Computer Science and Engineering, MNIT, Jaipur. Her areas of expertise are wireless sensor networks, information and network security, software-defined networks, and Internet of things. She has supervised 5 Ph. D. scholars and more than 15 master's thesis. She is handling Department of Science and Technology, sponsored R&D projects. Her 20+ publications are in peer-reviewed international journal and conferences. She is the principal investigator in DST projects having around 60 lakhs of funding.

Tweaking Big Data Analysis for Malicious Intents

Talks Abstract

Governments of various nations and major industries working in the world have started relying on big data analysis for future predictions of their workflow and making critical decisions based on the results. With the indefinite scope of big data in today's scenario, assessing the risks pertaining to data breach, data tampering, and all related security factors is at a very minimal level. Understanding all the malicious activities and intents from the tweaking of big data analysis by individuals, organizations, or governments that could lead to catastrophic outcomes should be treated with utmost priority.

Mr. Prajal Mishra President of Graduate Student Senate at University of Texas at Arlington, Arlington, TX, USA.

Prajal is a graduate student at University of Texas, Arlington, pursuing his Master of Science in Computer Science, specializing in the field of databases and software engineering. He also focuses on secure programming paradigms and will be graduating in Spring 2018. He is working as the President of Graduate Student Senate at University of Texas at Arlington, which is the official representative body of all graduate students on campus. During his undergraduate, he was a Microsoft Student Partner for 2 years and co-organized "Hour of Code, Indore" in the year 2014 and 2015 with a team of 60+ volunteers who taught and motivated 5000+ school students from 25+ schools about coding.

Big Data and Web Mining

Talks Abstract

The era of growing Web users on the Internet, i.e., the world of the World Wide Web, has led to the generation of massive data with such a high speed and variety that it has given a scope to analyze the behavior of Web user. Almost all the streams existing today have become almost dependent on Web technologies, and thus, we have the opportunity to analyze the user data which can help us in improvising services and the products offered, utilizing the likes and dislikes of the user to ensure a better user experience and by providing better quality products. Predicting the trends of user can be helpful in deriving his future patterns. The user's past data can be helpful in drawing out this pattern. This is one of the paradigms where software products can be harnessed to provide more pleasing experience to the users.

Big data tools and technologies help to mine patterns, association, prediction for improving quality of service, applying personalization, i.e., recommendation of product or promotional offers, placing advertise, to handle business-specific issues like customer attraction, retention, cross sales, and departure. Web mining uses Web content, hyperlinks, server logs, and other Web data to retrieve useful information and draw the patterns. Big data challenges first deals with the massive volume of data which includes Web logs, content, their social media data like Facebook, Twitter, images, video. Fortunately, advance data management like cloud computing and virtualization provides effective capture, storage, and manipulation of the large volume of data. Secondly, it helps in dealing with a variety which includes structured data like the product information, unstructured data like the tweet by the user, Web page content, and the semi-structured data. Big data tools and techniques have the potential to convert, store, access, analyze, and visualize a variety of data. The continuous flow of massive new data, i.e., the data velocity, happens to be a bigger challenge today. So, the volume and variety of data that are collected and stored have changed. Real-time data is necessary to retrieve, analyze, compare, and decide the user trends and shall totally depend on the new findings offered by the big data technologies.

Dr. Bhawna Nigam Assistant Professor, IET, Devi Ahilya University, Indore, India.

Dr. Bhawna Nigam received her B.E. and M.E. degrees with honors in Computer Engineering Institute of Engineering and Technology (IET), Devi Ahilya University, Indore, India, in 2003 and 2008, respectively. In 2017, she obtained her Ph.D. degree in Computing Engineering. She is currently with Institute of Engineering and Technology (IET), Devi Ahilya University, Indore, India, as an assistant professor in Information Technology Department since 2007. Her current research interests include data mining, Web mining, big data analytics, machine learning. She has published 20+ papers.

Big Data and Data Science Trends: Challenges and Opportunities

Talks Abstract

Due to rapid growth in data, our society is facing a paradigm shift in Information and Communication Technology (ICT). These trends are going to change the industry and business process and impact the society. The basis of this paradigm shift is data science which gives meaning to the big data. Data science framework consists of three basic elements: people, data, and technology. To deal with data science framework, data scientist should possess few important characteristics such as data-driven approach, computational intelligence, and latest technological knowledge for better analysis, visualization, and preservation of huge amount of data. Here, we will discuss various digital challenges and opportunities ahead of us in big data and data science.

Keywords Data science, Big data, Trends, Challenges, Opportunity, ICT

Dr. Samiksha Shukla Assistant Professor, Department of Computer Science and Engineering, Christ University, Bengaluru, Karnataka, India.

Dr. Samiksha Shukla received her M.Tech. degree in Computer Science and Engineering from RGTU, Bhopal, and Ph.D. degree in Computer Science Engineering from Christ University, Bengaluru, India. She is currently working as an assistant professor in Department of Computer Science and Engineering, Christ University. She has published 35 papers in peer-reviewed conferences and journals in the area of data and computation security. She served as advisory board member for ICTBIG'16, CSIBIG'14, Technical Program Committee Member ICRITO'14, ICRITO'16, Organizing Committee Member ICERDS'18, and Conference Chair NCCOCE'16, NCCOCE'17.

A Study of the Correlation Between Internet Addiction and Aggressive Behaviour Among the Namibian University Students

Poonam Dhaka and Cynthia Naris

Abstract The explosion of online Social Networking Sites over time has its benefits as well as its risks. A potential risk is the fact that so many individuals have become victims of aggressive and cyber-bullying acts via Online Social Networking Sites. In the paper, the aim of this study is to analyse the correlation between Internet addiction and Aggressive Behavior Among the Namibian University Students. Based on statistical analysis the paper concluded that there is a worthwhile correlation between Internet addiction and Aggressive Behaviour and a sizable majority of the students who participated in the study suffer from moderate addiction problems due to their Internet usage. Also, the results indicate that the two most prevalent forms of aggression among the majority of the students are hostility and Physical Aggression.

Keywords Inferential statistics · Internet addiction · Descriptive statistics
Psychological wellbeing · Social desirability · Aggression

1 Introduction

The use and accessibility of the Internet and communication devices such as iPad, smartphone, computers, and laptops have increased drastically throughout the world in last two decades and have changed university students' daily life drastically. These days despite the positive impact the Internet has brought up, there has been a growth of literature about the negative impact of excessive and compulsive Internet usage [1].

A research concern of this research is that aggression is influenced by multiple factors, namely biological, psychosocial and the environmental factors [2]. Thus,

P. Dhaka (✉) · C. Naris
Department of Human Sciences, University of Namibia, Windhoek, Namibia
e-mail: pdhaka@unam.na

D. K. Mishra et al. (eds.), *Data Science and Big Data Analytics*,
Lecture Notes on Data Engineering and Communications Technologies 16,
https://doi.org/10.1007/978-981-10-7641-1_1

using Internet addiction as a dominant factor for electronic aggression and other forms of aggression prevalent in this study would be irrelevant. The explosion of Online Social Networking Sites and Internet facilities over time has its benefits as well as its risks. Due to the fact that Internet Addiction is a relatively modern phenomenon, there is little information available on electronic aggression, Internet addiction and the impact it has on aggression levels in students. Thus, this study will mainly focus on how Internet addiction affects aggressive behaviours among University of Namibia students.

2 Literature Review

Its been reported that a co-morbid relationship exists between psychological disorders such as suicidal ideation, social phobia, obsessive-compulsive compulsive disorders and problematic alcohol usage [3]. This correlation can be related to the bidirectional causality of Internet Addiction, namely environmental and biological factors. It is thus very important to explore the relationship between Internet addiction (addiction to social networking) and aggression levels in students at the University of Namibia.

Internet Addiction
According to Young [4], Internet Addiction is the compulsive use of the Internet by an individual. Widyanuto et al. [5] made a significant contribution to Internet addiction research, he identified underlying factors that collectively identify Internet Addiction based on the Internet Addiction Test (IAT) developed by Young in 1998 [6].

The three major underlying factors identified were; emotional conflicts, time management and mood modification. The emotional conflicts factor refers to an individual's preference to being online over social engagements such as spending time with family members and friends. The time management factor refers and individuals preferences to spend more time online, this leads to them neglecting important tasks and decreased productivity in school, work, etc. The third factor is mood modification, this factor suggests that individuals with Internet addition develop emotional problems such as depression or excessive levels of stress to escape their reality [5]. Griffiths [7, 8] made another significant contribution regarding Internet Addiction.

As per ITU report, 31% Namibian population are using the Internet and 20.24% Internet users are Facebook subscribers as on June 2017. Most literature regarding Internet Addiction is Euro-centric and U.S.-centric, but according to Nath et al. [9] who studied Internet addiction among University students in Uganda and Namibia. A sizable majority of the students in both Namibia and Uganda suffer from frequent problems due to their Internet usage. Results also showed that the Internet Addiction psychometric constructs in the African context differ from those in the Western context. According to recent studies using Young's Internet Addiction Test [6], adolescents are more prone to developing Internet addiction because of their variability of developing cognitive skills as well as self-control [10]. Using Young's

Internet Addiction Test, 4% of U.S. adolescents [10], 8% of Chinese adolescents [11] and 11.3% of adolescents in Germany [12] were found to be addicted to using the Internet.

Addiction Correlates

According to Beard and Wolf [13], Internet Addiction can disrupt of an individual's psychological state (mental and emotional), academic performances, occupational and social interactions. Internet Addiction correlates such as aggression, academic performance problems, enhanced anxiety and aggression can be implemented in the Namibian context.

The study [14] constructed a correlative model between the Internet addiction and mobile phone addiction of 448 college students from two colleges on an island in Taiwan, with 38.8% females and 61.2% males.

This study used the Mobile Phone Addiction Scale (MPAS) and the Chinese Internet Addiction Scale to serve as the measurement tools to conduct investigations on the subjects and adopted the structural equation model (SEM) to investigate the collected data from the survey. The study concluded that Internet addiction and mobile phone addiction are positively related. Online interview conducted by Choi et al. [15] with college students and reported that the major physical complaints like a backache, headaches and sore shoulders by students stem from their Internet use. Internet addiction was also reported to have a correlation with anxiety disorders, depression and hyperactivity disorder [16]. Online Social Networking sites have also been regarded as one of the major contributors to Internet addiction disorder. Internet Addiction studies include studies by Kim et al. [17], on 1573 Korean high school students and a study by Johanasson and Gotestam [18], they studied the prevalence of Internet addiction in 3237 Norwegen learners between the ages of 12–18.

Aggression

According to Social Psychologists, aggression is defined as a behaviour which intends to harm another individual who does not wish to be harmed. Social psychologists frequently use the term "violence" as a subdivision of aggression. Violence is a physical way of expressing one's aggression by inflicting harm on yourself or another individual. The aim of this study is to examine how an external stimulus such as Social media and Internet usage triggers aggressive behaviours in University of Namibia main campus students.

Aggression and Aggression Measures

According to Bushman and Anderson [19], there are four main types of aggression, namely; Emotional aggression, Instrumental or Hostile aggression, Physical aggression and lastly Verbal aggression. Emotional aggression is also known as impulsive aggression. This form of aggression is mainly classified as the result of extreme levels of negative emotions. It occurs as an impulsive response to an internal or external threat. Instrumental or Hostile aggression is a form of aggression that is intentionally instigated or done. It is aimed at accomplishing a certain goal or satisfying a need. Physical aggression is a form of aggression that involves inflicting physical harm to another individual as and attempted to deal with one's anger. The last main type of anger is Non-physical or Verbal anger, this entails harming another individual without using any physical force, this includes verbal as well as psychological forms of violence. This study will use the self-report measures of aggression developed by Buss and Perry in 1992 [20].

3 Methodology

This study is an exploratory descriptive study following a quantitative approach. According to Kothari [21] exploratory studies aim to gain familiarity with or gain insight into an unknown phenomenon, while descriptive research aims to accurately portray an individual, group or situation.

3.1 Study Design and Population

The study was conducted in the University of Namibia during 2016 academic year. With the increase in smart devices including smartphone, iPad, laptop and computers and the Internet of thing, multimedia and other universal applications of the Internet play the important role. The university students in Namibia are using the Internet for study, research and entertaining purposes.

The population to which findings of this research can be generalised is the University of Namibia, Windhoek main campus. The sampling technique used in this study was the simple random sampling technique. A total of 100 students were selected to participate.

The male and female ratio in the sample was 50: 50. 45% of the participants were Namibians, whilst 55% of the participants were foreign nationals. The male and female breakdown in the sample was: Namibian 44.4% males and 55.6% females; foreign nationals: 54.5% males and 45.5% females. The minimum age recorded in the study was age 18, whilst the maximum age was age 33. The results indicate that 20% of the respondents were in their first year, 21% in their second year of study, and an overwhelming 30% in their third year, whilst 29% of the respondents were in their fourth year of study.

3.2 Measures and Data Collection

Research instruments and measures that were used to collect data was a self-constructed demographic questionnaire, the Internet Addiction Test (IAT) as well as the Aggression Questionnaire. The demographical questionnaire form was developed by the researcher to determine age, gender, nationality and the year of study of the participants. Three paper and pen questionnaires (demographic questionnaire, the

Internet Addiction Test (IAT) and the Buss Perry Aggression Questionnaire (BPQA)) were handed out to randomly selected students from the University of Namibia, Windhoek, main campus. It took approximately 5–10 minutes for the participants to complete all three questionnaires.

The Internet Addiction Test (IAT)
It is a 20-item scale developed by Young [6] used to measure the level of addiction to the Internet by experienced Internet users who utilises this technology on a frequent basis. Each of the 20-items is measured using a 5-point Likert scale (1-Never, 5-Very Frequently). There are three underlying psychometric constructs based on the statements of the IAT: emotional conflict (1–9); time management issues (10–14); and mood modification (15–20). The questionnaire also asked respondents to provide information about their age, years online, whether they use the Internet for and lastly in school, daily time spent online, and lastly an estimation of the percentage of time they spend on 15 online activities. Further, a cut-off score of 20–49 points implies that the participant is experiencing a mild level of Internet Addiction. A cut-off score of 50–79 points implies that the participant is experiencing moderate Internet Addiction, with frequent problems in their personal as well as social life due to their Internet usage. Whilst a cut-off score of 80–100 points imply that the participant is experiencing severe Internet Addiction, with extreme problems in their personal as well as social lives due to their Internet usage.

The Buss Perry Aggression Questionnaire (BPAQ)
It is a 29-item, 5-point Likert scale constructed by Buss and Perry in 1992 [20]. It is used to measure the different levels and factors of Aggression present among the participants in the study. The minimum score obtainable is 29 whilst the maximum score obtainable is 145. The Buss Perry Aggression Questionnaire (BPAQ) consists of four factors represented by different statements in the questionnaire namely, Physical Aggression (PA), Verbal Aggression (VA), Anger (A) and Hostility (H). Physical Aggression (2, 5, 8, 11, 13, 16, 22, 25 and 29), Anger (1, 9, 12, 18, 19, 23 and 28), Verbal Aggression (4, 6, 14, 21 and 27) and lastly Hostility (3, 7, 10, 15, 17, 20, 24 and 26). The total sum of aggression is the sum of the factor scores.

3.3 Data Analysis

In this study to investigate the collected data, the Statistical Package for Social Sciences (SPSS) was used. Pearson's correlation coefficient (the data ranges from −1 to +1) was used to measure the significance of a correlation between Internet Addiction levels and Aggression levels.

Table 1 Internet addiction and aggression (N = 100)

Research instruments	Mean	Standard deviation (SD)
The Internet addiction test (IAT)	59.58	16.637
The Buss Perry aggression questionnaire (BPAQ)	84.04	18.015

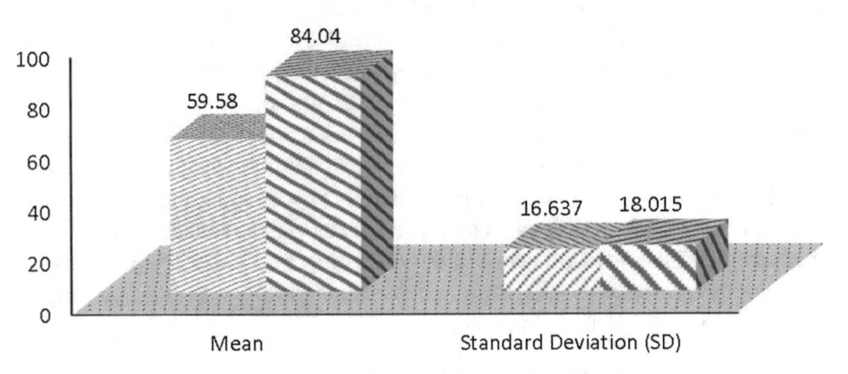

Fig. 1 Mean and SD of IAT and BPAQ

4 Results and Analysis

4.1 Descriptive Statistics

The study adopts descriptive statistics to analyse the university students' self-reported situations of Internet addiction and aggression. Table 1 shows the mean score of Internet addiction and Aggression among all the participants in the study was 59.58 and 84.04 respectively (Fig. 1).

The results reveal that students exert moderate levels of Aggression. The most prevalent form of aggression found in this study was Physical Aggression (27.8%) followed by Hostility (27.7%), Anger (25.2%) and lastly Verbal Aggression (19.3%).

4.2 Correlations Between Variables

The results of the correlation between Internet Addiction and Aggression are presented in Table 2, 3, and 4. A total positive linear correlation ($r = 0.460$) was found between Internet addiction and Aggression. The results depicted in Table 3 and

Table 2 Correlations between Internet addiction and aggression

		IAT	BPAQ
IAT	Pearson correlation	1	0.460
	Sig. (2-tailed)		0.000
	N	100	100
BPAQ	Pearson correlation	0.460	1
	Sig. (2-tailed)	0.000	
	N	100	100

Table 3 Internet addiction and aggression in male participants

Level of Internet addiction	N	Average %		Sum: aggression	Most prevalent aggression factor	Least prevalent aggression factor
		Internet addiction	Average aggression			
Severe	2	81.5	95	190	Physical	Verbal
Moderate	26	62.2	96	2498	Physical	Hostility
Mild	22	34.2	74.2	1632	Hostility	Hostility

Table 4 Internet addiction and aggression in female participants

Level of Internet	N	Average %		Sum: aggression	Most prevalent aggression factor	Least prevalent aggression factor
		Internet addiction	Average aggression			
Severe	3	84	95.7	287	Hostility	Verbal
Moderate	31	62	81	2511	Hostility	Verbal
Mild	16	35.3	74.1	1186	Hostility	Anger

4, indicate that the most prevalent form of Aggression among male participants is physical aggression, whilst the most prevalent for aggression in female participants is hostility. It also indicates that more females (n = 35) suffer from severe and moderate Internet addiction than males (n = 28). The results of the correlation between Internet Addiction and Aggression are presented in Table 2. A total positive linear correlation was found between both the variables, which is understandable (the higher the degree of addiction, the higher the Aggression).

The results depicted in Table 3 and 4, indicate that the least prevalent form of Aggression among male participants is verbal aggression, whilst the least prevalent for of aggression in female participants is verbal.

The finding reveals that 88% of the participants in the study use the Internet for purposes can easily be explained. According to Choi et al. [15], the academic use of the Internet for learning and research purposes has increased drastically over the past decade. This is related to the fact that many universities and colleges have Internet services available to their students on and campus 24 h a day. Results from the study indicate that the majority of the students experience psychological conflict due to their Internet usage.

In this study, it was found that 38% of the participants had low levels of Internet addiction, whilst 57% had moderate levels of Internet Addiction and 5% had severe levels of Internet addiction. It also indicates that more females (n = 34) suffer from severe and mild Internet addiction than males (n = 28) and study concluded that Internet addiction was more common among undergraduate university students. According to Choi et al. [15] young people should be monitored as they have a more common Internet addiction. The finding of this paper is consistent with the literature data Choi et al. [15].

5 Conclusion

This study explored the impact of level of Internet addiction and aggression as well as the prevalence of the different forms of aggression in students registered at the University of Namibia's Windhoek main campus. Three research objectives addressing the relationship between these variables were formed. A survey questionnaire based on three questionnaires (demographic questionnaire, the Internet Addiction Test (IAT) and the Buss Perry Aggression Questionnaire (BPQA)) was developed and conducted. Results were analysed using SPSS version 24. Results from the study indicate the level of Internet addiction in students is reciprocally related to their aggression levels. This course of research in Internet addiction reveals a promising and productive avenue for numerous future research opportunities.

Declaration and Acknowledgements We thank the Research Ethical Committe, Department of Human Science for granting permission to conduct the research and also thankful to the participants who participated in the study. Authors have obtained approval from the ethics committee and consent by all participants (university students).

References

1. Frangos CC, Frangos CC (2009). Internet dependence in college students from Greece. Eur Psychiatry 24(Suppl 1):S419
2. Carpenter DO, Nevin R (2010) Environmental causesof violence. Physiol Behav 99(2):260–268
3. Cheung LM, Wong WS (2011) The effects of Insomnia and internet addiction on depression in Hong Kong Chinese based adolescents: an exploratory cross-sectional analysis. J Sleep Res 20(2):311–317
4. Young KS (2010) Clinical assessment of Internet addictedclients. In: Young K, Nabuco de Abreu C (eds), Internet addiction: A handbook and guide for evaluation and treatment. Wiley, New York, pp. 19–34

5. Widyanto L, McMurran M (2004) The psychometric properties of the internet addiction test. CyberPsychol Behav 7(4):443–450
6. Young KS (1998) Internet addiction: the emergence of a new clinical disorder. CyberPsychol Behav 1(3):237-244
7. Griffiths MD (2000) Internet abuse and internet addiction in the workplace. J Workplace Learn 22(7):463–472. https://doi.org/10.1108/13665621011071127
8. Griffiths MD (2000) Internet addiction: time to be taken seriously? Addict Res 8(5):413-418
9. Nath R, Chen L, Muyingi HL, Lubega JT (2013) Internet addiction in Africa: A study of Namibian and Uganda college students. Int J Comput ICT Res 7(2):9–22
10. Liu T, Potenza MN (2007) Problematic Internet use: Clinical implication. CNS Spectrum 12(6):453–466. https://doi.org/10.1017/S1092852900015339
11. Leung L (2007) Stressful life events, motives for internet use, and social support among digital kids. CyberPsychol Behav 10(2). https://doi.org/10.1089/cpb.2006.9967. Retrieved on https://www.liebertpub.com/doi/abs/10.1089/cpb.2006.9967
12. Wolfling K, Muller K, Beutel M (2010) Diagnostic measures: scale for the assessment of Internet and gaming addiction. In: Young K (ed) Internet addiction: the emergence of a new clinical disorder. Cyberpsychol Behav 1(3):237–244
13. Beard KW, Wolf EM (2001) Modification in the proposed diagnostic criteria for Internet addiction. CyberPsychol Behav J 4:377–383
14. Chiu S-I, Hong F-Y, Chiu S-L (2013) An analysis on the correlation and gender difference between college students' Internet addiction and mobile phone addiction in Taiwan. ISRN Addiction, vol 2013, Article ID 360607, 10 pp. https://doi.org/10.1155/2013/360607
15. Choi J, Hwang SY, Jang KE (2008) Internet addiction and psychiatric symptoms among Korean adolescents. J Sch Health 78(3):168–171
16. Mehroof M, Md G (2010) Online gaming addiction: The role of sensation seeking, self-control, neuroticism, aggression, state anxiety, and trait anxiety. Cyberpsychol Behav Soc Netw 13(3):313–316
17. Kim K, Ryu E, Chon MY et al (2006) Internet addiction in Korean adolescents and its relation to depression and suicidal ideation: a questionnaire survey. Int J Nurs 43:185–192
18. Johansson A, Götestam KG (2004) Internet addiction: characteristics of a questionnaire and prevalence in Norwegian youth (12–18 years). Scand J Psychol 45(3):223–229
19. Bushman BJ, Anderson CA (2001) Is it time to pull the plug on the hostile versus instrumental aggression dichotomy? Psychol Rev 108(1) 273–279
20. Buss AH, Perry M (1992) The aggression questionnaire. J Pers Soc Psychol 63:452–459
21. Kothari CR (2004) Research methodology. Methods and techniques, 2nd ed. New Age Publishers, New Delhi

Genetic Algorithm Approach for Optimization of Biomass Estimation at LiDAR

Sonika and Aditi Jain

Abstract Estimation of Biomass at ICESat/GLAS footprint level was finished by incorporating information from various sensors viz., spaceborne LiDAR (ICESat/GLAS). The biomass estimation accuracies of Genetic Algorithm were studied by optimizing the waveform parameters. Multiple linear regression equation was generated using the most important variables found by Genetic Algorithm. The results of the study were very encouraging. Optimum results were obtained using the top 18 parameters derived from GLAS waveform. The biomass was predicted at small area by Genetic Algorithm with an R^2 63% of and RMSE of 18.94 t/ha using the best six variables, viz. wdistance, wpdistance, R50, Ecanopy, Home., wcentroid over to 18 independent variables. The same methodology can be used for biomass estimation over a large area. The estimation is done at Tripura area. The study finally established that Genetic approach is produced the better result to predicting AGB. The best outcome of the study was the formulation of an approach that could result in higher biomass estimation accuracies.

Keywords ICESat/GLAS · Footprint · Waveform · AGB · LiDAR
Genetic algorithm

Sonika (✉) · A. Jain
Jayoti Vidyapeeth Women's University, Jaipur, Rajasthan, India
e-mail: saxenasonalika12@gmail.com

A. Jain
e-mail: aditijain.csi@gmail.com

D. K. Mishra et al. (eds.), *Data Science and Big Data Analytics*,
Lecture Notes on Data Engineering and Communications Technologies 16,
https://doi.org/10.1007/978-981-10-7641-1_2

1 Introduction

It is notable that the forest ecosystem acts as worldwide carbon sink [1] approximately 80% of all aboveground and 40% of underground terrestrial organic carbon are stored in forest [2]. An accurate and precise measurement of carbon sequestered and released is very important for biomass and to understand the role of carbon cycle on global climate change. Above Ground Biomass symbolizes the major portion of aggregate forest biomass. Estimation of biomass of forest is one of the best critical regions of research that makes AGB as forest structural feature to be discovered.

The technology of remote detecting has been generally utilized for biological system and vegetation contemplates. Typically remote sensing refers to acquiring information about vegetation and ecosystem using sensor on board passive optical systems and active systems like airborne system (e.g., aircraft, balloons) or space-borne system (e.g., space shuttles, satellite) platforms (remote sensing application). An active system such as radar provides sequence of ecological applications. Passive remote sensing exploits spectral reflectance properties of forest region to basic characteristic of forest ecosystem. LiDAR based instruments from space-borne, airborne, and terrestrial stages give an immediate method for measuring forest qualities.

For evaluating biomass/carbon in vegetation utilizing remotely sensed information, one approach is field review based biomass appraisal which is dreary and more time consuming at huge scale. Radar data is beneficial for accessing cloud-free datasets.

Numerous studies have endeavored to active microwave remote sensing and latent passive optical device to access forest-dependent parameters for estimation of AGB and also carbon storage in forest [3]. The major advantage of LiDAR is its potential to estimate in forest when LiDAR is combined with advanced statistical models with sample plots. While LiDAR ranges strike into densest forest and the output data are without being in any obstacles like scapegoat by clouds and shadows, hence producing more accurate outputs in comparison of other remote sensing techniques [4].

Presently study is mainly focused on the study of large footprints in full waveform signals that is produced by GLAS (Geo-Science Laser Altimeter System), which is mounted at the ICESat (Ice, Cloud and Land Elevation Satellite System) [5]. This study suggested optimizing parameters using sensor for estimating biomass at ICESat GLAS footprint using Genetic Algorithm.

Previous literature provides an overview of LiDAR [6, 7] and evaluates specific sensors and techniques [8] for estimating biomass of forest. Present study has broadly aimed to utilize the precise measurements of forest stand heights from LiDAR and type density parameters from optical imagery to estimate biomass [9].

The most well-known direct techniques utilized for optimization are evolutionary algorithms, for example, Genetic Algorithm (GA). A Genetic Algorithm is utilized in present paper. This paper is based on probability to locate the best variable which are gives the important role to predict biomass. Genetic Algorithms are Random search strategies; it is based on that strategy to select the natural selection and natural genetics. It can discover the global optimization in complex huge data for search spaces. As of late, GA has been utilized broadly used for numerous optimization searching issues including numerical optimization [10].

2 Review of Literature

Carbon stock and biomass estimation is playing the most important role in forestry and climate change. The accuracy of biomass and carbon stock assessment is thus most important. The field data combined with the remote sensing has more potential to measure the accurate biomass and it is highly reliable model. It is very difficult to estimate accurate biomass using single data type instrument like only field cannot be sufficient to measure the accurate biomass.

2.1 Biomass Estimation

The biomass evaluation is a strong component to research. Biomass is the dry weight or total quantity of living organism of one plant species [11]. Biomass can be measured in the terms of g/cm^2, kg/m^2 or ton/ha units. Vegetation biomass includes leaf, stem, root, fruits, and flower.

2.2 Review of Global Forest Biomass Assessment

The word "LiDAR" means "Light Detecting and Ranging". For collecting accurate and dense elevation data across landscapes, shallow water areas, and other sites that is why LiDAR has become a conventional method. A fundamental LiDAR system involves a laser beam. This laser beam detected by a range finder with the help of inbuilt rotating or oscillating mirror. For gathering distance measurements at particular angle intervals in one or two dimensions, after that the laser scans around the scene being digitized,

LiDAR can be classified into two based on: systems and data acquisition. First is system containing three different systems space-borne, airborne, and terrestrial systems and also their sensors related to the biomass estimation.

2.3 Spaceborne LiDAR (ICESat/GLAS)

The principle negative mark of airborne LiDAR is its comparing low horizontal coverage and high cost. In January 2003, the Ice cloud and land elevation satellite (ICESat) was introduced. Its purpose was to measure atmospheric profiles of cloud and aerosol properties, polar ice-sheet elevation change, land topography profiles and height of vegetation canopies [12]. These purposes are refined with Geo-Sciences Laser Altimeter System (GLAS).

GLAS (Geo-science Laser Altimeter System) on ICESat (Ice Cloud and land Elevation Satellite) is a sensor of the spaceborne LiDAR. This is the initial space-borne tool which can digitize the backscattered waveform and offer coverage. GLAS instrument provides a facility to measure ice-sheet topography and associated with temporal changes, cloud, and atmospheric details. It also provides information about height and its thickness of important cloud layers.

For measuring distance GLAS contains a laser system, a GPS (Global Positioning System) receiver, and a star tracker attitude determination system. The laser transmits small pulses (approximate 4 ns) of infrared light (1064 nm wavelength) and visible green light (532 nm wavelength). Photons returned back to the spacecraft from the ground surface of the Earth and from the atmosphere, including the interior of clouds, will be collected in a 1 m diameter telescope. Laser pulses at 40 times/s will illuminate spots (called footprints) meter intervals along Earth's surface [13]. In this research we will used basically two GLAS product GLA 01 and GLA 14.

2.4 ICESat/GLAS Data Processing

The process of this methodology initially started with GLA 01 and GLA 14 product data. GLA 01 product for global elevation and GLA 14 product for transmitted as well as received waveform. The GLA 14 product data is stored in file format include accurate geo-location of the footprint center from the corresponding surface elevations. The GLA 14 also contain UTC (coordinated universal time) time at that waveform from a footprints are documented and the UTC (coordinated universal time) time is used to fetch the raw wave form with the help of GLA 01 file. Data processing tools used plus a description.

3 Study Area and Materials

The study area of "Tripura" (Fig. 1) was large for biomass prediction that is why for garneting biomass equation and testing the methodology a small area of "Doon Valley" was chosen. The study area selected for the presented study is Tripura (22° 56 N–24° 32 N and 91° 09 E–92° 20 E).

4 Methodology

See Fig. 2.

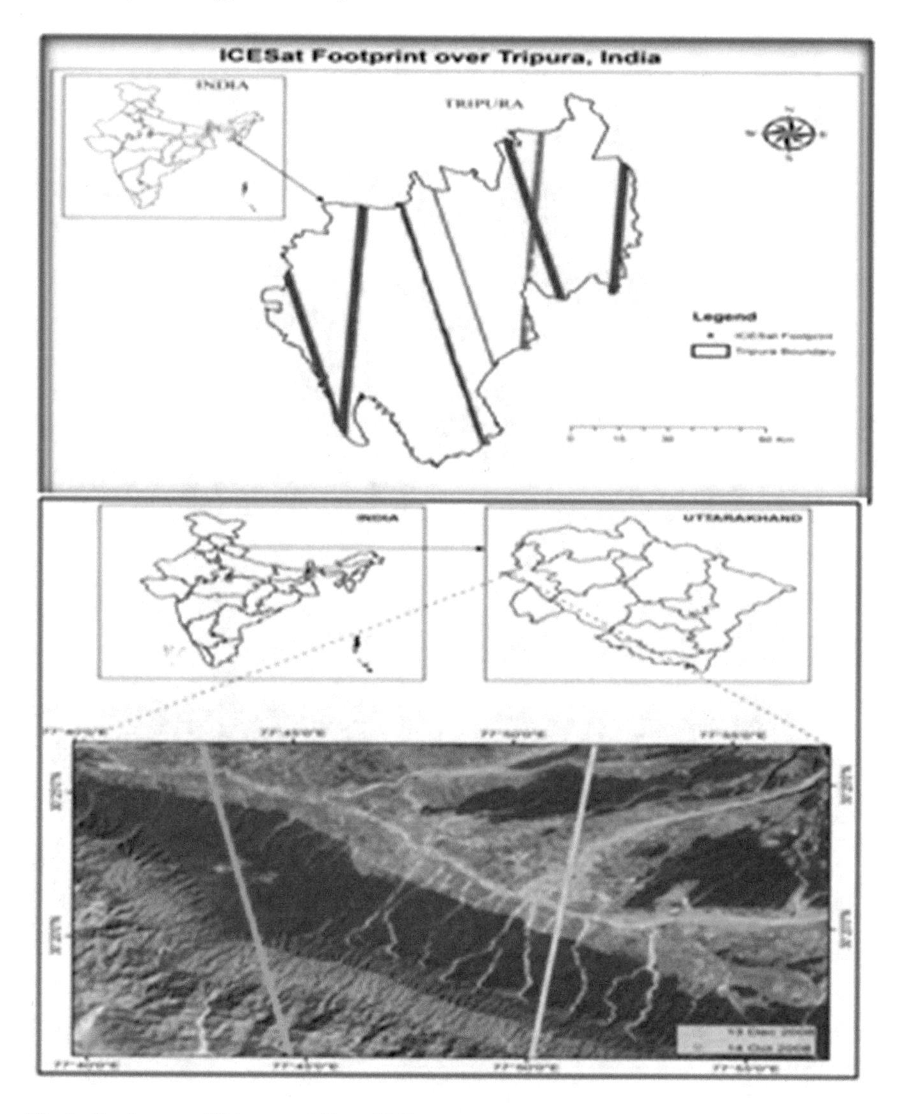

Fig. 1 Study area of Tripura and Doon Valley area

5 Description of Datasets

5.1 Satellite Data

The Geosciences Laser Altimeter System (GLAS) is a Spaceborne LiDAR system. It is the individual scientific tool on ICESat. Its goal is to register polar ice-sheet

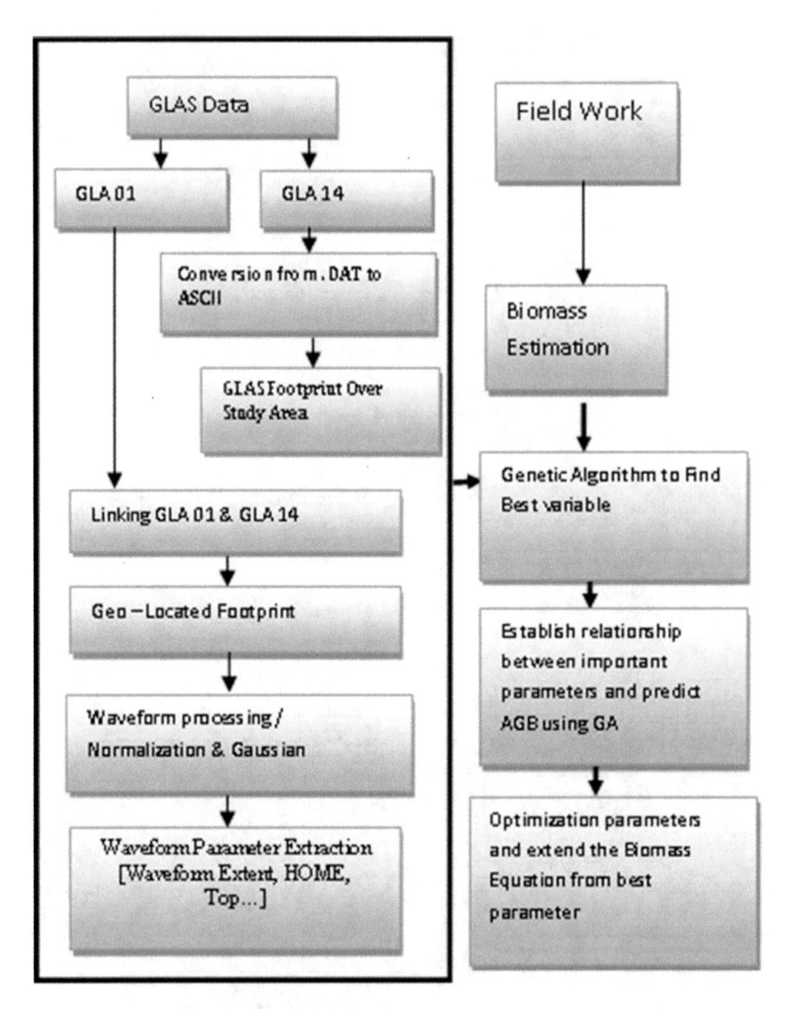

Fig. 2 Methodology

elevation change, barometrical outlines of aerosol properties, cloud properties and land topography profiles.

Satellite laser altimetry has an exclusive capability for assessing canopy height of the forest, that has a straight and increasingly well understood correlation to above-ground carbon storage. ICESat/GLAS data product (GLAS 01 and GLAS 14), Land sat data (7 bands). GLA 01 is Level 1 GLAS (Global Land Surface Altimetry) data product it includes broadcasted and received waveforms from the altimetry and GLA 14 (Global Land Surface Altimetry, level 2) data product it offers surface elevations for ground. It as well contains the laser footprints geo-location and reflectance, as well as instrument, geodetic, and atmospheric alterations for range measurements.

5.2 *Field Data*

The field data sampling procedure is the systematic study which is estimated at each footprint located by ICESat/GLAS data. The geo-location coordinate of the footprint were acquired from GLA 14 data products. In this study, 20 m radius circular plots were selected for height and biomass assessment. The biomass estimation from field data, footprints at the plot area was matched from GLA 14 results that were already plotted at Google Earth map during at GLA 14 data process. For every plot visited by, interested variables measured were tree species, tree height, and DBH.

6 Genetic Approach for Optimization of LiDAR Biomass Estimation

Genetic Algorithms are a variable pursuit methodology which depends upon the rule (Algorithm steps) of evolution by natural selection. This technique works on advancing arrangements of chromosomes (variables) that fit certain criteria. This is underlying arbitrary populace by means of rotations of differential repetition, recombination of chromosomes, and change of the best fit chromosome.

A Genetic algorithm (GA) is a searching process used for finding extract or approximate result to optimization and find exact problems. The GA produced the optimal solution after some iterative computation. GA work with the initial population which is depends on the problem structure and its data. GA generates Successive Population of another solution that is represented by a chromosome that gives the solution of that problem [14].

There is the procedure to optimize GA.

```
Simple GA ()
 {
     Initial-population;
     Evaluation-chromosome;
     while (Stopping standard has not been reached)
          {

               Selection (do Reproduction);
               Crossover;
               Mutation;
               Evaluation;
          }
 }
```

Step1: The method at first takes various values (chromosomes). These various variable sets called chromosomes.

Step2: Initialized the chromosome set for evaluation process. In this step the GA tests the precision of the prediction.

Step3: Each of the chromosomes must be evaluated for the selection method. In this step calculate the value of each gene in the chromosome, adding, and averaging the value for the chromosome. Third step evaluate the elite chromosome of the generation is determined.

Step4: At this step there are some selections procedures evaluate next generation of chromosome. The number of inhabitants in chromosomes is replicated. With the help of fitness function calculate chromosomes higher fitness score will create an increasingly various offspring.

Step5: The genetic information enclosed in the replicated parent chromosomes are joined through crossover method. Two messily chose parent chromosomes are used to make two new chromosome combinations. This crossover permits a improved exploration of possible arrangements recombining great chromosomes (Fig. 3).

Step6: After the crossover chromosome population generation mutate the population will then obtainable in the chromosome haphazardly. The changes of this population produce that new genes are utilized as a part of chromosomes.

Step7: This procedure is repeated from step2 until a precise chromosome is acquired. The cycle of repetition (step3), crossover (step4) and mutation (step5) is called generation.

7 Optimization Genetic Algorithm Using R Package

R is open source data analysis software for statisticians, data scientists, analysts and others who need for statistical analysis, data visualization, and predictive modeling. R is a programming language and environment for statistical computing and graphics. R provides a wide variety of statistical (linear and nonlinear modeling, classical statistical tests, time-series analysis, classification, clustering, etc.) and graphical techniques, and is highly extensible. R provides GALGO package to optimization for a large no of data. It is a generic algorithm optimization software package that is used for Genetic Algorithm to solve optimization problems concerning the selection of variable. GALGO is used in the statistical encoding environment R platform using object-oriented programming under the S3 methods package. R also provides glmulti package to extract R^2 and RMSE [13].

7.1 Calculate of RMSE

The principle criteria used to decide show fitness level were certainty and prediction difference, Root mean square Error (RMSE). The value of model's R^2 was used to calculate the model strength. The RMSE is used to show the error (in) the value of biomass which would be expected in using the Genetic algorithm model.

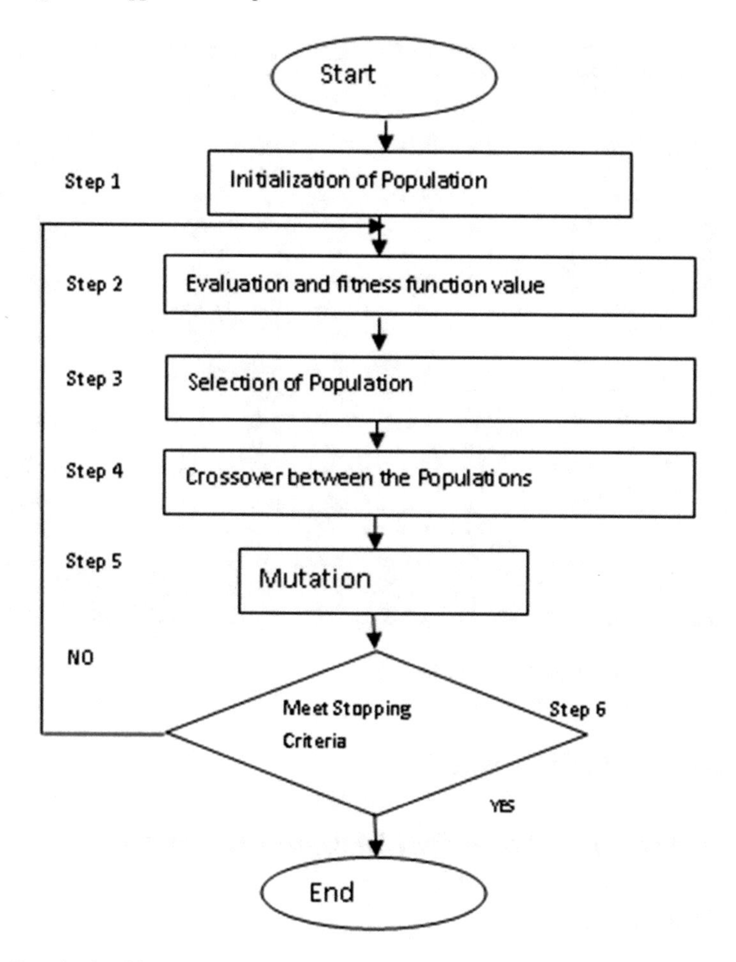

Fig. 3 Genetic algorithm procedure

$$\mathrm{RMSE} = \sqrt{\frac{\sum_{i=1}^{n}(o_i - E_i)^2}{n-1}}, \tag{1}$$

where

O_i Original value of *i*th perception.
E_i Expected value of *i*th perception.
N The total number of perception.

So RMSE is computed to find out the amount of error that is predicting biomass using the 18 independent variable when field data biomass is used for this.

Index number	Date of Acquisition	Time of Acquisition	Latitude	Longitude	Altitude	Geoids	Saturation	Gain	UTC Time
126E15241	10/21/2003	17:28.4	24.307532	93.299839	479.613	-50.31385	0	74	120014248.386
126E15241	10/21/2003	17:28.4	24.310663	93.299381	420.986	-50.31641	0	99	120014248.436
126E15241	10/21/2003	17:28.5	24.312228	93.299153	396.762	-50.31769	0	123	120014248.461
126E15241	10/21/2003	17:28.5	24.313791	93.298926	380.379	-50.31897	0	129	120014248.486
126E15241	10/21/2003	17:28.5	24.315349	93.298698	397.472	-50.32026	0	85	120014248.511
126E15241	10/21/2003	17:28.5	24.316903	93.298471	409.157	-50.32154	0	65	120014248.536
126E15241	10/21/2003	17:28.6	24.318457	93.298243	388.209	-50.32282	0	49	120014248.561
126E15241	10/21/2003	17:28.6	24.320007	93.298016	386.98	-50.3241	0	63	120014248.586
126E15241	10/21/2003	17:28.6	24.321554	93.297788	392.245	-50.32539	0	116	120014248.611
126E15241	10/21/2003	17:28.7	24.324642	93.297331	454.7	-50.32795	0	108	120014248.661
126E15241	10/21/2003	17:28.7	24.326186	93.297102	508.656	-50.32923	0	97	120014248.686
126E15241	10/21/2003	17:28.7	24.327735	93.29687	566.595	-50.33051	0	94	120014248.711
126E15241	10/21/2003	17:28.7	24.329291	93.296639	513.541	-50.3318	0	77	120014248.736
126E15241	10/21/2003	17:28.8	24.330844	93.296405	562.379	-50.33308	0	85	120014248.761
126E15241	10/21/2003	17:28.8	24.33396	93.295937	497.663	-50.33564	0	95	120014248.811
126E15241	10/21/2003	17:28.8	24.335518	93.295702	464.43	-50.33692	0	72	120014248.836
126E15241	10/21/2003	17:29.0	24.344832	93.294264	391.367	-50.34462	0	86	120014248.986
126E15246	10/21/2003	17:29.3	24.36668	93.290946	280.837	-50.39231	0	112	120014249.336

Fig. 4 .out file convert in the Excel File format

8 Results and Discussions

8.1 Analysis of GLAS Derived Waveform Parameters

8.1.1 GLA 14 Outcomes

The Final result obtained from the processing of GLA 14 data is shown in Fig. 5-2. These Geo-located footprints are used in field data estimation for calculating the Biomass of the respective area.

(a) **Footprint Location**

After the processing of GLA 14 we have found latitude longitude and UTC time, etc., values. The details are shown in Fig. 4.

After the completion GLA 01 it produced the 18 independent variables (Fig. 5).

8.1.2 GLA 01 Outcomes

After the processing of GLA 14 in Tripura, we have found 65 lines in whole area. In this process forest and non-forest footprints were separated. At every line have approximately 250 points (Fig. 6).

	A	B	C	D	E	F	G	H	I	J	K	L	M	N	O	P	Q	R
1	[illegible]	[illegible]	[illegible]	[illegible]	[illegible]	[illegible]	[illegible]	[illegible]	[illegible]	[illegible]	Peaks	[illegible]	roughness	[illegible]	[illegible]	[illegible]	[illegible]	[illegible]
2	18.2	18.9	21.45	0.38	-0.71	0.81	29.7	26.55	20.85	804	3	0.0083	5.7	19.85	3.82	15.83	0.24	0.81
3	17.4	20.85	23.55	0.6	-0.72	0.81	30.9	29.1	24.75	768	3	0.0103	4.35	18.74	3.22	15.53	0.21	0.83
4	19.2	22.05	25.05	0.63	-0.75	0.83	32.4	30.3	23.85	757	3	0.0089	8.45	17.81	3.27	14.54	0.23	0.82
5	19.35	22.85	24.45	0.69	-0.81	0.88	30.6	27.9	25.35	749	6	0.0085	2.55	17.72	1.4	18.33	0.09	0.92
6	13.8	19.5	22.35	0.5	-0.71	0.81	33.75	27.6	20.1	770	5	0.0082	1.5	15.73	3.13	12.6	0.25	0.8
7	17.85	21.3	24	0.6	-0.72	0.81	31.35	29.7	24.6	772	3	0.0065	5.1	17.79	2.23	15.35	0.14	0.87
8	14.4	17.25	19.8	0.55	-0.66	0.78	31.5	28.1	17.85	789	2	0.0047	8.25	17.18	1.89	15.28	0.12	0.88
9	20.85	24.8	26.55	0.68	0.8	0.86	32.4	30.75	20.1	745	4	0.0111	4.85	17.27	1.44	15.83	0.09	0.92
10	23.55	27.15	28.7	0.7	0.8	0.88	35.55	33.75	32.4	663	5	0.017	3.35	18.85	2.85	15.98	0.18	0.85
11	18.55	25.8	28.95	0.28	-0.77	0.87	35.1	33.45	30.8	770	5	0.0138	2.85	17.69	1.51	18.18	0.09	0.91
12	10.45	8.45	25.35	0.01	-0.28	0.76	35.7	33.45	25.8	880	3	0.0024	7.85	15.73	3.79	11.94	0.32	0.76
13	8.5	22.5	27	0.01	-0.69	0.83	35.25	32.7	28.35	786	3	0.0065	4.35	18.05	7.46	10.59	0.7	0.59
14	17.35	23.4	25.95	0.57	-0.75	0.84	30.45	31.05	28.5	677	4	0.0089	2.55	22.82	5.29	17.52	0.3	0.77
15	8.9	22.05	24.75	0.03	-0.73	0.82	32.55	30.3	23.7	748	2	0.0067	6.8	20.01	5.72	14.29	0.4	0.71
16	17.25	20.85	23.25	0.58	-0.71	0.79	31.8	29.4	21.9	763	3	0.0089	7.5	20.84	2.65	18.19	0.15	0.87
17	13.95	21.75	24.9	0.44	-0.69	0.79	33.3	31.5	24.8	773	3	0.0049	6.9	18.42	1.38	17.04	0.08	0.92
18	18.75	22.2	24.45	0.85	-0.77	0.85	31.35	28.8	24.6	750	3	0.0026	4.2	20.14	3.52	16.62	0.21	0.83
19	18	22.8	24.9	0.81	-0.77	0.84	31.05	29.55	24.3	754	3	0.011	5.25	19.84	2.82	17.02	0.17	0.86
20	14.1	18.95	22.35	0.51	-0.73	0.81	28.4	27.45	23.25	786	5	0.0106	4.2	18.83	2.4	16.43	0.15	0.87
21	10.85	17.25	20.55	0.38	-0.65	0.78	28.65	28.4	21.9	794	5	0.0089	4.5	17.48	3.21	16.25	0.17	0.93
22	0.75	2.1	8.3	0.03	-0.08	0.34	29.85	27.75	21.3	827	4	0.0019	6.45	19.04	3.58	15.45	0.23	0.81
23	0.6	20.25	23.7	0.02	-0.69	0.81	32.1	29.25	25.8	757	5	0.0081	3.45	20.25	5.6	14.04	0.38	0.72
24	1.05	9.9	19.8	0.06	-0.34	0.69	31.05	28.8	22.5	877	5	0.003	6.3	21.75	2.03	18.82	0.16	0.87
25	-0.3	1.35	12.75	-0.01	-0.05	0.5	30.3	25.5	20.75	922	4	0.0051	3.75	21.32	8.44	12.88	0.66	0.8
26	1.35	4.05	16.2	0.04	-0.15	0.52	34.2	31.05	22.2	898	6	0.0011	8.85	20.15	1.74	18.42	0.06	0.91
27	0.15	1.35	8.6	0.01	-0.05	0.25	30.6	26.85	13.8	824	4	0.0007	13.05	24.25	8.67	15.58	0.56	0.64
28	0.6	1.8	4.2	0.04	-0.12	0.28	18.3	15	7.95	925	3	0.0029	7.05	28.86	4.15	22.71	0.18	0.85
29	1.05	2.4	17.7	0.04	-0.08	0.61	31.05	28.8	23.55	923	6	0.0027	5.25	20.08	3.78	16.3	0.23	0.81
30	-0.15	0.9	20.85	-0.01	-0.05	0.79	28.35	26.4	21.9	931	3	0.0018	4.5	5.12	2.09	2.43	1.11	0.47
31	-0.3	0.3	20.4	-0.01	-0.01	0.87	32.25	30.6	22.8	941	2	0.0004	7.8	3.7	2.4	1.3	1.85	0.35

Sheet1 Sheet2 Sheet3

Fig. 5 Results from genetic algorithm

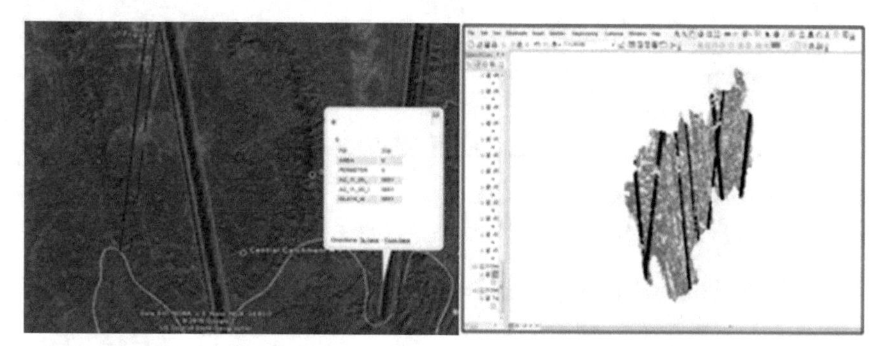

Fig. 6 **a** Footprints on Google Earth Map **b** clip forest area of Tripura with footprints

8.2 *Results from Genetic Algorithm*

The Genetic Algorithm is able to optimize all the parameters for biomass estimation. According to this algorithm the ranking of parameters were done which is the most participant important for prediction of biomass. Optimization using the genetic algorithm was done, in R package.

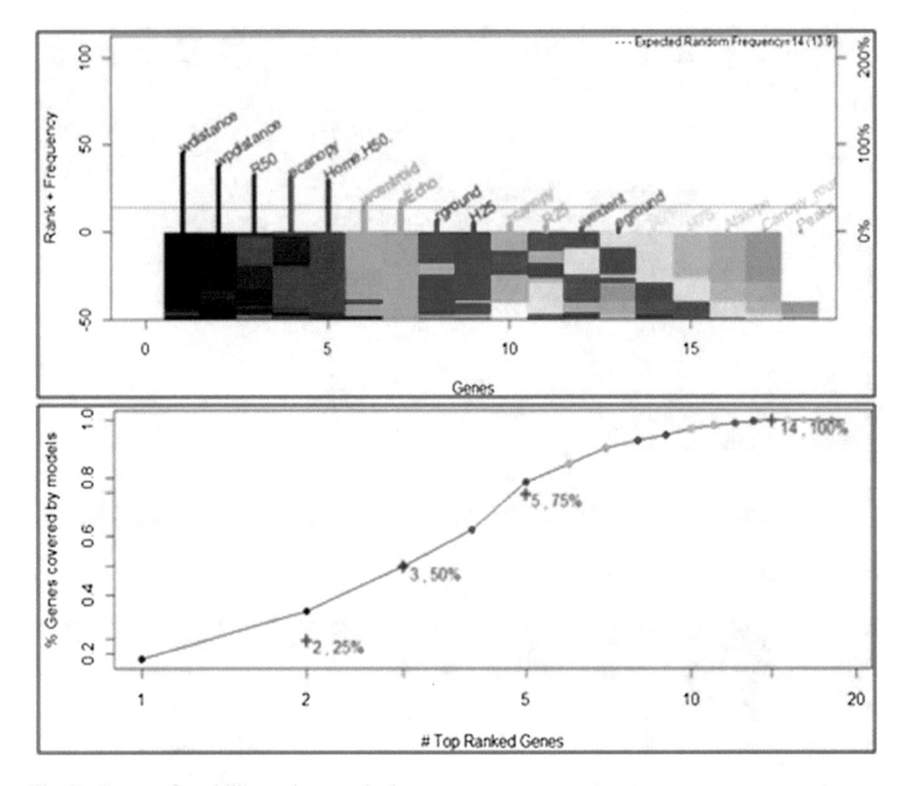

Fig. 7 Gene rank stability and top ranked genes

8.2.1 Using Single Sensor (GLAS Data)

R programming provides the many packages and libraries of genetic algorithm. For optimization it uses the GALGO package. GALGO plays out the component selection from extensive scale information. Genetic Algorithms used for Multivariate Statistical Models from Large scale Functional Genomics Data. GALGO gives the ranking (Importance) of independent 18 variables, shows by the stability of gene ranking. GALGO gives the coupling six grouping strategies. These are mainly extreme probability likehood discriminate functions (MLHD Method), k-nearest neighbours (KNN Method), nearest centroid (NEARCENT Method), classification trees (RPART Method), Support vector machines (SVM Method), neural networks (NNET package) and random forest (RANFOREST package) (Fig. 7).

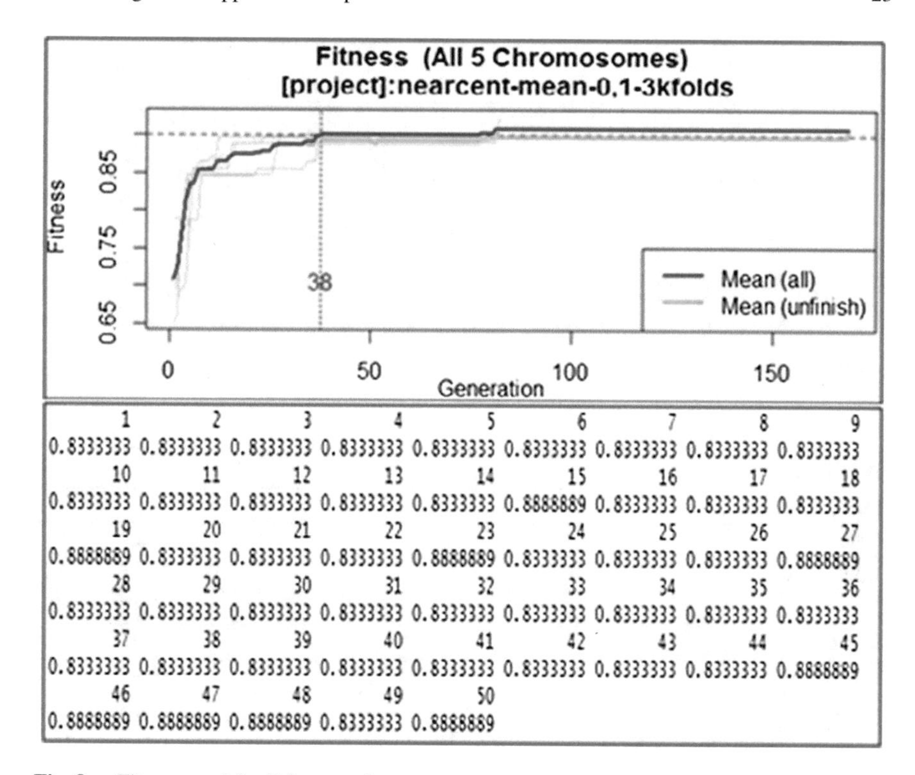

1	2	3	4	5	6	7	8	9
0.8333333	0.8333333	0.8333333	0.8333333	0.8333333	0.8333333	0.8333333	0.8333333	0.8333333
10	11	12	13	14	15	16	17	18
0.8333333	0.8333333	0.8333333	0.8333333	0.8333333	0.8888889	0.8333333	0.8333333	0.8333333
19	20	21	22	23	24	25	26	27
0.8888889	0.8333333	0.8333333	0.8333333	0.8888889	0.8333333	0.8333333	0.8333333	0.8888889
28	29	30	31	32	33	34	35	36
0.8333333	0.8333333	0.8333333	0.8333333	0.8333333	0.8333333	0.8333333	0.8333333	0.8333333
37	38	39	40	41	42	43	44	45
0.8333333	0.8333333	0.8333333	0.8333333	0.8333333	0.8333333	0.8333333	0.8333333	0.8888889
46	47	48	49	50				
0.8888889	0.8888889	0.8888889	0.8333333	0.8888889				

Fig. 8 **a** Fitness graph **b** all fitness value

Ordinarily the main seven "Black" genes are settled rapidly, while low positioned "Dim black" genes are appeared in some different colors with around 2–3 genes for every color. Horizontal axis demonstrates the genes arranged by the Y axis Vertical pivot shows the gene predicted frequency and the color coded rank of every gene in earlier developments. Therefore green and red genes are not yet stable; this is on behalf of a few chromosomes are insufficient to stabilized these genes. Different chromosomes would create more steady outcomes; be that as it may, chromosomes should as much as possible. This algorithm represents the fitness graph of these genes and the best fittest value is 0.8433333 (Fig. 8).

```
> predict.Biomass
            1        2        3        4        5       6        7        8       9       10
[1,] 290.2535 311.1008 316.7832 306.9477 306.1416 314.057 281.8089 323.3805 349.487 344.4978
           11       12       13       14      15       16       17       18       19       20
[1,] 330.8313 337.3163 327.7319 316.2947 306.846 322.8017 309.7111 313.0926 299.5095 290.1624
           21       22       23       24       25       26      27      28       29       30
[1,] 287.4358 313.4958 298.3799 276.0654 307.4596 268.7345 186.878 296.857 285.0199 310.1686
           31       32      33       34       35      36       37       38       39       40
[1,] 317.6026 312.5607 299.668 298.8756 312.0725 302.088 308.8429 302.8994 286.5875 300.9687
>
```

Fig. 9 Predicted biomass

Table 1 Table of parameter

Parameters	R^2
LiDAR	
wdistance	0.57
wpdistance	0.57
R50	0.19
Ecanopy	0.13
Home.H50	0.07
wcentroid	0.03
wdistance + wpdistance + R50 + ecanopy + Home.H50 + wcentroid	0.63

Genetic Algorithm used one of the Libraries (glmulti) for prediction of Biomass. glmulti is a generic function that acts of a wrapper to functions that actually fit statistical models to a dataset (such as lm, glm or gls). glmulti works out of the box with several types of function (such as lm, glm). Using this GA package can be predicted the biomass (Fig. 9).

Using the linear model generates the coefficient of best variable and created the formula of biomass prediction with help of important variables "wdistance, wpdistance, R50, ecanopy, Home.H50.wcentroid, eEcho". Predicted value generates the R^2 0.63 and RMSE 18.94 t/ha (Table 1).

8.2.2 Predicted Versus Observed Values of Biomass from Genetic Algorithm

See Fig. 10

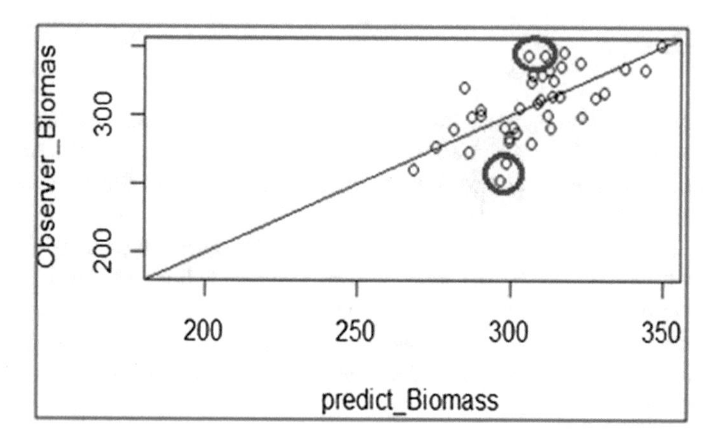

Fig. 10 Using ICESat derived parameters

8.3 *Biomass Equation Generated*

$$\text{Biomass} = 5.0417 * \text{wdistance} + 2.8720 * \text{wpdistance} + 16.1316 * \text{R50} - 0.6016 * \text{Ecanopy} - 4.1925 * \text{Home.H50} - 5.1938 * \text{wcentroid} + 16.6187$$

The biomass estimation along with an important method of data collection from ICESat data footprints on data sat LiDAR and field data (Table 2).

This table shows that the single sensor parameters predict Biomass equation using the GA and, it has r^2 value of LiDAR data is 0.63. It is much accurate value.

Table 2 Table of R^2 and RMSE

Dataset	R^2	RMSE
LiDAR data	0.63	18.94

9 Conclusions

In this research study, biomass inference at ICESat/GLAS footprint Single sensor approach was studied. The estimation was done by making use of Genetic Algorithm. The important variables were extracted using genetic Algorithm and linear equations were generated using Multiple Linear Regression algorithm. This research study was successful in achieving its objectives. The estimations were done using Genetic Algorithm for predicting biomass on concept of optimization. The most important variables were chosen using genetic algorithm and linear equation was developed.

The above research has been done for small area of Doon Valley. The same methodology can be applied for a large area of Tripura and generate a biomass equation to predict biomass.

Acknowledgements I would like to express my profounder gratitude towards Dr. Subrata Nandy (Scientist/Engr. SD, FED, IIRS) who guided me throughout this paper. He supervised me for this thought and inspired me to complete it.

References

1. Anaya J, Chuvieco E, Palaciosorueta A (2009) Aboveground biomass assessment in Colombia: a remote sensing approach. For Ecol Manag 257:1237–1246
2. IPCC (2000) Good practice guidance and uncertainty management in national greenhouse gas inventories
3. Viergever KM, Woodhouse IH, Marino A, Brolley M, Stuart N (2008) SAR interferometry for estimating above-ground biomass of Savanna Woodlands in Belize. In: Geoscience and remote sensing symposium, 2008, IGARSS 2008, IEEE International, pp V-290–V-293
4. Parmar AK (2012) A neuro-genetic approach for rapid assessment of above ground biomass: an improved tool for monitoring the impact of forest degradation, geo-information science and earth observation of the University of Twente
5. Duong HV (2010) Processing and application of ICES at large footprint full waveform laser range data. Doctoral thesis, Delft University of Technology, Netherlands
6. Lu D (2006) The potential and challenge of remote sensing-based biomass estimation. Int J Remote Sens 27(7):1297–1328
7. Eisfelder C, Kuenzer C, Dech S (2011) Derivation of biomass information for semi-arid areas using remote-sensing data. Int J Remote Sens 33:2937–2984
8. Hall FG, Bergen K, Blair JB et al (2011) "Characterizing 3D vegetation structure from space": mission requirements. Remote Sens Environ 115(11):2753–2775
9. Dhanda (2013) Optimising parameters obtained from multiple sensors for biomass estimation at icesat footprint level using different regression algorithms
10. Roeva O (2005) Genetic algorithms for a parameter estimation of a fermentation process model: a comparison
11. Jensen JR (2007) Prentice Hall series in geographic information science. Pearson Prentice Hall, 592 pp
12. Zwally HJ, Schutz B, Abdalati W, Abshirre J, Bentley C, Brenner A, Bufton J, Dezio J, Hancock D, Harding D, Herring T, Minster B, Quinn K, Palm S, Spinhirne J, Thomas R (2002) ICESat's laser measurements of polar ice, atmosphere, ocean and land. J Geodyn 34(3–4):405–445

13. GALGO (2006) An R package for multivariate variable selection using genetic algorithms. Victor Trevino and Francesco Falciani School of Biosciences, University of Birmingham, Edgbaston, UK Bioinformatics
14. Upadhyay D (2014) An ethno-botanical study of plants found in Timli Forest Range, District Dehradun, Uttarakhand, India. Cloud Publ Int J Adv Herb Sci Technol 1(1):13–19, Article ID Med-157

E-alive: An Integrated Platform Based on Machine Learning Techniques to Aware and Educate Common People with the Current Statistics of Maternal and Child Health Care

Garima Malik, Sonakshi Vij, Devendra Tayal and Amita Jain

Abstract Data science finds a variety of applications in day-to-day life. Its practical uses can cater to needs of improving the lifestyle and health standards of the individuals of the society. This paper proposes an intelligent tool, called E-alive, build to encourage people towards the sensitivity of maternal and child health care. This tool serves as an integrated platform for rural and urban people, government officials and policy makers to actively participate and analyse the current statistics of various parameters such as infant mortality rate, life expectancy ratios for females and males individually, female and male sterilization rates and maternal mortality rates for the next subsequent years. This can help them in taking quality decisions in order to improve upon the predicted values. Further this tool can assist in classifying the educational status of an individual, community and state on the basis of total fertility rates. This implies that the awareness factor among the people of respective community or state and total fertility rate can be predicted by this tool for the future years. The current analysis analyses the two government schemes in detail: Swadhar Scheme and Janani Suraksha Yojana. Other analysis factors include Life Expectancy Ratio, Education Details, Maternal Mortality Rate and the Contraceptive Methods used by people in major cities.

Keywords Data science · Data mining · Health care

G. Malik (✉) · S. Vij · D. Tayal
Indira Gandhi Delhi Technical University for Women, New Delhi, India
e-mail: annu.2353@gmail.com

A. Jain
Ambedkar Institute of Advanced Communication Technologies and Research, New Delhi, India

D. K. Mishra et al. (eds.), *Data Science and Big Data Analytics*,
Lecture Notes on Data Engineering and Communications Technologies 16,
https://doi.org/10.1007/978-981-10-7641-1_3

1 Introduction

Health and well-being are the two most fundamental need of a human being. The art of Health can be described as the absence of social and mental illness from the life standards. Child and maternal care is also a key issue of the community in healthcare system. Every individual needs basic facilities to derive their lives; health management plays a significant role in it. During the recent past years Indian health system has grown significantly well but still we are in a process of becoming an independent nation which can offer world class health facilities to the citizen of India. In order to tackle the present day problems in health stream, this paper suggests an online recommendation system based on the important and key parameters of child and maternal health care. This system will try to curb out the rooted problems of rural as well as urban population regarding the basic healthcare facilities and tries to educate the people about the importance of good health to achieve wellness in life. The consistency of the system can be explained as it uses various data mining algorithms to classify and predict the values of parameters such as maternal mortality rate and total fertility rate. It clearly shows the extensive application of Data science and machine learning algorithms which are robust in nature.

The biggest challenge in building this system is the data variations in Maternal Health and child health care both are very sensitive issue and the education level in India is not uniform, as we know the literacy rate in rural areas is low than in urban areas so as to build a recommendation system the cases used should be consistent. After facing all the hurdles the system is developed with the supervised algorithms to provide a flexible and adaptable environment to the user.

Previously in the literature, an intelligent tutor system is built for the maternal and child health care specifically the IOT device which senses the body conditions of a pregnant women and provides the solution. This system uses sensors to recognize the variations in body internal body movements and the system is a great success in rural areas where the facilities of hospitals are not easily available. For educating people or to aware people about the real healthcare scenarios we have developed this system so that people can assume the risk factors or they can choose their preventive measures according to the respective situation.

The proposed system fulfils all the basic requirements in addition to that we have provided the knowledge centre portal which is a kind of blog in which recent or updated Government schemes related to direct benefit transfers for people is shown so that people will know their fundamental rights and get benefits from government schemes. Along with those articles on basic hygiene and health care is provided in local language so that rural people can understand well.

This system can become a powerful source which can inspire people and help our country to sustainable development goal as early as possible. E-alive is the new paradigm which can be intelligent tool to encourage people towards the sensitivity of maternal and child health care. This tool also provide an integrated platform where rural and urban people, government officials and policy makers actively participate and analyse the current statistics of various parameters such as infant mortality rate, life expectancy ratios for females and males individually, female and male sterilization rates and maternal mortality rates for the next subsequent years so that they can take quality decision making in order to improve the predicted values.

E-alive also shows state-wise statistics of health parameters so that government officials and people can see that which state is progressing fast and which state needs financial assistance for the health care facilities.

Further this tool can classify the educational status of an individual, community and state on the basis of total fertility rates which will implies the awareness factor among the people of respective community or state and Total fertility rate can easily be predicted by this tool for the future years.

2 Motivation

To develop a country we must develop the human resources associated with it. India is one of the leading developing countries in the world and progressive towards the developed countries but the fundamental barrier is the health and awareness statistics of India. Maternal and child health has been always one important issue for India. During the past years, India has shown significant development in maternal and child health care such as reduction of infant mortality rate to 35 by 2017 or reduction of maternal mortality rate to 127 by 2017. India has progressed a lot on all the fronts of health-related issues but still it lacks behind all the developed countries to achieve the best results. Health, nutrition and family are the key prospects of a nation and "real development cannot ultimately take place in one corner of India while the other is neglected"—by Jawaharlal Nehru.

The problems underlying with the health sector need to be addressed such as structural problems like inadequate funding in health sector, lack of implementation of social policies and programmes, weak regulatory authorities and poor management of resources. To improve the current statistics technology can play a vital role, with the advancement of data science and machine learning we can integrate all the resources and information which can be provided to common masses which will aware all the citizens.

Rural health care is also one of the major issues for health ministry as 70% of the rural population is living in lower level of healthcare facilities. Rural hospitals or health centres are unable to provide the better medication because of less number of doctors and lab serviceman. 66% of rural people do not have access to the medicines and tribal people situation is more critical [1].

Recent union budget shows the advancement in the funds allocated to health sector but we need to educate the people about the maternal and child health care norms so that people can act accordingly or can take preventive measures in the respective situation. Rashtiya Swasthiya BimaYojana (RSBY) and Central Government Health Schemes (CGHS) are the examples of Government initiatives in the direction of development of health sector [2].

India has 900 million users who are using Internet facilities and increasing day by day and technology can bring smile to those faces which are struggling for the basic facilities. Our system also trying to address these above mentioned issues, providing a recommendation system which can solve the millions of people by educating them about the recent trends in health industry.

This system will provide an overview of healthcare facilities as user can predict the education status of a community, predict the values of life expectancy ratios of male and female for the subsequent years and system will also provide suggestions to improve the ratios so that we can reach the sustainable development goals. Maternal mortality rates prediction can also be done to see the real picture for urban and rural areas. A blog or forum is also provided for all users which will be updated with all the recent schemes organized by government for the common people. For the government authorities analysis of various policies is provided so that they can improve the implementation standards.

This paper aims to provide an integrated system which will help common people as well as government authorities to look upon all the current situations in India regarding the basic health issues so that best methods or ways of implementation can be suggested to improve the conditions.

3 Preliminaries

This section describes the preliminary concepts that would be dealt with in the paper.

3.1 Maternal and Child Health Care Parameters

These are described as follows:

(a) **Total Fertility Rate (TFR)**—It can be defined as the average number of children would be born per woman or per 1000 of women in the lifetime. It is calculated as the sum of age specific fertility rates.

$$\mathbf{TFR} = \sum \text{AGSR(sum of all the age specific fertility rate of a woman in life time)}$$

(b) **Maternal Mortality Rate (MMR)**—MMR can be described as the total number of maternal deaths under the 42 days of pregnancy termination because of complications in pregnancy, childbirth or any other reason in a particular geographical area (country, state, district, etc.) divided by the total number of live birth associated with the same geographical area for a particular time period such as a calendar year and multiplied by 100,000.

$$\textbf{MMR} = \frac{\text{Total no of maternal deaths} \times 100{,}000}{\text{Total no of live birth}}$$

(c) **Infant Mortality Rate (IMR)**—This term refers to deaths of an infant child within one year of live birth. It is measured by the total number of death of children in one year per 1000 live births.
(d) **Life Expectancy Ratio**—It can be described as the statistical measure which is calculated as the average time of an individual is expected to live and it depends on the year of birth, current age of an individual and other demographic factor.
(e) **Steralization Rate of Male**—It is the measure of any number of medical techniques that promotes the growth of infertility in a living being. It basically describes the inability of an individual to reproduce.
(f) **Contraceptive Methods**—The methods which are used to avoid and terminate the pregnancy are called contraceptive methods. It can be categorized in four major parts—

i. Modern methods such as barriers like condoms, cervical caps, vaults, etc.
ii. Natural Methods such as periodic abstinence
iii. Oral pills—these are oral contraceptives used to avoid pregnancy.
iv. IUD—Intra uterine devices such as copper-T and copper-7

(g) **HIV Prevalence**—This term refers to the number of people which are tested as HIV positive in a given geographical area over a period of time.

3.2 Government Schemes

These are described as follows:

A. **Swadhar Scheme**—This scheme was started by the Government of India under the union ministry of Women and child development in 2002. It aims to provide food, shelter, medication, clothing and care for the women or marginalized girls which are facing difficult situations and it also includes widows, women prisoners released from jail, women victims of violence and women survivors of natural calamities [3, 4]. The funds under these schemes are given to Government affiliated NGO's which are working for the women survivors or Government has also created the swadhar homes through which women can get direct benefits of this scheme. In this paper we have analysed the progress of this scheme over the years in different states such as which states are top ranked in providing swadhar homes to women or how many number of women received grants from this scheme, these questions will be answered by the tool E-alive.
B. **Janani Suraksha Yojana (JSY)**—This initiative promotes safe motherhood and child health care. It was started under the Ministry of health and family welfare and National Rural Health Mission (NRHM). The main objective of this scheme is to reduce the maternal and infant mortality rates by promoting institutional child delivery among the under privileged women. This scheme was initiated in

12 April 2015 by our honourable Prime minster and it is implemented in all the working states of India [5]. It basically tracks the pregnant women and inspire them to get register under thus scheme and give them a special card so that they can receive their benefits from this centrally sponsored scheme. In this paper we have analysed the progress of this scheme for different state and number of benefices under this scheme.

3.3 Techniques of Data Mining

These are described as follows:

(a) **Decision Trees**—It is a machine learning tool used to classify textual and numerical data based on the attributes of data. It is also called the conditional inference tree as the internal nodes of the tree represents a test or an attribute and each branch denotes the results of the test performed on the attribute of data. The leaf nodes of the tree represent uniform class or class distribution. In this paper we have used ID3 algorithm to classify the data into two classes according to the given Total fertility rate provided by the user. This algorithm basically works on entropy concept as the decision nodes will be selected on basis of highest entropy attribute means whichever attribute has higher number of variations in the values that attribute will have the greater entropy than other attributes. This algorithm is selected for the classification as the number of classes which needs to be classified are small and it builds faster and accurate decision tree with minimum number of nodes. Data which is classified is in the form of textual information so we have chosen decision tree algorithm but if the data is in the numerical format we can also select fuzzy decision tree technique for greater precision in numerical values.

(b) **K Means Clustering**—Clustering techniques is also one of the most popular technique for the classification. It basically works on different distance measures such as Euclidean distance, Manhattan Distance and Chebychev distance [6]. It primarily calculates the two farthest data points and then according to the respective distance measures it assigns the cluster to new data points but the catch is analyst should know the value of K initially for the smooth working of algorithm [7]. In this paper we have used this technique to classify the data of number of people who have adopted institutional deliveries or home deliveries and advantages of both the techniques in context with the maternal health care is discussed.

(c) **Linear Regression**—Simple linear Regression model is a statistical method which is used to analyse the linear relationship between two variables. This machine learning technique is extensively used in predictions of numerical data for future use [8]. In Linear regression theirs is one dependent and one independent variable and on the basis of values of R^2 which can be described as the linear associations between two variables we select the best curve fitting line such as log, exponential or polynomial, etc. E-Alive also used linear regression in the prediction of maternal mortality rate and life expectancy ratio of an individual, community, district or state on the basis of one calendar year.

(d) **Multiple Regression**—Multiple regression is one of the types of regression modelling in which there are more than one independent variable such as if the equation has two or more than two independent variables which can be used to identify the value of dependent variable.
The general equation of regression model is—

$$Y_i = F(X, B)$$

Here

Y_i is the dependent variable
X is independent variable
B represents unknown parameters

$$Y_i = B_0 + B_1X_i + B_2X_i^2 + \cdots + B_iX_i^n + €_i$$

This represents general equation of multiple regression, where Y_i is dependent on all the X_i variables present in the equation. E-alive also supports multiple regression as it has predicted the female and male sterility rates using this method.

(e) **Support Vector Machine (SVM)**—In machine learning SVM's are categorized as supervised learning algorithm that is used to analyse the data used for regression and classification. It trains the model based on probabilistic binary linear classifier. For unlabelled data it uses support vector clustering for unsupervised approach [9]. In this paper we have classified and analysed the different contraceptive methods used by women to avoid and terminate the pregnancy. The trained model predicted the best method for an individual to adopt and function accordingly. It also analysed the importance of each method in the various states of India for example—A female user can see the growth and various statistics related to contraceptive methods and decide which method is best for the respective situation.

(f) **Time Series Analysis**—A time series can be seen as the series of data points arranged with the time order. Time series data usually plotted in line graphs and in order to derive meaningful statistics from time series data this technique is used with linear regression [10]. Initially we plot the time series data with the help of line graph and then curve fitting method is used to fit the mathematical equation in correspondence with the line graph. In this paper Time series Analysis technique is used majorly in predicting various health parameters.

4 System Implementation

This section will explain the details of E-alive system and graphical user interface of the system. E-alive is developed in r studio as a platform and R as a programming language.

Figure 1 shows the basic architecture of the system and the left side menu in the system shows the bunch of tabs which signifies the flow of the system. Current status in the tab is initial or first GUI available for the user which works on the live data feeding as in Fig. 1 some graphs are shown which shows the latest growth of different states of India. Bar graphs present in the graph shows the top five states in India which has highest number of maternal mortality rate and total life expectancy ratio. These graphs can be interpreted as in which states government needs to do more work in order to improve the situations in health sector. This tab also shows pie chart which represents the total number of people which are HIV prevalent or HIV positive in a state in percentage. This tab primarily shows the latest trends of various health parameters in different states and a brief description of system working in orange box. Figure 1 also shows the list of tabs which are as follows:

- **Life Expectancy Ratio**—To predict the life expectancy ratio for male and female individually for a given year as user will provide us the year for which life expectancy ratio needs to be calculated and the E-alive system will show the ratio on the screen. To calculate the life expectancy ratio this system applies linear regression and for the curve fitting it uses polynomial equation and user will provide the year which will be consider as one of the parameter of the equation [11] (Table 1).

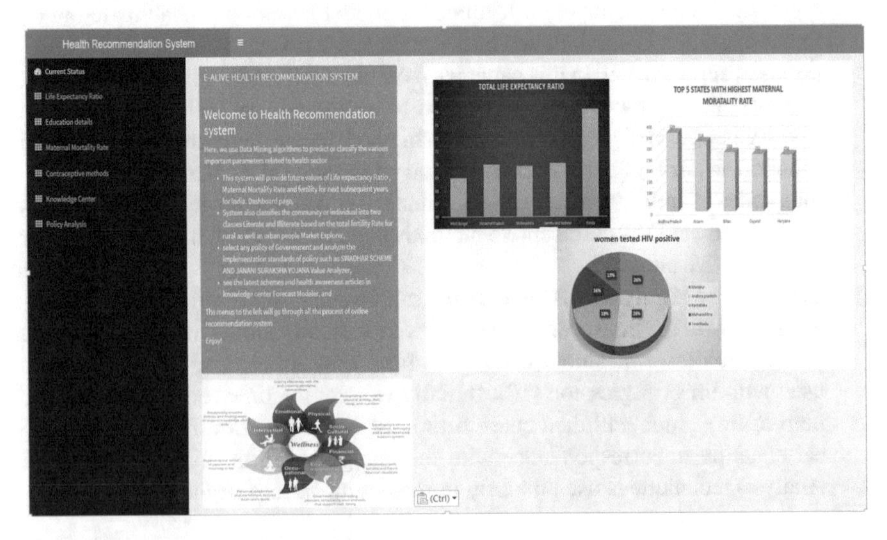

Fig. 1 Showing the initial web page of the system with latest and live trends for the states of India

Table 1 Showing the various mathematical equations and corresponding linear association value

Mathematical equation type	Value of R^2
Linear equation	0.9868
Exponential equation	0.9869
Polynomial equation	0.9995

Table 2 Showing the various mathematical equations and corresponding linear association value

Mathematical equation type	Value of R^2
Exponential equation	0.9876
Polynomial equation	0.9861
Linear equation	0.9812

- **Education Details**—This tab is interestingly used to classify an individual, community, district and state as literate and illiterate on the basis of total fertility rate such as user will provide the value of fertility rate and system will predict as whether the given value of fertility rate lies in which class. To classify the education details on the basis of fertility rates it applies conditional inference trees or decision trees []. It basically calculates the entropy of each attribute in data set and on the basis of that total fertility rate attribute possesses highest entropy and become the deciding node for this conditional inference trees.
- **Maternal Mortality Rate**—To predict the maternal mortality rate for the future years and for particular state. It also shows the future trends of all the states of India with the combination of bar graph and line graph. It uses data mining technique as multiple linear regression and uses exponential equation to calculate the maternal mortality rate. This system uses exponential equation to predict the values as described in the table the value of R^2 for exponential equation is close to 1 [12] (Table 2).
- **Contraceptive Methods**—This section of E-alive completely belongs to maternal health as it will provide female user a platform to judge various contraceptive method and user can choose the best method and see the rate of use of contraceptive methods for all the major states of India. It uses support vector machine, initially all the three methods named as IUD, Oral pills and modern methods are labelled as $-1, +1$ and 0 respectively then support vector clustering is performed to classify the training set [13]. After building support vector classifier it used for testing of remaining data set.
- **Knowledge Centre**—This is a unique tab for all the users as it will provide all the articles related to health and hygiene and user can also post their experience while using the system and report any issue. We will also provide the latest government benefit schemes to the users to increase the awareness so that more number of poor people can get direct benefit from government.
- **Policy Analysis**—This section is developed for government officials especially for the policy makers so that they can analyse the current implementation standards of

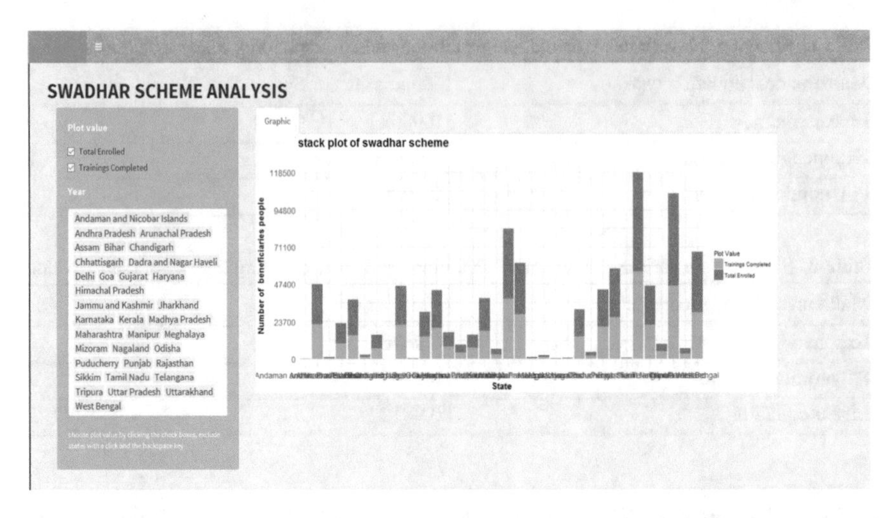

Fig. 2 Showing stack plot for Swadhar scheme as the comparison of funds released by government and real expenditure

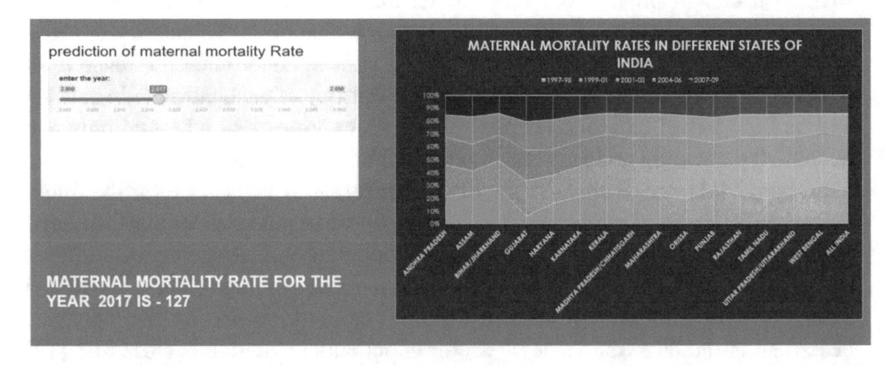

Fig. 3 Showing the maternal mortality tab in which user can predict MMR for a year from slider input and can see the variation in 3D planar graph

the government schemes. We have only analysed two schemes Swadhar scheme and Janani Suraksha Yojana. Analysis is done by planar and bar graphs for various states in India and separate graphs has been plotted for the best performing and worst performing states.

E-alive has no signup and login feature so it is freely available to user and the knowledge centre tab of the system is very unique in terms of that it will provide all the content in local Indian language so that user can experience or gain knowledge from the system (Figs. 2, 3, 4, 5, 6 and 7).

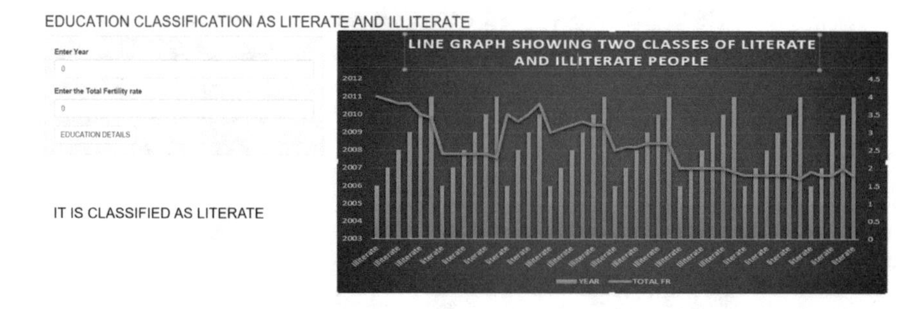

Fig. 4 Shows the education details tab in which user can predict the class of education according to the total fertility rate and year

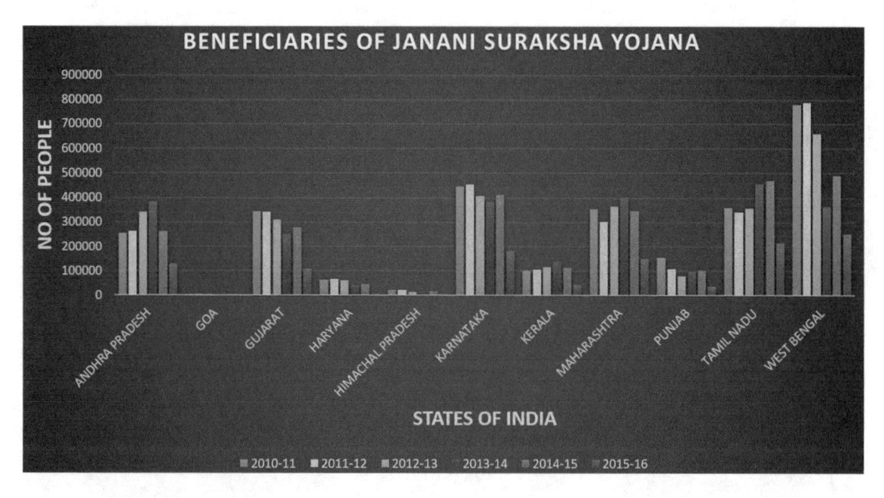

Fig. 5 Depicting the beneficiaries of Janani Suraksha Yojana (high performing states)

5 Results and Discussions

This section will describe the details of accuracy of our system in predictions and classification of various parameters. We have tested this system for 250 users and developed these following results. In these results we are showing the sample datasets derived from the user experiences (Tables 3, 4 and 5).

6 Conclusion and Future Scope

E-alive is an efficient system to predict and classify important health parameters using machine learning techniques. It can be seen as the intervention of technology with health and it has a power to educate and aware common people. Results discussed

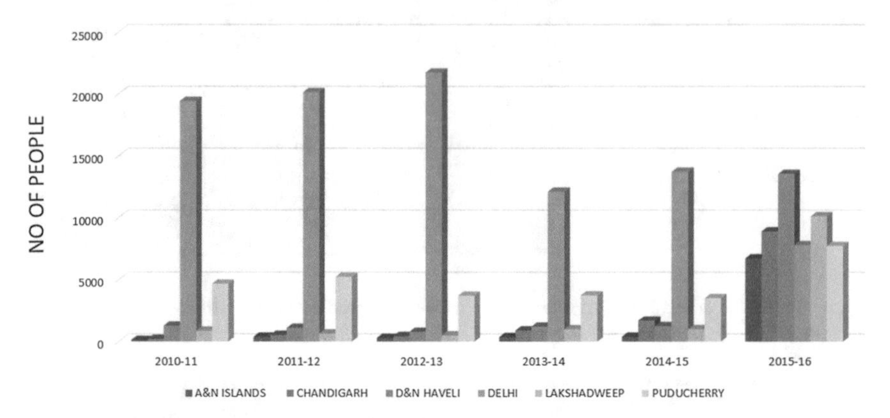

Fig. 6 Depicting the beneficiaries of Janani Suraksha Yojana (union territories)

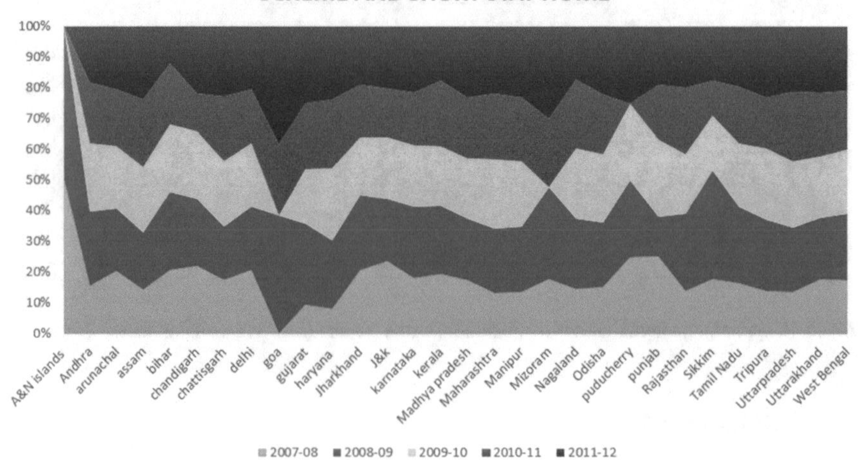

Fig. 7 Showing the planar graph for Swadhar scheme analysis for different states of India

in above section clearly signify the precision rate of predictions done in the system. In this system an upper bound to all the important parameters is provided such as life expectancy ratio and maternal mortality rate. For classifications of textual data unsupervised and supervised learning techniques were used to provide user a live experience. To expand this system, recommendation can be provided through artificial agents or bots. Fuzzy logics can also be incorporated in the system to improve the accuracy of the system. We can also consider more number of government policies for analysis such as policies implemented by state governments and union territories.

Table 3 Classification of education status comparison chart

S. no.	Total fertility rate	Observed class of education	Predicted class of education	People
1	3.9	Illiterate	Illiterate	Rural
2	3.7	Illiterate	Illiterate	Rural
3	3.5	Illiterate	Illiterate	Rural
4	2.3	Literate	Literate	Rural
5	1.6	Literate	Literate	Rural
6	2.1	Illiterate	Illiterate	Urban
7	2.3	Illiterate	Illiterate	Urban
8	1.5	Literate	Literate	Urban
9	1.2	Literate	Literate	Urban
10	1.9	Literate	Literate	Urban

Table 4 Life expectancy ratio prediction data comparison chart

S. no.	Year interval	Life expectancy ratio observed	Life expectancy ratio predicted	Sex
1	2012	67.1	67.3	Male
2	2017	69.2	68.8	Male
3	2024	69.0	69.8	Male
4	2030	73.78	71.09	Male
5	2035	74.68	71.31	Male
6	2012	68.9	69.3	Female
7	2017	71.40	71.30	Female
8	2024	73.5	72.3	Female
9	2030	76.07	73.99	Female
10	2035	78.09	74.51	Female

Table 5 Maternal mortality rate prediction data comparison chart

S. no.	Year interval	Maternal mortality rate observed	Maternal mortality rate predicted
1	2017	135	132.50
2	2019	110	113.45
3	2025	85	82.7
4	2029	71	70.81
5	2037	46	44.18
6	2040	38	37.76
7	2043	36	32.26
8	2026	85	82.7
9	2015	178	178
10	2056	17	17.2

References

1. http://www.in.undp.org/content/india/en/home/post-2015/mdgoverview.html
2. http://pib.nic.in/newsite/PrintRelease.aspx?relid=123683
3. https://www.nhp.gov.in/national-health-insurance-schemes_pg
4. http://planningcommission.gov.in/sectors/health.php?sectors=hea
5. http://nrhm.gov.in/nrhm-components/rmnch-a/maternal-health/janani-suraksha-yojana/background.html
6. Singh A, Yadav A, Rana A (2013) K-means with three different distance metrics. Int J Comput Appl 67(10)
7. Kanungo T, Mount DM, Netanyahu NS, Piatko CD, Silverman R, Wu AY (2002) An efficient k-means clustering algorithm: analysis and implementation. IEEE Trans Pattern Anal Mach Intell 24(7):881–892
8. Asai HTSUK (1982) Linear regression analysis with fuzzy model. IEEE Trans Syst Man Cybern 12:903–907
9. Hua S, Sun Z (2001) A novel method of protein secondary structure prediction with high segment overlap measure: support vector machine approach. J Mol Biol 308(2):397–407
10. Tong H (2011) Nonlinear time series analysis. In: International encyclopedia of statistical science. Springer, Berlin, Heidelberg, pp 955–958
11. Aiken LS, West SG, Pitts SC (2003) Multiple linear regression. In: Handbook of psychology
12. Daniel WW, Wayne WD (1995) Biostatistics: a foundation for analysis in the health sciences
13. Ben-Hur A, Horn D, Siegelmann HT, Vapnik V (2001) Support vector clustering. J Mach Learn Res 2(Dec):125–137
14. Issac R, Sahasranamam S (2014) Tele-consulting through rural health centres for tribal community—a case study from Wayanad. In: 2014 IEEE global humanitarian technology conference (GHTC). IEEE, pp 57–61
15. Al Nuaimi N, AlShamsi A, Mohamed N, Al-Jaroodi J (2015) e-Health cloud implementation issues and efforts. In: 2015 international conference on industrial engineering and operations management (IEOM). IEEE, pp 1–10
16. Kirchner K, Tölle KH, Krieter J (2004) Decision tree technique applied to pig farming datasets. Livestock Prod Sci 90(2):191–200
17. Olaru C, Wehenkel L (2003) A complete fuzzy decision tree technique. Fuzzy Sets Syst 138(2):221–254
18. Mendel JM (2001) Uncertain rule-based fuzzy logic systems: introduction and new directions. Prentice Hall PTR, Upper Saddle River, pp 131–184
19. Malik G, Tayal DK, Singh A, Vij S (in press) Applying data analytics to agricultural sector for upcoming smart cities. In: Proceedings of the 11th INDIACom, 4th international conference on computing for sustainable global development, INDIACom 2017. IEEE
20. Agrawal R, Gehrke J, Gunopulos D, Raghavan P (1998) Automatic subspace clustering of high dimensional data for data mining applications, vol 27, no 2. ACM, pp 94–105
21. Chen MS, Han J, Yu PS (1996) Data mining: an overview from a database perspective. IEEE Trans Knowl Data Eng 8(6):866–883
22. Kamber M, Winstone L, Gong W, Cheng S, Han J (1997) Generalization and decision tree induction: efficient classification in data mining. In: Seventh international workshop on research issues in data engineering, 1997, Proceedings. IEEE, pp 111–120

An Effective TCP's Congestion Control Approach for Cross-Layer Design in MANET

Pratik Gite, Saurav Singh Dhakad and Aditya Dubey

Abstract The fast expansion in correspondence innovation has offered ascend to strong research enthusiasm on Wireless Networks. As of late numerous scientists have concentrated on planning steering plans which would effectively work on the continuous condition of remote systems, e.g., MANETs. Every one of the hubs in the mobile ad hoc network (MANET) agreeably keep up the system network. Abusing the conditions and collaborations between layers has been appeared to expand execution in specific situations of remote systems administration. Albeit layered structures have served well for wired systems, they are not reasonable for remote systems. The standard TCP clog control system is not ready to deal with the unique properties of a common remote channel. TCP clog control works exceptionally well on the Internet. In any case, versatile specially appointed systems display some properties that extraordinarily influence the plan of fitting conventions and convention stacks when all is said in done, and of clog control component specifically. This paper proposes an outline approach, going amiss from the customary system plan, toward improving the cross-layer association among various layers, to be specific physical, MAC and system. The Cross-Layer configuration approach for blockage control utilizing TCP-Friendly Rate Control (TFRC) mechanism is to maintain congestion control between the nodes what's more, to locate a successful course between the source and the goal. This cross-layer configuration approach was tried by reproduction (NS2 test system) and its execution over AODV was observed to be better.

P. Gite (✉) · S. S. Dhakad
IES IPS Academy, Indore, India
e-mail: Pratikgite135@gmail.com

S. S. Dhakad
e-mail: sourabhdhakad@ipsacademy.org

A. Dubey
SGSITS, Indore, India
e-mail: Adi.jerry99@gmail.com

D. K. Mishra et al. (eds.), *Data Science and Big Data Analytics*,
Lecture Notes on Data Engineering and Communications Technologies 16,
https://doi.org/10.1007/978-981-10-7641-1_4

Keywords TCP · Congestion control · TFRC · AODV · Cross-layer · Routing

1 Introduction

Wireless networks [1] are naturally restricted by battery power and transfer speed imperatives. They are described by versatility, irregular changes in network, vacillations in channel and impedance because of neighboring hubs and so on. Because of these components, bundle loss of a remote system is considerably more than that of a wired system, in which parcel misfortune happens essentially because of clog in the system. Clog in a system is portrayed by deferral and parcel misfortune in the system. Transport Control Protocol (TCP) is utilized as a solid transport layer convention in the customary best exertion (wired) system and manages blockage adequately. The blockage control system of different forms of TCP gives better throughput in a wired system, where the parcel misfortune is primarily because of clog at different hubs and switches. Nonetheless, this system may not be reasonable in a remote system, where bundle misfortune because of time-fluctuating nature of channel and impedance of different hubs are extensively high [2–4].

1.1 Mobile Ad Hoc Network

In these days, wireless systems are ending up exceptionally well known innovation on the planet. Henceforth it is imperative to comprehend the engineering for this sort of systems before sending it in any application. In mobile ad hoc networks, correspondence among portable hubs happens through remote medium. Mobile Ad hoc network is an accumulation of cell phones that are self-arranging and speak with each other without utilizing brought together framework. In such a domain, hubs likewise go about as switch and forward parcels to the following bounce to convey it to the last goal through numerous jumps. Keeping in mind the end goal to be versatile, untethered availability utilizing remote interfaces should be available with each hub in the system. Normally versatile hubs will rely upon battery control for their operations. It is attractive to limit the power utilization in these hubs. Further, this issue is imperative as once the battery of the hub is depleted, it can't transmit and additionally get any information. It bites the dust bringing about effect on arrange availability since in impromptu systems, even middle of the road hubs are essential to keep up network [5]. When one of the middle of the road hubs passes on, the entire connection must be framed again this prompts substantial measure of postponement, misuse of rare hub assets like battery control consequently hampering the throughput of the entire framework. Further, portability introduces the difficulties as constantly factor topology and in this manner requiring a complex and vitality effective steering instruments [6] (Fig. 1).

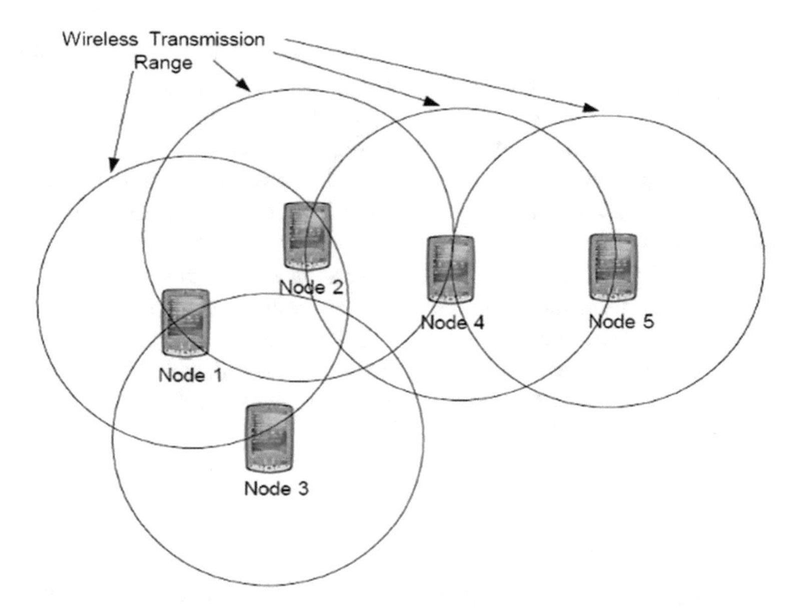

Fig. 1 Review of the versatile specially appointed system

1.2 Congestion Control

At the point when the total interest for assets (e.g., transmission capacity) surpasses the limit of the connection at that point comes about is clog. Clog is portrayed by deferral and loss of parcels in conveyance. In TCP, clog is said to have happened when the sender gets three copy affirmations or when a timeout (parcel misfortune) happens, bringing about wastage of assets. Clog Control and Congestion Avoidance are two known arrangements which address the above issue. In clog control [7], framework controls the system parameters in the wake of acknowledging blockage (receptive); while, in blockage shirking, framework controls the system parameters before clog (proactive).

The fundamental issue in MANET is clog control. It is completely related to control the approaching movement into a media transmission arrange. The arrangement is to send a decreased parcel sending rate to defeat the impact of blockage successfully. All in all, the Transmission Control Protocol chiefly consolidates the clog control and trustworthiness system without unequivocal input about the blockage position. The clog control standards incorporate bundle protection, Additive Increase and Multiplicative Decrease (AIMD) in sending rate, stable system. Alternate methods for clog control are end framework stream control, arrange blockage control, organize based clog evasion, and asset distribution [8, 9].

The packet misfortune can be dense by including clog control over a portability and disappointment versatile steering convention at the system layer. The blockage causes following troubles:

More delay: A few Congestion control instrument sets aside much time for distinguishing blockage. It prompts substantial deferral. At times the utilization of new courses in some basic circumstances is prudent. The primary issue is the postpone moving for course looking in on-request steering convention.

High Overhead: Congestion control instrument requires exertion for handling and correspondence in new courses for finding it. It likewise requires exertion in multi-path steering for keeping up the multi-ways, however there is another convention.

More packet losses: If the congestion is recognized, the parcels might be lost. Blockage control arrangement is connected either by diminishing the sending rate at the sender, or dropping bundles at the middle of the road hubs or by the two strategies to diminish the activity stack. On the off chance that the high bundle misfortune rate happens, little throughput might be happened in the system.

1.3 Cross-Layer Design

Cross-layer input implies cooperation of a layer with some other layers in the convention stack. A layer may collaborate with layers above or beneath it. The layered convention stack configuration is exceptionally unbending and firm and each layer just takes think about the layer straightforwardly above it or the one specifically underneath it. This outcome in non-coordinated effort which exists between various layers, apparently in light of the fact that nobody around then observed any requirement for such a non-shared plan known as the cross-layer outline.

To completely streamline remote broadband systems, both the difficulties from the physical medium and the QoS-requests from the applications must be considered [10]. Rate, power and coding at the physical layer can be adjusted to meet the necessities of the applications given the present channel and system conditions. Information must be shared between (all) layers to get the most elevated conceivable adaptively.

In [10–12], a general meaning of a cross-layer configuration is given as any infringement or change of the layered reference design. The aim of CLD, basically expressed, is to misuse data from various layers to mutually advance execution of those layers. The breaking of various leveled layers or the infringement of reference design incorporates converging of layers, making of new interfaces, or giving extra interdependencies between any two layers as appeared in Fig. 2.

This paper has been organized in four sections. Section 1 introduces some related work in this domain. Section 2 discusses the details of TCP congestion control proposed work. Section 3 illustrates the implementation view of work. Section 4 includes simulation results and analysis. Section 5 summarizes the work, i.e., conclusion.

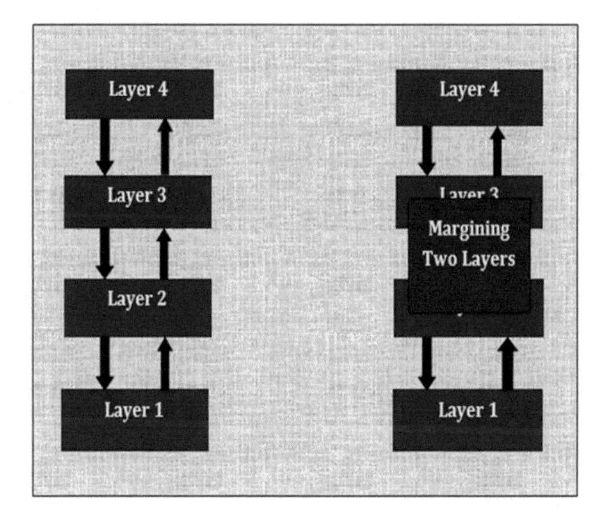

Fig. 2 Reference architecture and its violation (cross layer)

2 Literature Survey

TCP's congestion control mechanism has evolved as a reliable mechanism over the years. Supporting wireless applications over wireless communication has been one of the main fields of attention in the networking and cross-layer design is effective for TCP congestion control from the last decade there are numerous research works going on in this area. Some the prior works are described here:

Khan et al. [13] talked about the significance of cross-layer security systems and steering conventions for multi-jump remote systems by basic examination. The choice of ideal way to rout and the discovery of multilayer security assaults cannot be accomplished with the customary methodologies. They recommend that cross-layer configuration is the main answer for adapt to these sorts of difficulties in multi-bounce remote systems.

Jain and Usturge [14] surveyed that flag quality based estimations used to enhance such parcel misfortunes and no compelling reason to retransmit bundles. In this way, the hub-based and connect-based flag quality can be measured. On the off chance that a connection flopped because of versatility, at that point flag quality estimation gives transitory higher transmission energy to keep interface alive. At the point when a course is probably going to flop because of frail flag quality of a hub, it will discover substitute way. Therefore, maintains a strategic distance from congestion.

Sreedhar and Damodaram [15] proposed Medium Access Level Multicast Routing for Congestion Avoidance in Multicast Mobile Ad Hoc Routing Protocol to keep away from the clog in systems. This convention infers a calculation that transmits the information in multicast way at assemble level not at all like other multicast conventions, concentrating of information transmission in a succession to each focused on hub. Being free, the proposed work was with gathering of either tree or work.

Jain and Tapkir [16] gives their profound understanding to TCP blockage control instrument and proposed course disappointment identification systems. It is recommended that the majority of the components are going for expanding the life time of the systems and enhancing the execution of the parameters like parcel conveyance proportion, end-to-end postpone and so forth.

Senthil Kumaran and Sankaranarayanan [17] displayed the Congestion Free Routing in Ad hoc arranges (CFR), in view of progressively evaluated instrument to screen organize blockage by ascertaining the normal line length at the hub level. While utilizing the normal line length, the hubs clog status partitioned into the three zones like safe zone, congested zone and congested zone. The plan uses the non-congested neighbors and starts course disclosure system to find a blockage free course amongst source and goal. This way turns into a center way amongst source and goal. To keep up the clog free status, the hubs which are helping information parcel transmission occasionally figure their blockage status at the hub level.

3 Proposed Work

The proposed work is intended to reduce the TCP's congestion using TFRC in cross-layer design framework by which the mobile ad hoc network that improvement of TCP congestion. Therefore, the given chapter provides the detail overview of the methodology that justifies the developed solution.

3.1 Problem Identification

TCP/IP tradition was planned for wired frameworks which offers end-to-end strong correspondence among centers and ensures asked for movement of packs. It moreover gives stream control and screw up control frameworks. As it is so far a productive tradition in wired frameworks, incidents are transcendently a result of blockage. However, in the event that there ought to be an event of uniquely named frameworks allocate are a direct result of obstruct in the framework and in light of ceaseless association dissatisfactions so when we change TCP to offhand frameworks it puzzles the package adversities in light of association frustration as package setbacks due to blockage and in the event of a timeout, backing-off its retransmission timeout (RTO). This realizes trivial reducing of transmission rate because of which throughput of the whole framework ruins. In this manner, course changes on account of host movability can adversely influence TCP execution. On-ask for Routing Protocols, for instance, AODV and DSR are used for this execution examination of TCP. These sorts of coordinating traditions make courses exactly when requested by a source center point. Exactly when a center point needs to set up a course to an objective, it begins a course divulgence process inside the framework. One the course has been set up, it is kept up until either objective breezes up doubtlessly far off or the course is never again needed.

3.2 Methodology

A system of broad standards or principles by which particular techniques or strategies might be inferred to translate or take care of various issues inside the extent of specific orders. Unlike an algorithm, a methodology is not a formula but a set of practices.

To solve the issue of the TCP congestion in network, we enlist basic three modules of our methodology that construct entire system that proving efficiency and effectiveness of this work.

1. Basic Assumption
2. Parameter Selection
3. Threshold Estimation

Basic Assumption
For securing network, the proposed algorithm is developing using different constraint. So that we basically we need to assume some constraints to progress further. In this, firstly we create a normal network where with different number of network, e.g., 20, 40, 60, 80, and 100.

Initialize the Network, with N nodes where N = 1, 2, 3 … in ideal condition S initiates a RREQ message with the following components:

- The IP addresses of S and D
- The current sequence number of S and the last known sequence number of D
- A broadcast ID from S. This broadcast ID is incremented each time S sends a RREQ message.

The pair of the source S forms a unique identifier for the RREQ.

For route discovery, we process a route request RREQ to all other node except the node which is generating request. Therefore, source node waits for the route reply, i.e., RREP which is coming from that node to its match broadcast ID and IP address.

Parameter Selection
In this we select criteria as a parameter for calculating Threshold value for initiating their condition to checks network traffic. For this we select two parameter, e.g., Buffer size and packet loss rate for each node.

Threshold Calculation
Buffer Length: the buffer or queue length of a node demonstrates the amount of workload which is processed by any node. In this context the amount of buffer length is free to use indicate the node if free and can able to serve better the packet rate. This here for the length of buffer the letter B is used.

TFRC for Loss Event Rate: TFRC enables an application to transmit at a relentless rate that is regularly inside a factor of two from the TCP rate in the comparable conditions. TFRC does not have the transmission rate after a solitary bundle misfortune, but on the other hand is ease back to build the rate without blockage. The TFRC beneficiary is in charge of detailing the misfortune occasion rate p and normal get rate X_rcv to the sender. Computing the misfortune occasion rate as opposed to just assuming the parcel misfortune rate is an essential piece of TFRC. The default technique that TFRC utilizes for ascertaining the misfortune occasion rate is known as the Average Loss Interval. In this situation we need to alter TFRC for figuring normal edge rate to legitimize there is sans blockage way. Here we illustrate, proposed calculation for blockage taking care of in cross-layer configuration utilizing AODV convention (Table 1).

4 Performance Analysis

4.1 Implementation

Simulation is "the process of designing a model of a real system and conducting experiments with this model for the purpose of understanding the behavior of the system and/or evaluating various strategies for the operation of the system". With the dynamic idea of PC systems, we in this manner really manage a dynamic model of a genuine dynamic framework.

System Simulator (NS2.34) is a discrete occasion test system coordinated at systems administration look into. NS gives critical help to recreation of steering, TCP, and multicast conventions on wired and remote systems [18]. A test system model of a true framework is basically an advancement of this present reality framework self. In our reproduction, 100 versatile hubs move in a 1000-m $\times$ 1000-m^2 district for various changing recreation time. All hubs have a similar transmission scope of 500 m. The reenactment settings and parameters are condensed in the Table 2.

The simulation process is demonstrated the required constraints for developing the TCP's congestion control where we simulate the process using TCP-friendly Rate control mechanism. Therefore, we set up the simulation table.

4.2 Result Comparisons

We examine routing overhead, end-to-end delay and packet delivery ratio, etc., for evaluating congestion control for TCP protocol using TFRC mechanism. This result termed as comparison of normal AODV protocol and proposed work is implemented using combination of TFRC and AODV protocol representing when there is network is congested then the network performance is affecting via different parameter.

Table 1 TFRC-based congestion avoidance

Input: Number of Nodes, Data Packet; Output: Improved Cross Layer Congestion Rate;
Process: 1: Initialize the Network, with N nodes where $N = 1, 2, 3, \ldots$, in ideal condition. 2: Initialize Route Discovery by Source Node N_s 3: N_s sends RREQ Packets to Destination N_d 4: Wait Until all Route Replies not received 5: For each routing in routing table: 6: Consider for each node buffer length is B_i where $i = 1, 2, \ldots, n$ 7: Find average buffer threshold for all node $Threshold_{buffer} = \frac{1}{N}\sum_{i=1}^{N} B_i$ 8: Find packet event Loss rate for TCP-Friendly Assign packet sending and receiving time i.e. t_s and t_r Integrating Loss Rate for a single packet $P_{LR} = \int_{t_r}^{t_s} P_i d(t)$ Calculate average threshold loss rate $Threshold_{LR} = \frac{1}{N}\sum_{i=1}^{N} P_{LR_i}$ 9: ModifiedRecvTFRCresponse (feedback_packet) 10: Check congestion availability and assign traffic free path $if\ (Threshold_{LR} < P_{LR})$ Stop the packet sending event $else\ (Threshold_{buffer} > 50\%)$ Select alternate path for packet transmission *endif* 11: adapt_transmission (destination_address, loss_event_rate) 12: *end process*

End-to-End delay

End-to-end delay on organize alludes to the time taken, for a video bundle to be transmitted over a system from source to goal gadget, this deferral is figured utilizing the beneath given recipe.

Figure 3 shows the comparative average End-to-End Delay of the traditional AODV and the proposed cross-layer-based AODV using TCP-friendly control Rate. In this Fig. 3 the X pivot contains the quantity of hubs in organize and the Y hub demonstrates the execution of system delay regarding milliseconds. As indicated

Table 2 Simulation scenarios

Parameters	Values
Antenna model	Omni antenna
Routing protocol	AODV
Radio-propagation	Two ray ground
Dimension	1000 × 1000
Number of nodes	100
Traffic model	CBR
Channel type	Wireless channel
Mobility model	Random waypoint

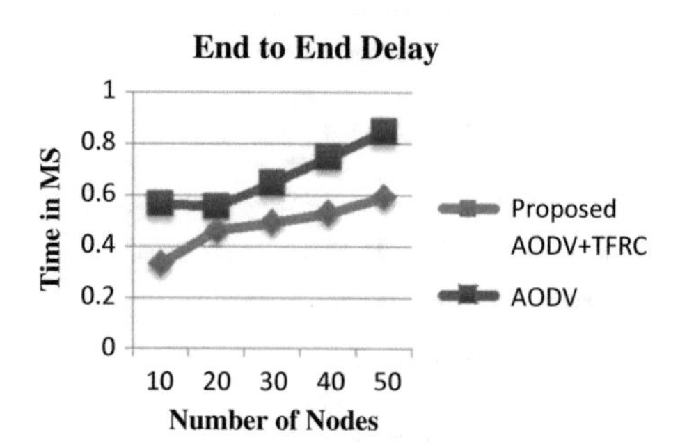

Fig. 3 End-to-end delay

by the acquired outcomes the proposed strategy is creates less end-to-end postpone when contrasted with conventional system under various hubs. We need to specify that the above change in normal end-to-end defer is critical for congestion control.

Packet Delivery Ratio

The performance parameter Packet delivery ratio now and then named as the PDR proportion gives data about the execution of any steering conventions by the effectively conveyed parcels to the goal, where PDR can be evaluated utilizing the recipe given

$$\text{Packet Delivery Ratio} = \frac{\text{Total Delivered Video packets}}{\text{Total sent Video packets}}$$

The comparative packet delivery proportion of the systems is given utilizing Fig. 4, in this outline the X pivot demonstrates the quantity of hubs in the system and the Y hub demonstrates the measure of information parcels effectively conveyed as far as the rate. The orange line of the diagram speaks to the execution of the customary AODV situation and the blue line demonstrates the execution of the proposed system. As per the got comes about the proposed system conveys more bundles when contrasted with the customary strategy notwithstanding when the system takes out the transmission blunder and creates exact information correspondence. This is critical parameter for dealing with movement on the course among various hubs.

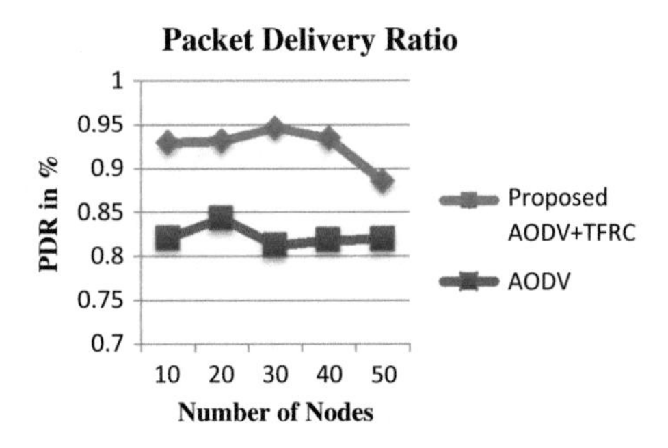

Fig. 4 Packet delivery ratios

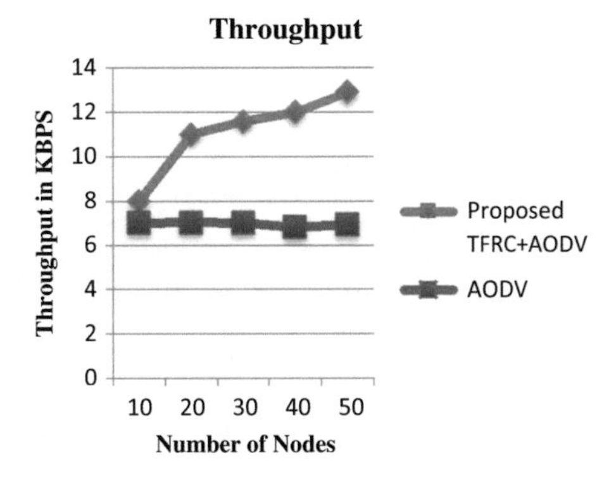

Fig. 5 Network throughput

Throughput

System throughput is the normal rate of effective message conveyance over a correspondence channel. This information might be conveyed over a physical or intelligent connection, or go through a specific system hub. The throughput is typically measured in bits every second (piece/s or bps), and now and then in information parcels every second or information bundles per availability.

The relative throughput of the system is exhibited utilizing Fig. 5. In this chart the X hub demonstrates the quantity of hubs in arrange and the Y pivot demonstrates the throughput rate of the system regarding KBPS. The blue line of the diagram demonstrates the execution of the proposed TFRC based approach and the orange line demonstrates the execution of the AODV steering situation. The cross-layer structure is exhibiting the diminishing blockage and expanding throughput information rate. We need to say that the change in throughput is imperative where clog is happened from the end client perspective in light of the fact that a little increment in throughput can prompt huge upgrade freeway.

Routing Overhead

Amid the correspondence situations it is required to trade the parcels for various following and checking reason. In this way the extra control messages in organize is named as the steering overhead of the system. The near steering overhead of both

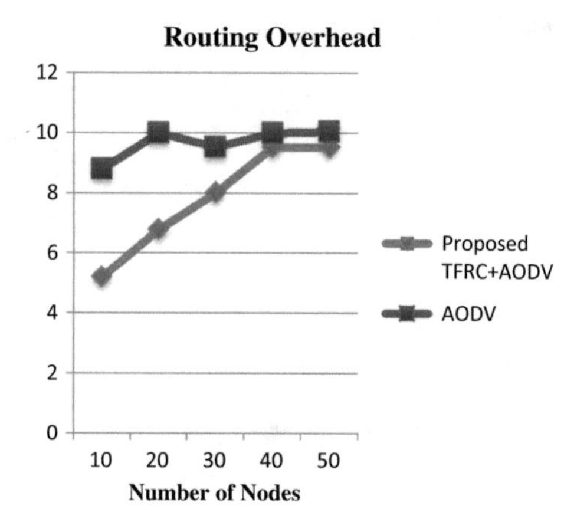

Fig. 6 Routing overhead

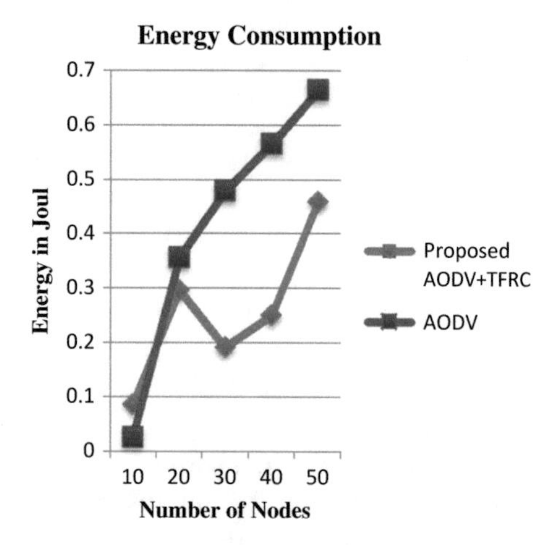

Fig. 7 Energy consumption

the cross-layer configuration is incorporate with the AODV and TFRC strategy for information bundle transmission. In this chart the X hub demonstrates the measure of system hubs exist amid the experimentation and the Y pivot demonstrates the steering overhead of the system. In this graph for showing the execution of the proposed method the blue line is utilized and for customary AODV the orange line is utilized. As indicated by the got execution of the systems the proposed procedure creates less steering overhead when contrasted with the conventional strategy. Along these lines the proposed strategy offers higher data transfer capacity conveyance when contrasted with the AODV steering convention (Fig. 6).

Energy Consumption

The measure of vitality devoured amid the system occasions is named as the vitality utilization or the vitality drop of the system. In systems administration for every individual occasion a lot of vitality is devoured. The given Fig. 7 demonstrates the vitality utilization.

Figure 7 indicates Energy Consumption of the system in both the reproduction situations. The orange line of the chart demonstrates the measure of vitality overwhelmed by the conventional AODV steering convention under furthermore the blue line demonstrates the measure of vitality devoured amid the proposed calculation based system. In the customary AODV the system vitality is as often as possible expended when contrasted with the proposed approach. In this manner the proposed system is compelling and ready to recuperate the system while diverse hubs are mimicked.

5 Conclusion

Enhancing TCP execution in the IEEE 802.11 specially appointed system is a confused cross-layer issue. Versatile hubs are moving haphazardly with no brought together organization. Because of essence of clog, the bundle misfortune happens pointlessly. So the hub needs to retransmit the parcel. It prompts greatest vitality utilization, top of the line to end deferral, and low bundle conveyance proportion. Hence, our plans to enhance this parameter productive system correspondence disposal or lessening clog overhead. In the main period of the plan, we need to characterize number of parameter to figure edge esteem. Here, data is shared between the distinctive layers of the convention stack. In clog recognition stage, we check the edge an incentive for discovering blockage rate utilizing TFRC component to decided activity rate which guarantees the identification of clog specifically interface. In blockage control stage, the course is found without clog. In this manner, proposed work is successful and adoptable to discover high activity/clog connect and enhances organize execution.

References

1. Schiller JH (2003) Mobile communications. Pearson Education
2. Ahmed AS (2015) Cross-layer design approach for power control in mobile ad hoc networks. Egypt Inform J 16(1):1–7 (2015)
3. Mourya AK, Singhal N (2014) Managing congestion control in mobile ad-hoc network using mobile agents. Int J Comput Eng Appl 4(3) (2014)
4. Nishimura K, Takahashi K (2007) A multi-agent routing protocol with congestion control for MANET. In: Proceedings of 21st European conference on modeling and simulation, pp 26–39
5. Seungjin P, Seong-Moo Y (2013) An efficient reliable one-hop broadcast in mobile ad hoc networks. Ad Hoc Netw 11(1):19–28
6. Bakshi A, Sharma AK, Mishra A (2013) Significance of mobile ad-hoc networks (MANETS). Int J Innov Technol Explor Eng (IJITEE) 2(4)
7. Ryu S, Rump C, Qiao C (2003) Advances in Internet congestion control. IEEE Commun Surv Tutor 3:28–39
8. Rajeswari S, Venkataramani Y (2012) Congestion control and QOS improvement for AEERG protocol in MANET. Int J Ad Hoc Netw Syst (IJANS) 2(1):13–21

9. Senthil Kumaran T, Sankaranarayanan V (2012) Congestion free routing in ad-hoc networks. J Comput Sci 8(6):971–977
10. Lin X, Shroff NB, Srikant R (2006) A tutorial on cross-layer optimization in wireless networks. IEEE J Sel Areas Commun 24(8):1452–1463
11. Srivastava V, Motani M (2005) Cross-layer design: a survey and the road ahead. IEEE Commun Mag 43(12):112–119
12. Shakkottai S, Rappaport TS, Karlsson PC (2003) Cross-layer design for wireless networks. IEEE Commun Mag 41(10):74–80
13. Khan S, Loo KK, Ud Din Z (2009) Cross layer design for routing and security in multi-hop wireless networks. J Inf Assur Secur 4:170–173
14. Jain S, Usturge SI (2011) Signal strength based congestion control in MANET. Adv Phys Theories Appl 1:26–36 (2011)
15. Sreedhar GS, Damodaram A (2012) MALMR: medium access level multicast routing for congestion avoidance in multicast mobile ad hoc routing protocol. Glob J Comput Sci Technol Netw Web Secur 12(13):22–30
16. Jain SA, Tapkir SK (2012) A review of improvement in TCP congestion control using route failure detection in MANET. Netw Complex Syst 2(2):9–13
17. Senthil Kumaran T, Sankaranarayanan V (2012) Congestion free routing in ad hoc networks. J Comput Sci 8(6):971–977 (2012)
18. The Network Simulator. http://www.isi.edu/nsnam/ns/. Accessed 12 July 2017

Baron-Cohen Model Based Personality Classification Using Ensemble Learning

Ashima Sood and Rekha Bhatia

Abstract These days intelligence is not the only factor for judging the personality of a human being. Rather, emotional quotient (EQ) as well as systemizing quotient (SQ) has a major role to play for classifying a human's personality in many areas. Using these quotients, we can foresee one's personality in the society. The broad classification of personality on the basis of EQ and SQ score has been well researched using machine learning techniques with varying degree of accuracy. In the present research work, the performance of different classification techniques have been enhanced using ensemble learning in which various combination of classification models with different permutations has been done. Ensemble learning technique increases the accuracy in most of the cases in the present work.

Keywords Baron-Cohen · Emotional quotient · Systemizing quotient Classification · Machine learning · Ensemble modeling

1 Introduction

Emotional intelligence (EI) is defined as "the ability to carry out accurate reasoning about emotions and the ability to use emotions and emotional knowledge to enhance thought" [1]. Systemizing quotient defines the ability to explore the system which is governed by some rules and regulations; this allows the person to think practically [2]. In the modern era of smart thinking success of the person depends 80% on EQ and 20% on IQ. People with high-score emotional quotient makes them more

A. Sood · R. Bhatia (✉)
Punjabi University Regional Centre, Mohali, India
e-mail: r.bhatia71@gmail.com

A. Sood
e-mail: soodashima91@gmail.com

D. K. Mishra et al. (eds.), *Data Science and Big Data Analytics*,
Lecture Notes on Data Engineering and Communications Technologies 16,
https://doi.org/10.1007/978-981-10-7641-1_5

emotionally strong than people with less score. On the same side high scorers in systemizing quotient are more practical or rational people those who explore the systems than the ones who are less scorers. In this research work personality has been predicted from the EQ and SQ level of the people. This level was achieved from the scores that people got in the respective EQ and SQ field from the test conducted by the Baron-Cohen model [3–5]. This test contains total 120 questions which are divided equally between emotional and systemizing quotient, i.e., 60 questions in each. These questions have been asked from the different people on the scale of 4 which is as strongly disagree, disagree, agree, and strongly agree. People rate their answers and their answers are saved for the future record from which data set has been generated [6]. E-S theory was proposed by Simon Baron-Cohen in which he gave the scores on EQ and SQ of the people and generating different types of brain. He also concluded that females are high on emotional quotient and males are high on systemizing quotient [3].

The research work is done using machine learning that is a novel approach in predicting the personality of the individual and classifying on the basis of EQ and SQ scores. These two parameters are themselves so wide and can be consider as necessary and sufficient conditions to classify the personality of the individual. The technique of predicting involves getting the secondary data and cleaning the data in excel which makes it viable to use [7]. Further hypothesis is developed taking some specific and standard range of the two parameters. After this data is subjected for training and testing of data in R programming. The main focus of the paper is predicting the category of the personality that a particular individual holds on the basis of scores. In testing and training of data various models are generated which give the accuracy of prediction. After this to increase the accuracy of the models ensemble modeling is subjected which involves combinations of different models and give better result when combined instead of working as a single unit or a model.

The in order organization of the paper which sets the clear and systematic work flow starts with the Sect. 2 related work which gives the literature review of the work done in the paper. Next Sect. 3 covers the implementation part which explains how research has been done and with which technique. It is further sub-divided into small sections namely data refining, hypothesis building, machine learning, k-fold validation, ensemble modeling. Following Sect. 4 covers the conclusion of the paper and last Sect. 5 giving the future scope of the current research work done.

2 Related Work

The thorough literature review on emotional quotient culminated that a lot of work has been done on the applications of the EQ and SQ. Baron-Cohen has evolved from his E-S theories about the autism disorders found in children and stated that the children with autism disorder are very low in EQ [2]. These children do not understand the emotions, feelings, or thoughts of other people and are poor in social interaction as they fail to put themselves in shoes of someone else's. There are many other areas where research emotional quotient has been done like in workforce, employment, medical, health disorders, creativity, and innovation [2, 8, 9]. In IT sector, IT professionals are trained with all the skill and technical aspects but they lack in emotional intelligence and it very important for them to be emotionally strong and with rational approach which makes them outstand from other organizations. The emotional quotient has played a major role in effective leadership which requires emotionally high and systemizing high person [10, 11]. A leader has to convey his message to his team which should touch their heart and connect with them and for that it is very important to understand the emotions and feelings of the teammates and involve them in the process. Emotional Intelligence includes capacities, for example, self-inspiration, diligence notwithstanding disappointment, state of mind administration and the capacity to understand, think and trust [12]. These variables are currently considered to have more prominent effect on individual and gathering execution than customary measures of insight. It is the capacity to sense, comprehend, and successfully apply the force and keenness of feelings as a wellspring of human vitality, data and impact. Enthusiastic Intelligence rises not from the thoughts of complex brains, yet from the workings of the human heart. In the previous study it was found that more is the emotional quotient more is the level of creativity in the individual [13]. Emotional intelligence is also very important for the managerial level. Managers should have good EQ to make decisions which are unbiased and are in favor of all [9]. This theory has been used to figure out the brain types in the individuals which results in empathy and sympathy category.

3 Implementation

The implementation part begins from the gathering of the raw data information. This information can be of primary or secondary type. In the primary-type analysts gather the information themselves through distinctive means, thus making their own particular dataset. While in the secondary type, the execution is done on officially gathered/looked after dataset. The exploration which we have done is on the secondary dataset. Considering the way that raw data comprises of noisy part, i.e., the information which is insignificant from the machine learning perspective. In this way, this information must be sifted through by cleaning the information in exceed expectations. Once the information has been cleaned, information handling including hypothesis is done to make the document suitable for machine learning. After this, bagging is done to make the K-Fold approval workable for checking the consistency of the models. When all the four models are tested separately, combinations of all the models are made to improve the accuracy through ensemble learning. At last the permutation which give the highest accuracy is recommended for the personality classification. This can be clearly abridged in Fig. 1.

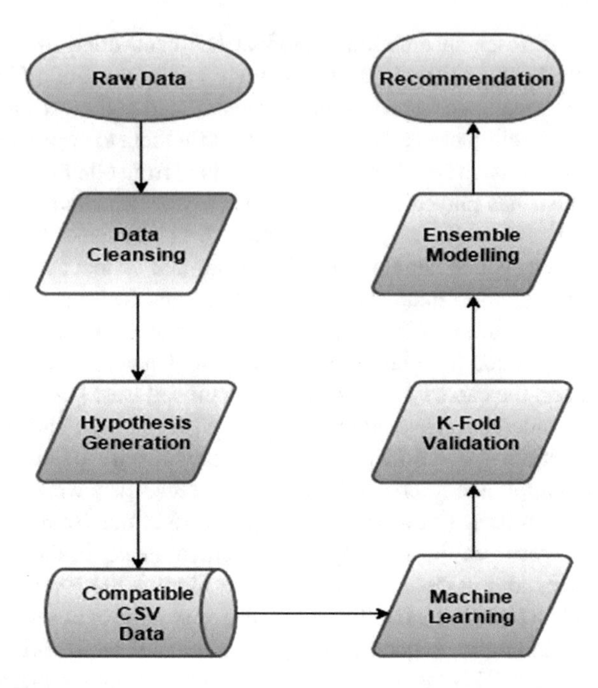

Fig. 1 Methodology

3.1 Data Set Details

The processed data that has been used to deploy in machine learning for testing and training is in the form of secondary data. Simon Baron-Cohen test is the empathizing-systemizing test in which there are total 120 questions, equally divided between EQ and SQ, i.e., 60 each [6]. People are supposed to answer these questions on the scale of 1–4. On the basis of these answers, data cleansing has been done for the acceptance of data for supervised machine learning after the hypothesis generation which is covered in the following sub-section. EQ and SQ scores has played the major role in determining the class of the person for the personality prediction.

3.2 Hypothesis

Once the questionnaire has been answered, the result is generated in the form of EQ and SQ scores. These scores are further used for hypothesis generation. Baron-Cohen has set the range of scores on the scale of 0–80 points. So, mid-value 40 has been used to divide the categories of the people and on the basis of less than and greater than EQ and SQ values classification of personality has been done [14]. Table 1 represents the classification of different categories of personality on the basis of EQ

Table 1 Hypothesis

S. No.	Range	Category of people	Category total # 13,256
1	$EQ \leq 40$ & $SQ \leq 40$	Amateurish	6529
2	$EQ \leq 40$ & $SQ > 40$	Practical	1992
3	$EQ > 40$ & $SQ \leq 40$	Emotional	3933
4	$EQ > 40$ & $SQ > 40$	Influencers	802

Table 2 Accuracy at seed value 822,938

S. No.	Model name	Method, Package	Accuracy (%)
1	Decision tree	rpart, rpart	61.2019
2	Random forest	rf, randomForest	71.6621
3	Support vector machine	svm, ksvm	75.3079
4	Linear model	lm, glm	73.1959

and SQ scores. Also, the count of people falling into each category out of total 13,256 participants has been given in the last column of the table.

3.3 Machine Learning

Machine learning is an effective approach in prediction through different techniques including clustering, classification, and regression. In our research work, we have used the supervised machine learning algorithms for the personality classification. We have used four classification algorithms namely Decision Tree model, Random Forest model, Support Vector Machine, and Linear Model [15, 16]. Machine learning is a technique where the user feeds the data to the machine in a supervised or unsupervised way [17]. In our case too, we fed the comma separated value (.csv) file to the different models. Training-Testing is the basic approach to supervised learning and we have used 70–30 standard for dividing the dataset into the training and testing of the machine. Seed value plays a key role in training-testing. When the training dataset is selected from the original .csv file, it is selected in such a way that every value after cycle of seed value iteration is included. In Table 2, different machine learning models with their method and package names are showing the accuracies with respect to seed value 822,938.

In the above table, column 2 represents the four different models that have been tested in machine learning algorithms.

Each one of the four models have their respective roles to play. Decision Tree model works like a segmented branch like structure which splits the data originating from root node which can be a record, file, or attribute. Random Forest model is itself a collection of various decision-making trees which are used in various classification problems. Support Vector Machine model classifies the data by constructing

Table 3 10-Fold check on the accuracies

Seed value	Model 1: decision tree	Model 2: random forest	Model 3: support vector machine	Model 4: linear model
42	56.7513	69.9522	73.3216	72.0392
834,845	58.6372	69.4242	73.5228	71.3352
822,938	61.2019	71.6621	75.3079	73.1959
1111	58.8383	70.9832	73.4725	72.0644
2222	59.2406	70.7819	73.9753	71.9135
3333	59.8441	70.2539	73.9250	71.7626
5555	58.5617	69.9019	73.6484	71.8380
7777	58.5365	69.7259	73.9502	71.9386
7711	58.5868	70.5556	74.0005	71.6620
8811	58.1594	69.8013	74.0507	71.6117

N-dimensional plane that divides the data into two halves. There are two types of SVM model—Linear SVM classifier and kernel-based SVM classifier. The working of Linear model is very much similar to mathematical linear equation in which relation of dependent and independent variables is explained. Column 3 gives the respective method and package names of the models used in R environment. Finally, the last column is showing that the SVM model is giving the best accuracy individually which can be improved to reach above 80% by ensemble learning technique. Moreover, the consistency of the model is tested through the bagging technique explained in the next sub-section.

3.4 K-Fold Validation (K = 10)

After the completion of machine learning phase, consistency is checked in every possible way via bagging technique, which means train-test-repeat. This is a method to ensure that the methods are consistent and if they are not then they will not be included in the ensemble permutations. Ensemble learning will be explained and results of it would be shown in next sub-section. In the following Table 3, 10 different random seed values have been used to check the consistency of different models as the result of their accuracies.

To show the results of the 10-fold validation done in Table 3 in more clear and representative way, accuracies can be shown in the graphical form which is shown in Fig. 2.

From the above figure, it is clear that all the models' performance is consistent. Hence, all of them will be included in the ensemble modeling.

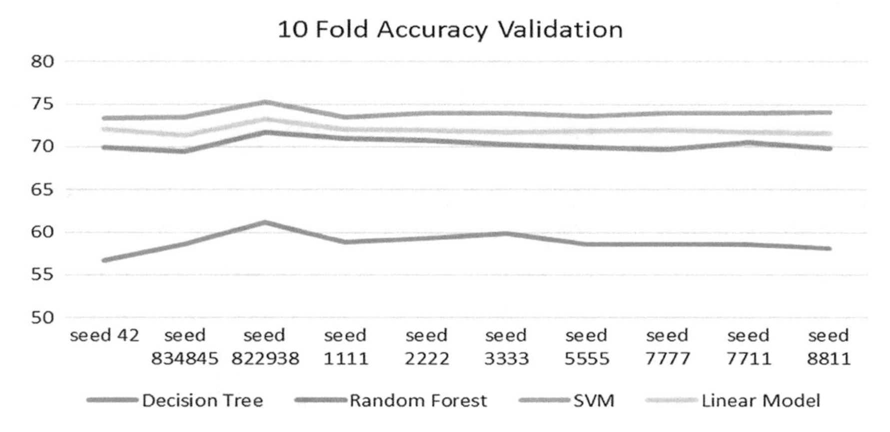

Fig. 2 10-Fold accuracy validation

3.5 *Ensemble Learning*

This is the final phase of implementation. Ensemble learning is the technique of finding the best optimized model which improves the performance of the models chosen. For example, in an organization there is a team of employees working on some particular task. Each one of the employee is expert in some specific domain; so this will give combined accuracy rate which will improve the decision-making of the organization. The accuracies have elevated to a great extent by trying all the possible combinations and permutations of the models which were earlier tested individually and resulted in low accuracy. In our work, we have obtained 11 different combinations of four different models and observed the various respective accuracies. As mentioned above, Table 4 represents all the different combinations checked for best ensemble recommendation. Referring to the combinations in the following table, 1 refers to Decision Tree model, 2 refers to Random Forest model, 3 refers to Support Vector Machine model and 4 refers to Linear Model.

It can be clearly observed in Table 4, that accuracy percentage is generally in increasing trend as more number of models are combined.

4 Results

The result is clear from the final phase of implementation, combination of all the four models that is, Decision Tree model, Random Forest model, Support Vector Machine model, and Linear model came out with the highest accuracy of 86.42193%.

Table 4 Ensemble Modeling

S. No.	Different combinations	Accuracy (%)
1	1 + 2	77.67161
2	1 + 3	83.00226
3	1 + 4	81.99648
4	2 + 3	80.23636
5	2 + 4	80.73925
6	3 + 4	79.48202
7	1 + 2 + 3	84.51094
8	1 + 2 + 4	84.76238
9	1 + 3 + 4	85.24013
10	2 + 3 + 4	82.90168
11	1 + 2 + 3 + 4	86.42193

5 Conclusion

The research findings concluded the overall classification of personality on the basis of EQ and SQ scores with considerable accuracy of 86.42193%. This work can be further used in some applications where emotional intelligence would be the deciding factor of one's personality.

References

1. Mayer JD, Roberts RD, Barsade SG (2007) Human abilities: emotional intelligence. Annu Rev Psychol 59:507–536
2. Groen Y, Fuermaier ABM, Den Heijer AE, Tucha O, Althaus M (2015) The empathy and systemizing quotient: the psychometric properties of the Dutch version and a review of the cross-cultural stability. J Autism Dev Disord (Springer) 45(9)
3. Baron-Cohen S, Richler J, Bisarya D, Gurunathan N, Wheelwright S (2013) The systemizing quotient: an investigation of adults with Asperger syndrome or high-functioning autism, and normal sex differences. In: Frith U, Hill E (eds) Autism: mind and brain. Oxford University Press, Oxford, pp 161–186
4. Baron-Cohen S, Wheelwright S (2004) The empathy quotient (EQ): an investigation of adults with Asperger syndrome and high-functioning autism, and normal sex differences. J Autism Dev Disord 34:163–175
5. Baron-Cohen Simon (2003) The essential difference. Basic Books, New York
6. http://goo.gl/zk1SH7
7. Tang H (2008) A simple approach of data mining in excel. In: 4th international conference on wireless communications network and mobile computing, pp 1–4
8. Tang HWV, Shang YM (2008) Emotional intelligence characteristics of IT professionals in Taiwan. IEEE Conference Publications
9. Syed F, Rafiq A, Ahsan B, Nadeem Majeed M (2013) An efficient framework based on emotional intelligence to improve team performance in developing countries. Int J Mod Educ Comput Sci

10. Grandey AA (2000) Emotion regulation in the workplace: a new way to conceptualize emotional labor. J Occup Health Psychol 5:95–110
11. Jordan PJ, Ashkanasy NM, Hartel CEJ, Hooper GS (2002) Workgroup emotional intelligence scale development and relationship to team process effectiveness and goal focus. Hum Resour Manag Rev 12:195–214
12. Wood LM, Parker JDA, Keefer KV (2009) Assessing emotional intelligence using the emotional quotient inventory (EQ-i) and related instruments
13. Chin STS, Raman K, Ai Yeow J, Eze UC (2013) The influence of emotional intelligence and spiritual intelligence in engineering entrepreneurial creativity and innovation. In: IEEE 5th conference on engineering education (ICEED)
14. Picard RW (2002) Toward machine emotional intelligence: analysis of affective physiological state. IEEE Trans Syst Pattern Anal Mach Intell 23(10)
15. Liaw. A, Wiener M (2003) Classification and regression by random forest. R News 2(3):18–22
16. Burges JC (1998) A tutorial on support vector machines for pattern recognition. Data Min Knowl Discov (Bell Laboratories, Lucent Technologies) 2(2):121–167
17. Alpaydin E (2010) Introduction to machine learning. The MIT Press

Investigation of MANET Routing Protocols via Quantitative Metrics

Sunil Kumar Jangir and Naveen Hemrajani

Abstract This paper aims at completing the analysis of the routing protocol of MANET with different categories using various parameters over the various scenarios. The MANET routing protocols are verified for different sets of data and on this basis the router which is best suited for data transmission among existing protocols is analyzed. To study the performance of a lot of routing protocols at the time of exchanging of data, we have generated assumed progress over a lots of MANET consisting of different pairs of source and destination node. The process of simulation has been done by using NS-3 which is an open-source simulator. Here we have successfully generated and analyzed the scenarios where the data communication effects can be analyzed over the rapid incrementation in network mobility and communication is evaluated and network data traffic is analyzed. The effort is beneficial for the candidate who is working on various problems of MANETs such as attacks, Quality-of-Service and effects of increasing number of nodes on various parameter etc. to know which protocol is best suitable for their effort towards a routing protocol.

Keywords MANET · Routing protocol · Security attack

1 Introduction

A ad Hoc network which is wireless is the combination of more than two devices that have the networking capability wireless communications [1]. This type network is deprived of a certain topology or central coordination point. So, forwarding and

S. K. Jangir (✉) · N. Hemrajani
Department of Computer Science & Engineering, JECRC University, Jaipur, Rajasthan, India
e-mail: sunil.jangir07@gmail.com

N. Hemrajani
e-mail: naven_h@yahoo.com

D. K. Mishra et al. (eds.), *Data Science and Big Data Analytics*,
Lecture Notes on Data Engineering and Communications Technologies 16,
https://doi.org/10.1007/978-981-10-7641-1_6

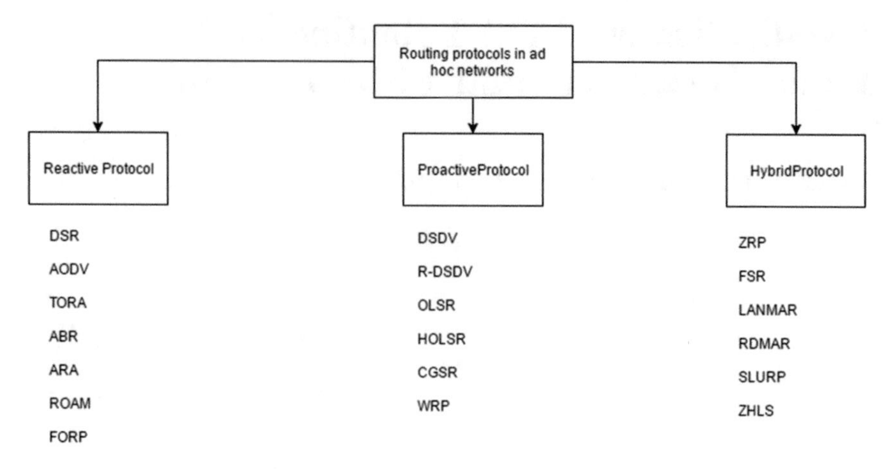

Fig. 1 Classification of routing protocol

collecting packets are very complex in comparison to that in infrastructure networks [2]. In the ad hoc network, the route data packets for the other nodes are forwarded by a node. Hence each node acts as both, as a router and the host, which makes it necessary for the use of routing protocols to perform routing decisions. Re-establishment and Establishment of paths in between the scenario of the moving network partitions and environment of network topology because of the node mobility are the biggest challenges which are faced by the routing protocols. On the basics of how the routing information is maintained and acquired by the mobile nodes, MANET routing protocols are classified as reactive and proactive.

MANET is an IP-based network for wireless machines and mobile devices having no central node and centralized administration and each node works as a router. The areas of WiNCS and MANETs [3–7] have received the attention up to the great extent in recent years. Many research outputs reveals that the ad hoc (mobile) networks known as MANET will be the dominant networks of the future wireless communications [8] (Fig. 1).

This paper aims at comparing the protocols which are for the purpose of routing in such infrastructure less networks. There are many distinctive of a lot of routing protocols that are used and investigated in the MANET Environment [9]. We have tried to compare two protocols from each (Reactive, Hybrid and Pro-Reactive).

The section below discusses the quantitative and qualitative comparison of proactive, hybrid, and reactive routing approach [10–30] (Table 1).

Table 1 Comparative study of MANET routing protocol

Key-constant	Reactive routing protocol	Table driven routing protocol	Hybrid routing protocol
Routing structure (RS)	level	Hierarchical	Hierarchical
Routing overhead (RO)	Low	High	Intermediate
Latency	High	Low	Low for local destination high for inter-zone
Control traffic	Low	High	Moderate
Routing information	Don't maintained	Keep maintained in table	Depends on requirement basis
Mobility support (MS)	Maintenance of route	Periodical reform	Combination of RRP and TDRP
Periodic route update (PRU)	Not entailed	Always entailed	Used inside each zone
Routing scheme	On demand	Table driven	Composition of both
Storage capacity (SP)	Low	High	Lean on the size of zone turn
Availability of routing information	Available when required	Always available stored in tables	Composition of both
Scalability level	Usually not fit for large networks	Low	Plan for large networks

2 Related Work

In [31], researchers have introduced their assessment comes about for execution-related correlation of AODV and DSR protocols which are two reacting protocols. With the expanded load on network mobility and network the creators have looked at two protocols.

In [32], the Researcher introduces the execution assessment of three routing protocols utilizing the outstanding open source test system known as ns-2. The protocols are changed in their packet size, inter-packet time, the node mobility and contrast. In numerous situations from result investigations we infer that the execution of AODV protocol is superior to both the DSR and DSDV protocols.

Besides, in [33] the researchers show the execution investigation report of DYMO (modified and improvised version of AODV protocol), AODV, ZRP and OLSR on different levels of mobility in network. Be that as it may, the creators did not check the execution in nearness of high network load. The measurements which have been utilized for the assessment procedure are PDR, EED, NT, and RO. Every one of recreations is done on a simulator with network scale called as Qualnet. As its recreation is done on a more modern simulator whose outcomes are near to the physical test bad comes about it is viewed as a decent work.

At a long last, in [34–36] likewise creators from various nations exhibit their execution assessment for MANET routing protocols by the use of the commercial simulators or free source on wide assortment of measurements and situations. As all the above work done on the appraisal of MANET directing conventions are either not surveying the most surely understood conventions or they are not utilizing all the normal estimation to display their feasibility or they does not use all the available or tasteful coordinating circumstances used by methods for several kinds of MANET applications.

3 Routing Protocols

AODV (Ad Hoc on-demand distance vector) [37]—It is the improved version for Destination-Sequenced Distance-Vector routing algorithm which was developed by Perkins and Bhagwat [38]. Rather keeping the up-to-date route data it discovers the route on-demand which decrease the number of broadcast message which is forwarded in the network. A source hub when it sends information parcels to a goal hub dependably checks its course table to check whether the course table has substantial goal hub. When a route present, it remit the packets to the next hop with the targeted node. On the off chance that there is no substantial goal course in the course table, the source hub initiates a course revelation process. Route request (RREQ) is broadcasted to the intermediate neighbor node and those nodes further broadcast their neighbor node until the destination is reached. The bundle which is sent to the hubs consists of the source hub (nodes) IP address, current grouping number, the sequence number. The neighbor node can only respond to the packet of request only when if it has the sequence number of destination contained which is same or more than the number contained in the route request of the packet header.

TORA (Temporally ordered routing algorithm)—TORA is a malleable and tensile algorithm that is routing working on the idea of reversal link, which was proposed by Park and Corson. It finds the numerous courses from the source to destination in exceedingly unique versatile systems network.

In TORA the control messages are set into the small sets of the nodes. Each node maintains routing information of their immediate one-hop neighbor.

The principal on which TORA protocol is being worked is to handle the link failure. Every link is reversed sequence searches for the alternate path to the destination. They use pass which is single of the distributed algorithm, which serves as the mechanism for the search of the alternate path as the tables are altered at the same time during search mechanism in the outward phase.

DSDV (Distance-sequenced distance vector)—DSDV is primeval ad hoc routing protocols which introduced by Perkins and Bhagwat [38]. As all the distance factor routing depends on the Bellman–Ford algorithm it is also based on the same algorithm. A routing table is maintained by every mobile network which along with the distance hop counts also contains all the possible destination of the network. Destination assigns a sequence number which is stored by each entry. Identification of entries that are stale and the shirking of loops are done through sequence number which is assigned by the target nodes. For maintaining the router consistency, router updates are forwarded periodically throughout the network.

OLSR (Optimized link state routing) [39]—This algorithm designed and developed by Clausen et al. OLSR is the changed rendition of the Classical link State Protocols on the premise of no. of improvements focused towards MANET Environment. These optimizations are unify on *multipoint relays* (MPR) which is the selected nodes. First, during flooding process of the route information, only MPR's forward message, resulting in decreasing the total messages forwarded. Also, MPRs is the only which generated link state information, the data is reduced as a result. Finally, the MPRs may report only to link between them and their respective MBR selectors.

The multipoint relay focuses on how to reduce retransmission within the equal region. All the nodes pick multipoint relays (MPR) for that node which is set of one-hop next. Since only MPRs forward the packets so the packets which are not MPRs system the packet but do not do not forward them. All two-hops neighbor must be covered in range while selecting multipoint relay.

ZRP (Zone routing protocol) [40]—Samar et al. designed ZRP protocol, which is used in very large-scale networks. Proactive mechanism is used by this protocol in node that is found within the immediate neighborhood of node while reactive approach carries the inter-zone communication By doing some modification ZRP which is a framework protocol allows any reactive/proactive protocols to use as its components [41, 42]. Zones are known as the local neighbor, which is defined for nodes. The zone size depends on the p-factor. p-factor is defined as the total number of hops to the total perimeter of that zone. There may be some zones that are overlapping which helps in the route optimization.

Neighbor discovery is done either by Intrazone Routing Protocol (IARP) or simple HELLO packet. The Intrazone Routing Protocol (IERP) uses an approach which is reactive in communication with different zones of node.

ZHLS (Zone-based hierarchical link state routing protocol)—Joa-Ng and Lu has been suggested ZHLS protocol [43]. They stated that the non-overlapping zones define the hierarchical structure by where every node has a unique node ID and also zone ID. Tools used are GPs for calculating ID's. The zone topology of a level and the node-level topology are two types of hierarchy division. ZHLS contains no cluster heads. Zone-level location appeal is broadcasted to every available zones when we required a route for a destination which is placed in another zone. When this location request is received at destination it replies with its path. We need only zone ID's and node ID's for the path discovery. There is no update required when a node continue to be in its boundary (Table 2).

Table 2 Comparison of routing protocol

Parameters	AODV	TORA	DSDV	ZHLS	OLSR	ZRP
Multiple routes	No	Yes	No	Yes	Yes	No
Route metric	Newest route and SP	SP or next available	SP	SP	SP	SP
Communication overhead	High	High	Low	Medium	High	Medium
Route repository	RT	RT	RT	IntraZ and InterZ RTs	RT	IntraZ and InterZ RTs
Routing philosophy	Reactive	Reactive	Proactive	Hybrid	Proactive	Hybrid
Loop freedom	Yes	Yes	Yes	Yes	Yes	Yes

RT routing table, *SP* shortest path

4 Simulation Table and Results

End-to-end delay can be called as the total time consumed by the individual packet to transmit over the network from the starting node to final node. In Fig. 2 Reactive Routing Protocol (AODV, TORA) the Average of End-to-End Delay should be minimum where as in Proactive Routing Protocol (DSDV, OLSR) it is average and in Hybrid Routing Protocol it is maximum. In Hybrid Routing Protocol (ZRP, ZHLS) as the quantity of hops are expanded the End to End defer values increments. This happens due the increase in load network increases, it also increases the number of packet of data in the middle node's IP output because of the overcrowding in the networks. As the overcrowding increases, it also increment the data packet waiting time in the IP output queue (Table 3).

We can find the PDR Fig. 3 and throughput of network Fig. 4 of all protocols that are routing ones which we have compared. Clearly from Fig. 3, it can be seen that

Table 3 Simulation table

Parameters	Values
Simulator	NS-3
Protocol studied	AODV, TORA, DSDV OLSR, ZHLS, ZRP
Simulation time	100 s
Simulation area	700 × 700
Transmission range	200
Node movement model	Random waypoint
Bandwidth	2 Mbps
Traffic type	CBR
Data payload	Bytes/packet

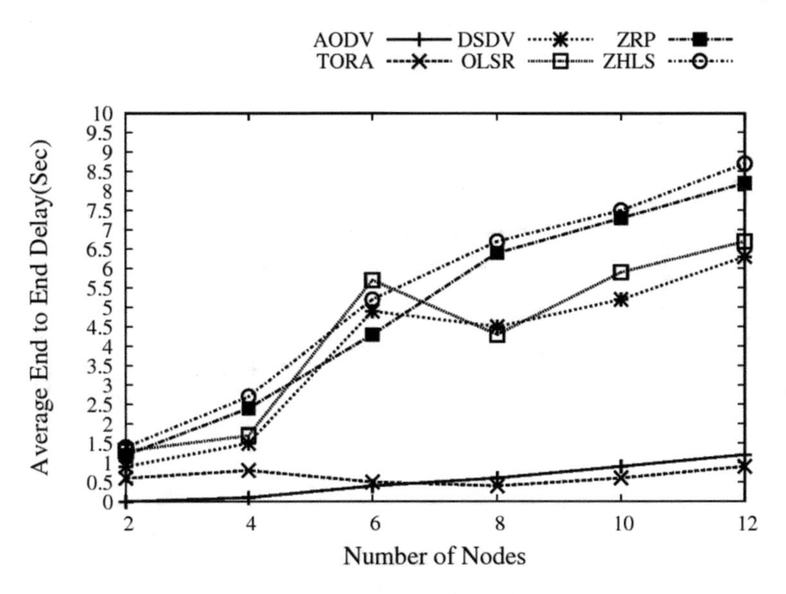

Fig. 2 Routing protocols equivalence on average end-to-end delay

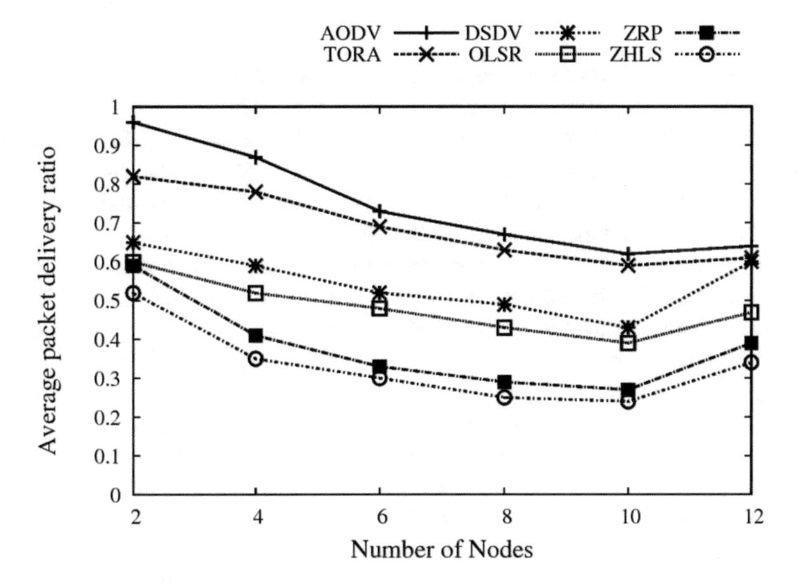

Fig. 3 Routing protocols equivalence on average packet delivery ratio

the both ratio of packet delivery and throughput of network of the routing protocol that we have compared, is large when the load of the data in the network is less as compared to created total bandwidth overloaded routes for the routing process. Therefore, it enlarges the PDR.

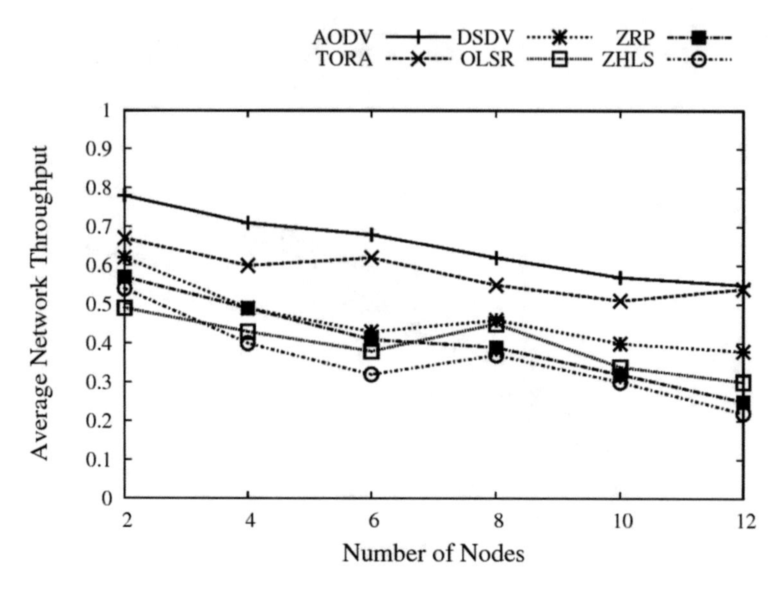

Fig. 4 Routing protocols equivalence on average network throughput

As the load of network enlarges the total number of packet which are diminished due to smashing and interference enlarges which is done by nearby flows. Moreover it can observed from Fig. 4 that this reactive routing protocol AODV is properly designed for the middle load networks if comparison done with the ZRP and OLSR protocols as their Ratio of packet delivery and throughput of network is higher and moderate environment networks in comparison to ZRP and OLSR. This is because of the reactive process of route discovery in the DSR and AODV which reduces the routes which are highly overloaded and locate or we can say discover the less.

Figure 5 is describing Transmission efficiency of routing of the routing correlating protocols. As it can be analyzed from Fig. 5 that Transmission efficiency of routing of the AODV is very large and it remains continual with increment in the network and the Transmission efficiency of routing of the OLSR is very low at the starting and then it increases slowly as the network load increases. This is mainly because in the AODV route discovery period will circumvent overloaded routes during its process of route discovery and carefully choose that route which is less overloaded therefore the total packets are in large quantity as compared to the AODV than other routing protocols. On the different side if we look, the Transmission efficiency of routing is persistent with the growth in the load of network because as this network load grows the number of packets that are transferred in the network are not increments fundamentally. This is on the grounds that the parcels are dropped because of no course or because of IP flood yield queue. So very less packets are forward therefore packets get less collided and because of this the Transmission efficiency of routing remains persistent with the increase of the load of network.

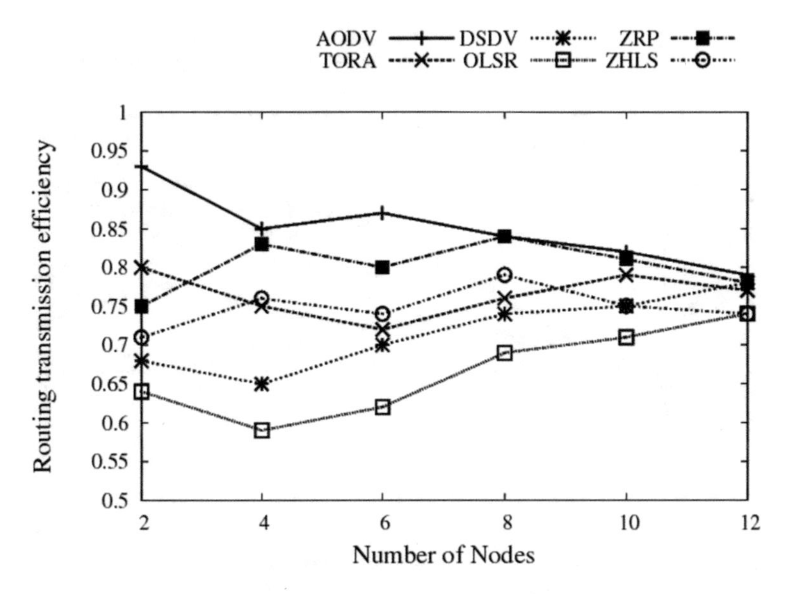

Fig. 5 Routing protocols equivalence on routing transmission efficiency

Normalized Routing Load (or also known as Normalized Routing Overhead) is considered aggregate number of routing packets in the systems that are transmitted per bundle of data. In Fig. 6 we conclude that there is increase in total transmission of data packet in all routing protocol as there is a growth in the node. As we start the number of nodes starts increasing, the control packets are transmitted to discover the routes. In On-demand routing, control overhead proportionately increments with the expansion in activity assorted variety as it is against proactive routing and is fundamentally normally versatile to movement decent variety control overhead proportionately increments with the expansion in activity decent variety as it is against proactive routing and is essentially normally versatile to movement diversity.

So, the routing which is on-demand is considered more reliable and efficient in terms of control overhead nonetheless of the relative node movement when the traffic diversity is very low. The routing overhead for on-demand routing could proceed towards that of proactive routing with high traffic diversity.

Path Optimality is defined as the difference the total hope required to achieve its goal and the length of the most limited way that exist in the system when the packets were produced. When all the protocols want to transfer packets then they choose the shortest path but in the case of AODV, it chooses the shortest and fresh path (Fig. 7).

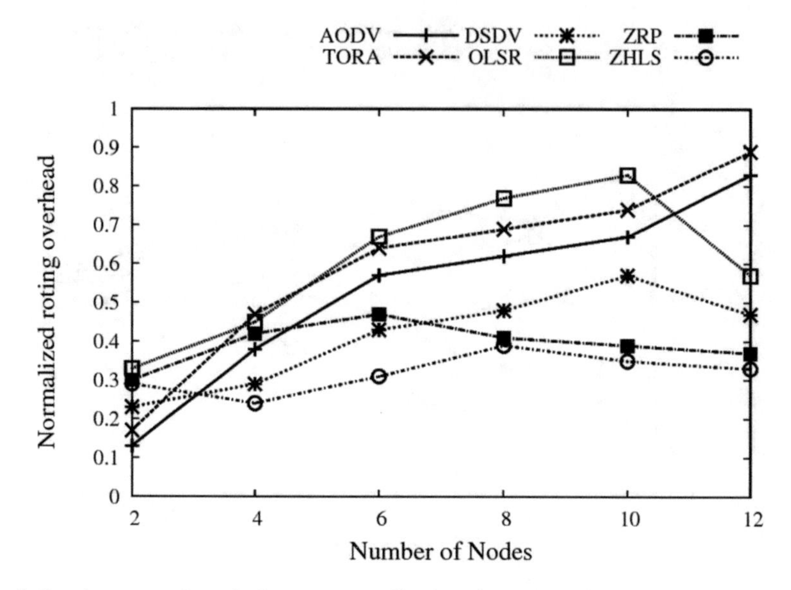

Fig. 6 Routing protocols equivalence on normalized routing overhead

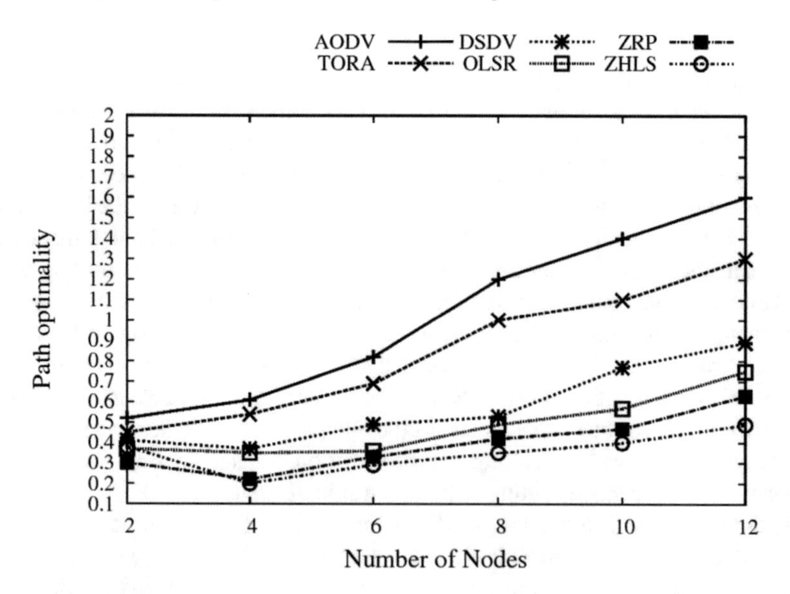

Fig. 7 Path optimality

5 Conclusions and Future Direction

In this paper, we have done very diligent execution for the assessment of different routing protocols. We have compared this assessment in a lot of situations which arrange load and versatility. The protocols examination is done by metrics which are picked in a way thus, to point that they can inhibit all the conceivable conduct of the routing protocols and give its uses and benefits in several cases.

If compared to the hybrid routing protocols, the reactive routing protocols are more perfect in the majority of the situations and problems in the MANETs. It is demonstrated by the outcome additionally that the protocols which are AODV routed are the most appropriate routing protocols as its route discovery stage can be adjusted in agreement to the dynamic states of the wireless network. Simulation results show that the routing protocols are just easy routing protocols that they can just offer help for a data correspondence on its exertion way.

In this manner, the MANETs routing protocols cannot send the information and data accurately and appropriately when the network load is increased and probability of network changes at a high recurrence. Due to the ease of the routing protocols of MANETs, further we are going to attempt to upgrade the elements of them so that they can ready to work in extensive networks with given confinements and furthermore ready to give application particular Services.

References

1. Elavarasipriyanka S, Latha V (2015) Design of routing protocol for MANET through energy optimization technique. Energy
2. Raj N, Bharti P, Thakur S (2015) Qualitative and quantitative based comparison of proactive and reactive routing approaches in MANET. Int J Comput Sci Mobile Comput 4(12):177–184
3. Chung WH (2004) Probabilistic analysis of routes on mobile ad hoc networks. IEEE Commun Lett 8(8):506–508
4. Floréen P, Kaski P, Kohonen J, Orponen P (2005) Lifetime maximization for multicasting in energy-constrained wireless networks. IEEE J Sel Areas Commun 23(1):117–126
5. Kwak BJ, Song NO, Miller LE (2003) A mobility measure for mobile ad hoc networks. IEEE Commun Lett 7(8):379–381
6. Ramanathan R, Redi J, Santivanez C, Wiggins D, Polit S (2005) Ad hoc networking with directional antennas: a complete system solution. IEEE J Sel Areas Commun 23(3):496–506
7. Weniger K, Zitterbart M (2004) Mobile ad hoc networks-current approaches and future directions. IEEE Netw 18(4):6–11
8. Velempini M, Dlodlo ME (2007) A virtual ad hoc routing algorithm for MANETs. In: 2007 the 2nd international conference on wireless broadband and ultra wideband communications, AusWireless 2007. IEEE, pp 22–22
9. Boukerche A, Turgut B, Aydin N, Ahmad MZ, Bölöni L, Turgut D (2011) Routing protocols in ad hoc networks: a survey. Comput Netw 55(13):3032–3080
10. Khatkar A, Singh Y (2012) Performance evaluation of hybrid routing protocols in mobile ad hoc networks. In: 2nd international conference on advance computing and communication technologies, pp 542–545
11. Patel B, Srivastava S (2010) Performance analysis of zone routing protocols in mobile ad hoc networks. In: 2010 national conference on communications (NCC). IEEE, pp 1–5

12. Michalareas T, Sacks L (2001) Reactive network management architectures and routing. In: International symposium on integrated network management, pp 811–824
13. Zhou J, Lin Y, Hu H (2007) Dynamic zone based multicast routing protocol for mobile ad hoc network. In: 2007 international conference on wireless communications, networking and mobile computing, WiCom 2007. IEEE, pp 1528–1532
14. Kathiravelu T, Sivasuthan S (2011) A hybrid reactive routing protocol for mobile ad-hoc networks. In: 2011 6th IEEE international conference on industrial and information systems (ICIIS). IEEE, pp 222–227
15. Chandra A, Thakur S (2015) Qualitative analysis of hybrid routing protocols against network layer attacks in MANET. IJCSMC 4(6):538–543
16. Garnepudi P, Damarla T, Gaddipati J, Veeraiah D (2013) Proactive, reactive, hybrid multicast routing protocols for wireless mess networks. In: International conference on computational intelligence and computing research (ICCIC). IEEE, pp 1–7
17. Mbarushimana C, Shahrabi A (2007) Comparative study of reactive and proactive routing protocols performance in mobile ad hoc networks. In: 2007 21st international conference on advanced information networking and applications workshops, AINAW'07, vol 2. IEEE, pp 679–684
18. Mahmood D, Javaid N, Qasim U, Khan ZA (2012) Routing load of route calculation and route maintenance in wireless proactive routing protocols. In: 2012 seventh international conference on broadband, wireless computing, communication and applications (BWCCA). IEEE, pp 149–155
19. Yang PL, Tian C, Yu Y (2005) Analysis on optimizing model for proactive ad hoc routing protocol. In: 2005 military communications conference, MILCOM 2005. IEEE, pp 2960–2966
20. Javaid N, Bibi A, Javaid A, Malik SA (2011) Modeling routing overhead generated by wireless reactive routing protocols. In 2011 17th Asia-Pacific conference on communications (APCC). IEEE, pp 631–636
21. Vanthana S, Prakash VSJ (2014) Comparative study of proactive and reactive ad hoc routing protocols using NS2. In: 2014 world congress on computing and communication technologies (WCCCT). IEEE, pp 275–279
22. Shenbagapriya R, Kumar N (2014) A survey on proactive routing protocols in MANETs. In: 2014 international conference on science engineering and management research (ICSEMR). IEEE, pp 1–7
23. Sholander P, Yankopolus A, Coccoli P, Tabrizi SS (2002) Experimental comparison of hybrid and proactive MANET routing protocols. In: Proceedings of the MILCOM 2002, vol 1. IEEE, pp 513–518
24. Rohankar R, Bhatia R, Shrivastava V, Sharma DK (2012) Performance analysis of various routing protocols (proactive and reactive) for random mobility models of ad hoc networks. In: 2012 1st international conference on recent advances in information technology (RAIT). IEEE, pp 331–335
25. Samar P, Haas ZJ (2002) Strategies for broadcasting updates by proactive routing protocols in mobile ad hoc networks. In: Proceedings MILCOM 2002, vol 2. IEEE, pp 873–878
26. Patel D, Patel S, Kothadiya R, Jethwa P, Jhaveri H (2014) A survey of reactive routing protocols in MANETs. In: International conference on information communication and embedded systems (ICICES), pp 1–6
27. Rajput M, Khatri P, Shastri A, Solanki K (2010) Comparison of ad-hoc reactive routing protocols using OPNET modeler. In: 2010 international conference on computer information systems and industrial management applications (CISIM). IEEE, pp 530–534
28. Rahman MA, Anwar F, Naeem J, Abedin MSM (2010) A simulation based performance comparison of routing protocol on mobile ad-hoc network (proactive, reactive and hybrid). In: 2010 international conference on computer and communication engineering (ICCCE). IEEE, pp 1–5
29. Hamma T, Katoh T, Bista BB, Takata T (2006) An efficient ZHLS routing protocol for mobile ad hoc networks. In: 2006 17th international workshop on database and expert systems applications, DEXA'06. IEEE, pp 66–70

30. Naserian M, Tepe KE, Tarique M (2005) Routing overhead analysis for reactive routing protocols in wireless ad hoc networks. In: 2005 IEEE international conference on wireless and mobile computing, networking and communications, (WiMob'2005), vol 3. IEEE, pp 87–92
31. Chaba Y, Singh Y, Joon M (2010) Notice of retraction simulation based performance analysis of on-demand routing protocols in MANETs. In: 2010 second international conference on computer modeling and simulation, ICCMS'10, vol 3. IEEE, pp 80–83
32. Tuteja A, Gujral R, Thalia S (2010) Comparative performance analysis of DSDV, AODV and DSR routing protocols in MANET using NS2. In: 2010 international conference on advances in computer engineering (ACE). IEEE, pp 330–333
33. Rahman MA, Anwar F, Naeem J, Abedin MSM (2010) A simulation based performance comparison of routing protocol on mobile ad-hoc network (proactive, reactive and hybrid). In: 2010 international conference on computer and communication engineering (ICCCE). IEEE, pp 1–5
34. Maan F, Mazhar N (2011) MANET routing protocols vs mobility models: a performance evaluation. In: 2011 third international conference on ubiquitous and future networks (ICUFN). IEEE, pp 179–184
35. Vir D, Agarwal SK, Imam SA (2012) Quantitative analyses and evaluation of MANET routing protocols in effect of varying mobility model using Qual-Net simulator. In: 2012 world congress on information and communication technologies (WICT). IEEE, pp 915–921
36. Shobana M, Karthik S (2013) A performance analysis and comparison of various routing protocols in MANET. In: International conference on pattern recognition, informatics and medical engineering (PRIME), pp 391–393
37. Perkins C, Belding-Royer E, Das S (2003). Ad hoc on-demand distance vector (AODV) routing, No. RFC 3561
38. Perkins CE, Bhagwat P (1994) Highly dynamic destination-sequenced distance-vector routing (DSDV) for mobile computers. In: ACM SIGCOMM computer communication review, vol. 24, no 4. ACM, pp 234–244
39. Jacquet P, Muhlethaler P, Clausen T, Laouiti A, Qayyum A, Viennot L (2001) Optimized link state routing protocol for ad hoc networks. In: 2001 technology for the 21st century, proceedings IEEE international multi topic conference, IEEE INMIC 2001. IEEE, pp 62–68
40. Samar P, Pearlman MR, Haas ZJ (2004) Independent zone routing: an adaptive hybrid routing framework for ad hoc wireless networks. IEEE/ACM Trans Netw (TON) 12(4):595–608
41. Haas ZJ, Pearlman MR (2000) Providing ad-hoc connectivity with the reconfigurable wireless networks. In: Ad Hoc networks. Addison Wesley Longman
42. Beijar N (2002) Zone routing protocol (ZRP). Networking Laboratory, Helsinki University of Technology, Finland, vol 9, pp 1–12
43. Joa-Ng M, Lu IT (1999) A peer-to-peer zone-based two-level link state routing for mobile ad hoc networks. IEEE J Sel Areas Commun 17(8):1415–1425

Review on Internet Traffic Sharing Using Markov Chain Model in Computer Network

Sarla More and Diwakar Shukla

Abstract The Internet traffic sharing is one of the major concerns in the dynamic field of Information and communications technology (ICT). In this scenario the concept of Big Data arises which defines the unstructured nature of data so that there is a strong need of efficient techniques to tackle this heterogeneous type of environment. Many things are dependent on Internet today, and a person has a lot of work to be done with the help of Internet. Due to this problems arise like congestion, disconnectivity, non-connectivity, call drop, and cyber crime. This review study is for the analysis purpose of all this type of problems. Various kinds of methods are discussed based upon the problem formation of Internet access and their respected solutions are discovered with the help of Markov chain model. This model is used to study about how the quality of service is obtained and the traffic share is distributed among the operators on the basis of state probability, share loss analysis, call-by-call attempt, two-call attempt, two market, disconnectivity, index, iso-share curve, elasticity, cyber crime, re-attempt, least square curve fitting, bounded area, area estimation and computation, Rest state, and multi-operator environment.

Keywords Internet traffic sharing · Markov chain model · ICT · Big data
Stochastic processes · Sampling techniques · Quality of service

S. More (✉) · D. Shukla
Department of Computer Science and Applications, Dr. Harisingh Gour University, Sagar, Madhya Pradesh, India
e-mail: sarlamore@gmail.com

D. Shukla
e-mail: diwakarshukla@rediffmail.com

D. K. Mishra et al. (eds.), *Data Science and Big Data Analytics*,
Lecture Notes on Data Engineering and Communications Technologies 16,
https://doi.org/10.1007/978-981-10-7641-1_7

1 Introduction

In computer networking data exchange is main objective and can be achieved by many approaches whose basics are wired and wireless media. Various services and applications are available such as fax machines, printers, digital video and audio, e-mail, and instant messaging, etc. Internet is the best-known computer network. Millions of computer networks are connected through this global network and around 3.2 million People are using this. Internet and www are different; a collection of computer networks is Internet, whereas for accessing and sharing this resources a bridge is used which is called World Wide Web. The occupancy of the server is defined as the traffic. The voice traffic and data traffic are served by the telecommunication systems. Traffic analysis serves the purpose of determinations of condition by which some services can be provided to the customers and the economical use of resources that provide the services.

2 Markov Chain in Big Data Environment

A mathematical model called Markov chain is used for a stochastic system, who is having states regulated by transition probability. In case of a first-order Markov chain the current state only depends on previous states and ignores all other states. There is a commonness between a Markov chain and the Monte Carlo casino that both are driven by random variables. For the generation of fair samples by random number in probability of high-dimensional space MCMC is used as a general purpose technique and drawn in certain range with uniform probability. Some examples of Markov chain process are birth death process, gambling and drunkard walk on a street.

In the environment of big data, unstructured form of abundant data comes into picture. It is a tedious task to deal with this type of data because form the fundamental level to the architectural level we need lot of effort to process the data to make it useful. Various examples can be seen of Markov chain such as the drunkard walk on street, weather condition model and Page Rank used by Google to determine order of search result is also an example of Markov chain. So in all these cases we can see that the big data is encountered to be processed. The heterogeneity property of big data requires efficient type of software to tackle any type of the difficulty and to provide solution of problems caused by big data environment.

2.1 Markov Chain Model

Naldi [1] proposed a phenomenon for two operator environment. Given below some of the hypotheses which are created to design the activity of user and the traffic sharing between two competitors:

- Initial choice of operator done by user as operators O_1 with probability p and O_2 with $1 - p$ probability in first step.
- P probability calculates the choice of operator with related factors, some relevant services and the required experiences.
- Switching to another operator or the termination with p_A probabilities done if the user has performed a failed attempt.
- Call-by-call attempt is the basis for the switching between the two operators and depends just on latest attempt.

The term L_1 and L_2 blocking probabilities and abandonment probability p_A remain constant while the repetition of call attempt (when operator O_1 or O_2 fails for call attempt process it is known as blocking probability). By these hypotheses a four state discrete time Markov chain can be modeled, in which state O_1 is the state telling to operator O_1 for placing a call and respectively the state O_2 is responsible for placing a call with operator O_2. A successful call attempt shown with the state Z and state A responsible for the termination of the user. The users attempt process terminates due to completion of a call or the abandonment of both absorbing states Z and A. To represent the transition probabilities arcs are used which also represent the states of chain. The actual state of the number of attempts represented by the Time X and indicated as

$$P\left[X^{(n)} = O_i\right] \; (i = 1, 2)$$

Probability of n + 1th call attempt is placed through operator O_i, $P[X^{(n)} = Z]$ is the probability that user has given up after no more than n attempts. Before attempting the first call attempt the states are distributed, so the starting conditions [1] are

$$\begin{aligned} P\left[X^{(0)} = O_1\right] &= p, \\ P\left[X^{(0)} = O_2\right] &= 1 - p, \\ P\left[X^{(0)} = Z\right] &= 0 \\ P\left[X^{(0)} = A\right] &= 0 \end{aligned}$$

In condition when a call fails, user switches between O_1 and O_2 and if abundance does not occur $L_1(1 - p_A)$ becomes blocking probability of the transition probability from O_1 to O_2. In the same way a call placed through O_1 generates $1 - L_1$ probability and able to be completed in a single attempt. The matrix of transition probabilities at a single step is represented as (Fig. 1)

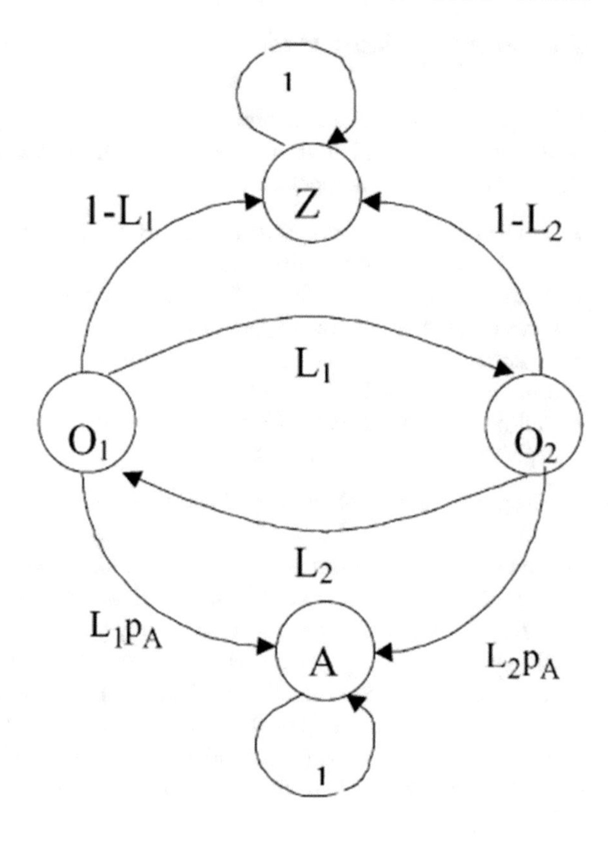

Fig. 1 The Markov chain model of the user's behavior in the two operator's case [1]

$$M = \begin{pmatrix} 0 & L_1(1 - pA) & 1 - L_1 & L_1 p_A \\ L_2(1 - p_A) & 0 & 1 - L_2 & L_2 p_A \\ 0 & 0 & 1 & 0 \\ 0 & 0 & 0 & 1 \end{pmatrix}$$

2.2 *General Internet Traffic Sharing*

An Internet connection allows us to connect to any of the person sitting locally or remotely in very quick and easy ways. The fast Internet connection can slow down due to heavy traffic in network. By identifying the causes of Internet traffic it will be helpful to overcome the congestion issues and make the required bandwidth available so that the data transfer can be done in easy ways.

- File Sharing
- Streaming Media
- Videoconferencing
- Malware.

3 Recent Contributions in Internet Traffic Sharing

3.1 Based on State Probability Analysis

In Internet traffic sharing the study of state probability is performed by considering two operator's competition environment. A Markov chain model by Shukla et al. [2] is represented in which network blocking probabilities are mutually compared and simulation study shows that some specific types of users are affected. For gain of Internet connection user makes some call attempt on which state probability is dependent. This probability reduces sharply as attempt increases. Faithful users (FU) have a tendency to stick with their favorite operators up to seven to eight attempts but partially impatient user (PIU) group has negative tendency in this regards. The Completely Impatient user (CIU) bears a better proportion of state probabilities. When blocking of network of ISP_1 is high, it gains state probabilities related to FU.

3.2 Based on Call-by-Call Attempt

More traffic gain by two-call attempt model rather than one-call attempt by Shukla et al. [3] has been proposed. In two systems it is represented that when L_1 probability of operator O_1 varies it presents the graphical pattern for operator O_1's traffic sharing P_1 probability. It also observed that system-I is faster than system-II, the traffic sharing goes down in system-II. The traffic share call blocking reaches nearly at zero level. Due to call difference both user behavior systems have some difference in traffic sharing. For operator O_1 two-call based system is not able to bear blocking more than 60%. System-II provides better traffic share than system-I if operant blocking is high for operator O_1. If initial traffic share is high the Internet traffic is more in system-II than in comparison to System-I.

3.3 Quality of Service Based

Two parameters blocking probability and initial preference by Shukla et al. [4] are used in this model in order to predict the traffic distribution between the operators. Some blocking competitors gain benefit by the call-by-call technique. The two-call attempt comes into picture and new mathematical results are derived. but it gives more traffic than one-call attempt. Due to call difference there is a little difference between traffic sharing by the systems of user behavior. For operator O_1 the blocking probability is less than two-call based system. So in system-II the operant blocking is high than system-I. If initial traffic share is high the Internet traffic is more in system-II than in comparison to System-I. The QoS is a function of blocking probabilities

(L_1, L_2) given by Internet service providers due to congestion in the network. User's high level blocking probability corresponds to lesser quality.

3.4 Share Loss Based

A share loss analysis by Shukla et al. [5] is presented in two market environment of Internet traffic where two operators are in competition for QoS.

Iso-share curve is drawn using Markov chain model and required analysis is performed. To obtain the effects over initial traffic share simulation study is done and observed that initial share of traffic is highly affected by network blocking probability for a network operator. The final share loss of traffic is linear function of self blocking probability of networks. Operator has to reduce blocking probabilities if final share loss goes high. With increment of r1 parameter proportion of FU improves. If *PR* and *r* both have increment then, FU proposition for operator O_1 uplifts. It seems the rest state has strong impact on upliftment of FU. To maintain the prefixed final share of PIU, operator O_1 has to reduce his blocking probability to keep the earlier initial share. P_{R1} probability related to rest state if high then operator O_1 have not to worry. The CIU users are high affected by opponent network blocking probabilities.

3.5 Index Based

Congestion is a cause of heavy usage of Internet, by service providers it can be managed at some extent. Shukla et al. [6] gives quality of service assessment for operators. To observe Internet access behavior of the user in two ISP environment a Markov chain model is applied and compared using simulation study. The probability index for CIU increases on comparing CIU with increasing L_2. Index increases for the lower values of L_2 when r is high. Variations of p_R and r have similar effects. A stable value for larger number of attempts has been given with small L_1, L_2, r and p_R CIU index. These views show that for FU there is a seamless variation for larger attempts for probability index. Contrary to this, PIU index has little variation preferably in range (0.3–0.5). For larger number of attempts stability pattern occurs in this same range but suggested here is the smaller parameter choice. In compare to FU there is a similarity between the Index fluctuation of PIU and CIU but there is a slight higher shifting for stability pattern over large attempt.

3.6 Disconnectivity Based

ISP is the source to manage calls. Call blocking is a problem for the Internet users. Blocking, congestions, delay, non-connectivity, Internet call termination and hard-

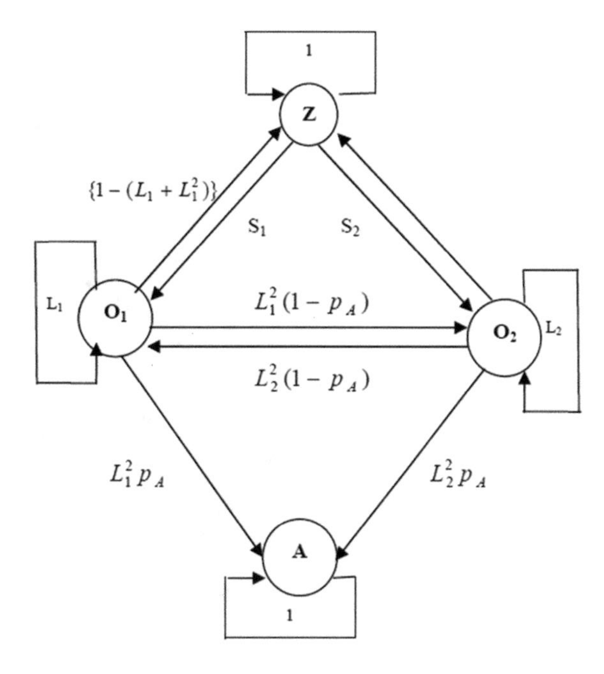

Fig. 2 Transition model (using Shukla et al. [2])

ware failures are the reasons for disconnectivity of call. To give better connectivity in minimum ISP blocking probability are properly managed. Shukla et al. [7] contributed a research where a stochastic model using Markov chain is represented to observe the characteristics and operations of Inter traffic sharing with unit step transition probability matrix. Various users perform the analysis of Internet traffic distribution in disconnectivity factor between two operators. In case of assumption of two-call basis the attempting number of calls depends on the operator's traffic share property. At time when self blocking is low the initial traffic share get in proportion to the traffic share amount. In case of $n=4$, $n=8$ the traffic share level decreases when self blocking of network is high. In two-call attempt model self-blocking and traffic sharing relationships are inversely proportional to one another. To enhance traffic share every network operator should keep lower level of network blocking than competitor (Fig. 2).

3.7 Iso-share Curve Based

With network establishment a user, as soon as connected the cyber crime may be happen. By Shukla et al. [8] to analyze the user's behavior during traffic sharing in the presence of cyber crime probability a Markov chain based model is used as a tool. Cyber criminals and non-cyber criminals are two groups of users. Simulation shows operators benefited by a better proportion of traffic who promote to cyber crimes as

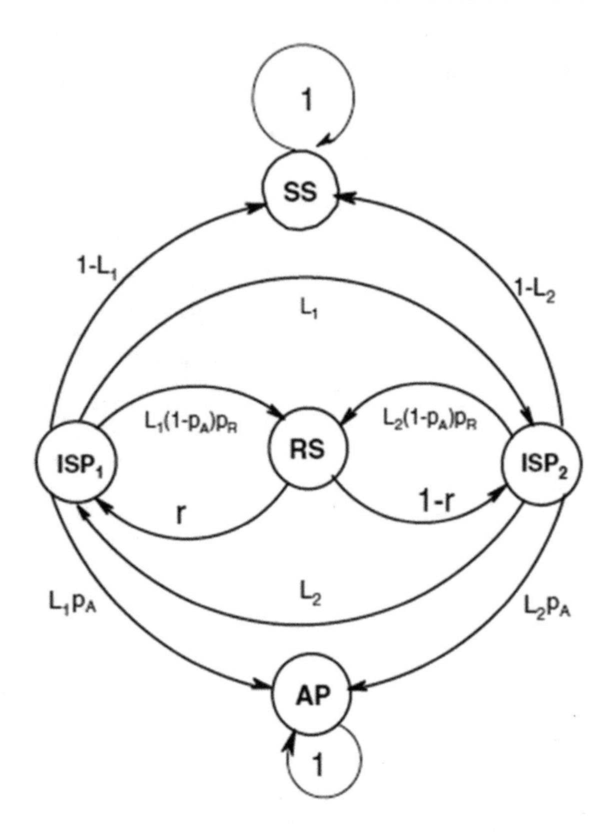

Fig. 3 Transition model (using Shukla et al. [3])

marketing strategy. In iso-share curves NC group of users and reflect growing trend under increasing pattern of opponent blocking L_2. For 50% final share, operator O_1 has to keep more initial share. If $L_2 = 0.2$ operator O_1 has to keep initial share nearly 60% to maintain 50% final share of NC group. The variation in c_1 probability does not affect the balance between initial and final share. In case of CC users, the reverse pattern exits in iso-share curves. O_1 has to maintain at least 20% value of initial share to keep final share 50% and blocking probability $L_2 = 0.2$. In the same way, when $L_2 = 0.3$, to maintain 50% final share, O_1 has to put in the initial share only on 10%. If opponent faces 10% increment in blocking probability O_1 gains 10% CC customer proportion (Fig. 3).

3.8 Two Market Based

In two market environment for analyzing the Internet traffic two operators are compared. Faithful, impatient, and completely impatient are three types of user's behavior which can be considered for analysis. By Shukla et al. [9] in the analysis of Markov

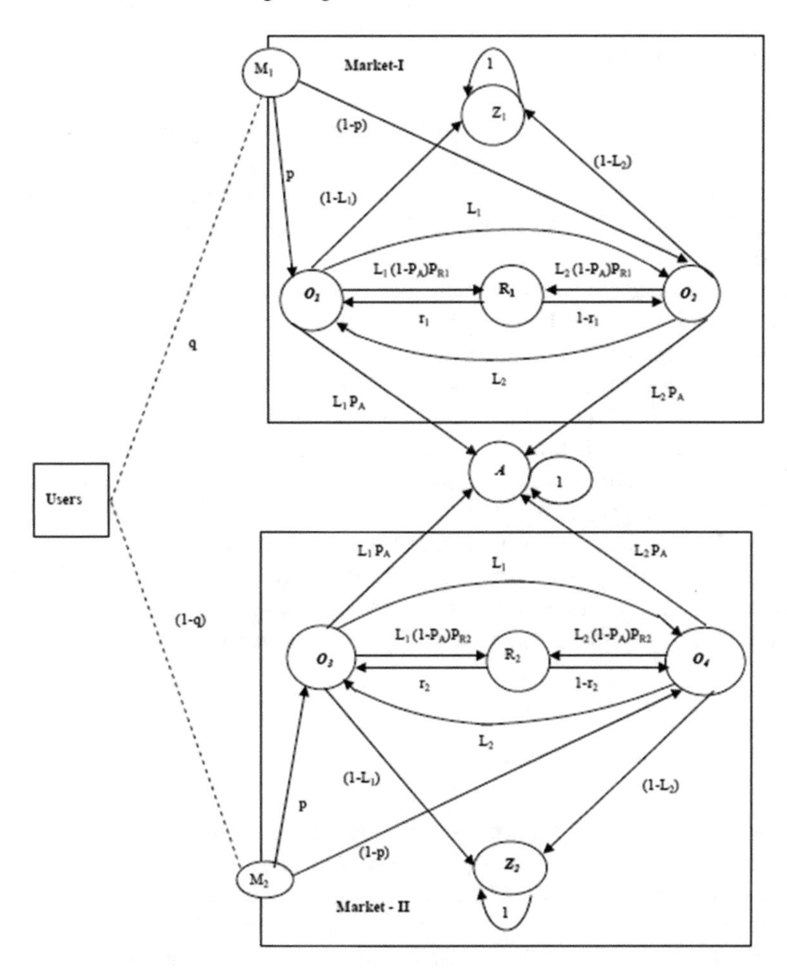

Fig. 4 Transition diagram of model (using Shukla et al. [9])

chain model, to obtain the user's behavior the network's blocking probability is considered and also useful in selection of an operator as an ISP. The operator bears the amount of faithful user which depends on the initial traffic share. The initial traffic share can be reduced by the high volume of self blocking probability. For operator O_1 the faithful user proportion is high in case the opponent's network blocking is high. For the purpose to increase the faithful user group in multi-market system its recommended to reduce the network blocking (Fig. 4).

3.9 Elasticity Based

The contribution by Shukla et al. [10] demonstrate blocking probabilities effects on the elasticity. For measurement of the effects of different model parameters on the elasticity variance, simulation study is performed. In this contribution operators O_1's blocking probability L_1 increases the level of elasticity to traffic sharing. This represents that related to network blocking operator has to be aware about the traffic share amount. Larger blocking get more fluctuation in traffic. Rest state probability p_R depends on this tendency. Elasticity trend is with negative side if opponent blocking probability L_2 increases. Due to increase of L_2 parameter for O_1 the elasticity level decreases. Over the variation of blocking probability L_1 and L_2 the elasticity level seems to be stable. The positive showing of Rest state takes importance due to effect of rest state probability $p_{R.}$. The L_1 level of network blocking induces that elasticity level has slight variation elasticity level is non-positive and independent to the variation of p_R probability for L_2 level variation. At blocking level of L_1 and L_2 the elasticity level becomes heterogeneous in nature shown by the simulation study. In case of L_2 elasticity level is negative and comparatively L_1 is positive and stable and traffic sharing is affected by the p_R rest state probability. It is concluded that network blocking levels affects to elasticity of traffic sharing.

3.10 Cyber Crime Based

The identification of cyber crime is performed by some known examples such as wrong letters, threaten to others, unlikely mails, hacking of data and phishing cyber attacks. The probabilistic model for the cyber crime evaluation has been proposed which examines the elasticity of Markov Chain model. In the two operators environment how the traffic sharing be calculated with the crime probability's effects is the main objective to be obtained. A simulation study is performed for the discrimination of effects in model parameters. In the study by Shukla et al. [11] it is observed that in case if cyber crime probability is high than to a specific network operator some high proportion of traffic collapses. In the sharing of traffic for cyber criminal, for the operator O_1 the elasticity becomes negative and for the operator O_2 the elasticity is positive and stable. From here onwards a conclusion is made that in traffic sharing for operator O_1 the fluctuation in L_1 is harmful as operator's blocking probability is a serious matter.

Network blocking probabilities is the main cause of high affection of elasticity's of traffic sharing. In case of larger blocking level of operator's network a highest collection of users and cyber criminals can be reduced. The operator's and cyber criminals achieve traffic share in case when other operator's network blocking is high. The cyber criminal shifting towards first operator is possible due to minimum network blocking and it should always be the case to lower down the blocking so that to attract maximum users in the competitive environment.

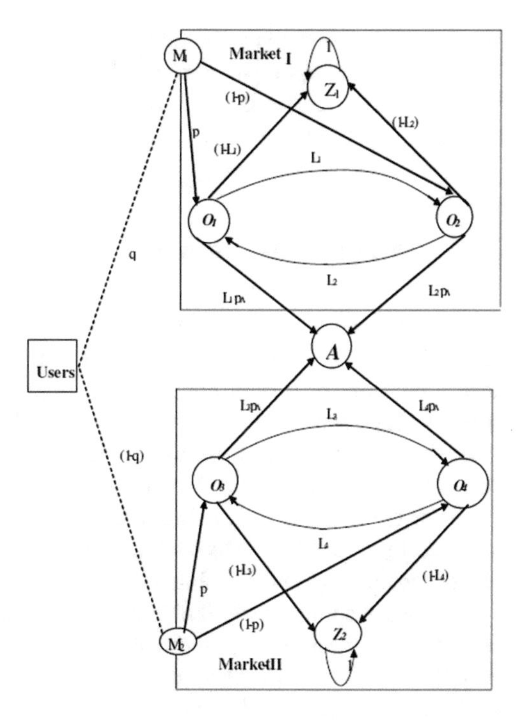

Fig. 5 Transition diagram of model (using Shukla et al. [12])

The contribution by Shukla et al. [12] says market position decides the value of elasticity. The highest level of market leads to priority position which represents analysis of elasticity between traffic sharing pattern of operators. For analysis of sharing of traffic of operators and the impact of elasticity on this a simulation study has been performed. Elasticity and blocking probability depends on one another. The in nature negative elasticity becomes high when the other operator's blocking is high. A highest elasticity level is achieved when market is of high priority. There is always a possibility that maximum abandon choice generates stable pattern of traffic share which are independent to the blocking probability (Fig. 5).

3.11 Re-attempt Based

In place of broadband people uses dial-up connection to avoid non-connectivity and call-disconnectivity. In the environment of traffic sharing of Internet and user's behavior the Markov chain model is used and the rejoining probability is observed by this contribution by Shukla et al. [13]. User's Re-attempt connectivity is the cause of getting affected to an attempt process. The success of connectivity depends on the re-attempt probability which must be high. To simplify cluster based Internet service management a new persistent Internet connectivity service is designed. The re-attempting connectivity for masses are successfully brings Internet services. With the help of re-attempting data structure the rejoin probability comes with the concept if ISP and connections are failed, we can connect and reconnect the Internet and service connectivity.

3.12 Specific Index Based

A computer and a network is responsible for all unlawful activities which refers to cyber crime. For the crime happening and for the target of crime computers are used. The main focus of criminal activity is to manipulate the Internet for intended purpose. The security of Nation and even its financial matters will be challenged. Once the confidential and useful information is accessed by unauthorized entity to do some modification in it or for the harm to the information, this leads to the breach problems of privacy policy. Issues related to Internet such as security and privacy are removed by rules of security and distribution of Internet traffic. Strengthening the security is the major need in Internet environment today.

Indexing techniques by Shukla et al. [14] represented which coincides various factors into one value to predict the Internet user's behavior. In computer network for cyber criminals how the index based analysis be presented is contributed by this work and in order to find some mathematical inferences the simulation analysis is performed. Two difference type of patterns shown by the indices at the different blocking stages, but in case if self network blocking is high the index reduces itself. Finally to achieve better quality of services the blocking conditions of operators must be removed as a concluding remark.

3.13 Least Square Curve Fitting Based

Various Internet service providers are in competition at various stages to provide the Internet services. Congestion is the major problem which leads to a barrier in order to extract the services provided by the ISP. This problem of congestion is suffered by network so user always demands for a network of low congestion. During the process of repetition in connection for attempts call-by-call effort is maintained by user. A Markov chain model by Shukla et al. [15] has been established a relationship between Internet access traffic sharing and network blocking probability which is simplified to linear relationship. The determination coefficient used to find accuracy of the fitting so that the complicated relationship of multiple parameter has been eased up to a much simplified form. By using least square method the linear relationships are represented through tables. In between P_1 and L_1 the value nearly equal to 1 shows the best fitting straight line for coefficients of determination (COD). The confidence interval for estimated value shows good fitting line. it defines

$$P_1 = a + b(L_1)$$

For prediction of best fitted linear relationship the average is calculated as under by the table where a and b are average estimates

$$P_1 = 0.6237 + 0.6029(L_1)$$

3.14 Curve Fitting and Cyber Crime Based

All round the world cyber crime is spreading so fast cyber crime correlated with Internet traffic growth. A Markov Chain Model by Shukla et al. [16] examines the inter-relationship between blocking probability and traffic sharing. Model based relations are derived and extended for the two-call basis analysis. Least square based curve analysis between network blocking and traffic sharing for Markov Chain model based relationship is presented. The computed determination coefficient shows high value towards unity. For expressing the complex relationship there is a thumb rule that a simplified linear relationship perform well between traffic sharing and network blocking probability. When defined

$$P_1 = a + b(L_1)$$

We found that

$$P_1 = 0.400915 - 0.381292(L_1)$$

For P_1 and L_1 there is no direct relation exists. Least square method and the model used together to simplify the relationship.

$$P_1 = 0.400915 - 0.381292(L_1)$$

This linear type of relationship is equal to a firsthand rule that the determination coefficient shows strength of fitting straight line which is nearly equal to 1.

3.15 Curve Fitting and Two Market Environment

Using Markov chain model by Shukla et al. [17] a numerical based relationship has been derived between system parameters and variables. The relationship should be simple enough to analyze the output variable in case of known input variables. A linear and simple relationship is presented by the least square curve fitting approach which makes the complicated relationship simpler and linear. This technique of two operators and two market environment is applied in traffic sharing of Markov chain

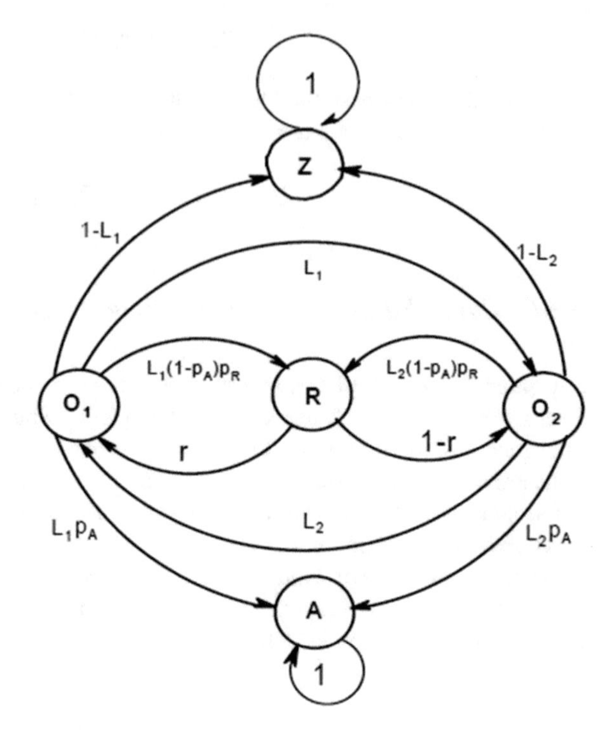

Fig. 6 Transition model (using Shukla and Thakur [6])

model. In between two prime system variables the determination's coefficients are used for line fitting judgment of accuracy (Fig. 6).

3.16 Bounded Area Based

Naldi [1] presented the problem of two operator's traffic sharing of Internet. Author has also developed numerical relationship in between network blocking probability and traffic share. A probabilistic function of quadratic nature is given by Gangele et al. [18] with some mathematical relationship and a fix bounded area. For estimation of bounded area a Trapezoidal rule of numerical quadrature is proposed. Bounded area is in direct proportion to choice of customer and network blocking. It explain the relationship of traffic share and network parameters of computer. In case when p_A, L_2 remain constants, initial choice p is in direct proposition to the bounded area A and this has only 5–10% increase if p is doubled and rate of increment is slow. 50% of the level of bounded area A is approached if p is at highest level. The network competitor's opponent blocking probability L_2 is in proportion to the bounded area A. In area A, the maximum area is 0.69 and the rate of increment is on the 0.9 maximum level. There exist inversely proportional relationship between bounded area A and p_A. Bounded area A will decrement faster to the larger p_A. There is a huge information about traffic sharing which is revealed by the observations in bounded area A. The two factors that is the initial preference p of consumer and the blocking probability of the owner's network is directly proportional to the bounded area A. The bounded area reduces when the blocking probability of network competitor increases.

3.17 Area Computation Based

From research point of view cybercrime is always an interesting area to be explored various criminal tasks are done by the use of Internet which threatens to the speed, convenience and Internet anonymity. Naldi [1] contributed the problem of Internet traffic share in a new style with the use of Markov chain model to generate a connection between cyber criminals traffic share and operator's blocking probability. In a definite bounded area a probability function is produced due to this relationship. This type of estimation is done for the bounded area with various network parameters. The numerical analysis applied method such as trapezoidal method by Gangele and Shukla [19] is used to attempt this kind of area. The observation says that other parameters and the cyber crime's probabilistic attraction are inversely proportional to one another. Area is inversely proportional to C_1 [$(1 - C_1)$ is cyber crime probabilistic attraction]. Another inversely propositional relationship exist between p and p_A for the other type of operator in condition of the owner operator's predefined inputted network parameters. It gives initial information of relationship between cybercrime based user behavior traffic share and network blocking operator's parameter.

3.18 Two-Call Index Based

In the era of computers and networks cyber crime generates many criminal acts. Heavy load of Internet traffic encompasses the competitors to gain much more share of Internet. By Gangele and Dongre [20] Index formation and Simulation study for mathematical findings are done with help of this approach for different kind of users. Indexing technique plays an important role for prediction of behavior of cyber criminals especially in two-call basis setup. Index formations of different categories indicate visibility of probability based study of user behavior. The conclusion is that in case of Internet traffic sharing in two-call based setup ratio based user index provides probability based evaluation of cyber crime behavior for different stage of network blocking.

3.19 Area Estimation Based

In the environment of Internet the congestion is a very big problem and various technical developments are in progress to avoid congestion problem in computer networks and now in a modern era to avoid congestion ISP's try to make congestion control and congestion avoidance techniques. Because of the heavy load and the fluctuation of operators Internet traffic sharing the disconnectivity appears on network. On the basis of Internet in Traffic share problem Naldi [1] explored a model for comparative operators mode for the organization of a relationship between probabilistic quadratic function and a fixed bounded area A. This area contains many network parameters to assess and estimate various properties of Internet traffic.

This is an appropriate approach using Simpson 1/3 rule usually applied in numerical quadrature and designed to estimate this type of bounded area by Gangele [21]. To tackle the problem of Internet congestion the share status of Internet traffic and network disconnectivity are two parameters onto which a mathematical relationship has to be maintained by using this approach. To provide a solution for the problem in networks Simpson's 1/3 rule of numerical analysis provides study and analysis of network parameters and the estimated bounded area and demonstrated an approach.

3.20 Numerical Analysis Based

The contribution by Gangele et al. [22] presents an analysis of numerical based method which compute probability and the bounded area. The two methods, i.e., Simpson 1/3 rule and Weddle rule are used for this analysis purpose and the comparison in both is performed. The probabilistic curve bounded area is a function of many input parameters. Most of cases, it bears the linearly increasing or decreasing trend on the variation of model parameters. The L2 and p, if increases, the corresponding area also increases but with the increase of pA the reverse pattern exists. Area computation by Waddles method is constantly higher than the Simpson 1/3 rule. By using Simpson procedure at $p = 0.9$ highest bounded area computation is 39% while at $p = 0.9$ the highest Weddle rule is 52%. Overall Weddle rule based computed values are constantly above than Simpson 1/3 rule.

3.21 Multi-operator Environment Based

By Gangele and Patil [23] some parameters like QoS, initial preferences and others has used to analyze the Internet traffic share of market based various types of operators have derived with simulation study to obtain traffic share. The correlation of traffic share is done by geographic position of market and network blocking probability. In two market based environment users transition behavior can be effectively obtained by Markov chain model. Different kinds of results can be observed at different stages of network blocking. In some cases there is overlapping of traffic share patterns and also some patterns of traffic share are in cubic form. Traffic share depends on market position, In two market based environment, in comparison to competitive service provider blocking probability of service provider is higher if market if of high priority.

3.22 *Rest State Based*

The contribution by Gangele and Patil [24] in the expression of Internet traffic share of market based has derived for multi-operator situation by using Markov chain model. In simulation studies it is found that marketing plan like rest state inclusion has the potential to enhance the traffic proportion of operators. User behavior analysis is an important task done by rest state based markov chain model. User's categories like FU, CIU, and PIU and network traffic share between operators also studied and observed that an exponential traffic pattern exist in market- I for second kind of operator in case when $r_1 = 5\%$, $L_1 = 35\%$, $q = 25\%$, $p = 30\%$ and $p_A = 15\%$. Similar traffic pattern also found in market-II in second kind of operator when $L_2 = 30\%$, $r_1 = 25\%$, $q = 35\%$, $p = 15\%$ and $p_A = 25\%$. In case of FU when $P_{R1} = 25\%$ and $P_{R2} = 35\%$ traffic pattern overlapped for I and II kind of market with constant network parameters. It is concluded that p_R and r if both have increases, then FU proportion for operator uplifts and they have hard core stuff for increasing Internet traffic of operators.

References

1. Naldi M (2002) Internet access traffic sharing in a multiuser environment. Comput Netw 38:809–824
2. Shukla D, Thakur S, Deshmukh AK (2009) State probability analysis of internet traffic sharing in computer network. Int J Adv Netw Appl 1(2):90–95
3. Shukla D, Tiwari V, Kareem PA (2009) All comparison analysis in internet traffic sharing using Markov chain model in computer networks. Georgian Electron Sci J CS Telecommun 23(6):108–115
4. Shukla D, Tiwari VK, Thakur S, Tiwari M (2009) A comparison of methods for internet traffic sharing in computer network. Int J Adv Netw Appl 01(03):164–169
5. Shukla D, Tiwari V, Thakur S, Deshmukh AK (2009) Share loss analysis of internet traffic distribution in computer networks. Int J Comput Sci Secur 3(4):414–427
6. Shukla D, Thakur S (2010) Index based internet traffic sharing analysis of users by a Markov chain probability model. Karpagam J Comput Sci 4(3):1539–1545
7. Shukla D, Tiwari V, Parchur AK, Thakur S (2010) Effects of disconnectivity analysis for congestion control in internet traffic sharing. Int J Comput Internet Manag 18(1):37–46
8. Shukla D, Thakur S, Tiwari V (2010) Stochastic modeling of internet traffic management. Int J Comput Internet Manag 18(2):48–54

9. Tiwari VK, Thakur S, Shukla D (2011) Analysis of internet traffic distribution for user behavior based probability in two-market environment. Int J Adv Netw Appl 30(8):44–51
10. Shukla D, Gangele S, Verma K, Trivedi M (2011) Elasticity variation under rest state environment in case of internet traffic sharing in computer network. Int J Comput Technol Appl 2(6):2052–2060. ISSN: 2229-6093
11. Shukla D, Gangele S, Verma K, Trivedi M (2011) Two-call based cyber crime elasticity analysis of internet traffic sharing in computer network. Int J Comput Appl 2(1): 27–38. ISSN: 2250-1797
12. Shukla D, Gangele S, Verma K, Singh P (2011) Elasticity of internet traffic distribution in computer network in two market environment. J Glob Res Comput Sci 2(6):06–12
13. Shukla D, Verma K, Gangele S (2011) Re-attempt connectivity to internet analysis of user by Markov chain model. Int J Res Comput Appl Manag 1(9):94–99. ISSN: 2231-1009
14. Shukla D, Gangele S, Verma K, Thakur S (2011) A study on index based analysis of users of internet traffic sharing in computer networking. World Appl Program 1(4):278–287. ISSN: 2222-2510
15. Shukla D, Verma K, Gangele S (2012) Least square based curve fitting in internet access traffic sharing in two operator environment. Int J Adv Netw Appl 43(12): 26–32. ISSN: 0975-8887
16. Shukla D, Verma K, Bhagwat S, Gangele S (2012) Curve fitting analysis of internet traffic sharing management in computer network under cyber crime. Int J Adv Netw Appl 47(24):36–43. ISSN: 0975-8887
17. Shukla D, Verma K, Gangele S (2012) Curve fitting approximation in internet traffic distribution in computer network in two market environment. Int J Comput Sci Inform Secur 10(4):71–78
18. Gangele S, Verma K, Shukla D (2014) Bounded area estimation of internet traffic share curve. Int J Comput Sci Bus Inform 10(1):54–67. ISSN: 1694-2108
19. Gangele S, Shukla D (2014) Area computation of internet traffic share problem with special reference to cyber crime environment. Int J Comput Netw Wirel Commun 4(3):208–219. ISSN: 2250-3501
20. Gangele S, Dongre A (2014) Two-call Index based Internet traffic sharing analysis in case of cyber crime environment of computer network. Int J Eng Trends Technol 13:271–280
21. Gangele S (2014) An approach for area estimation towards conjunction control of internet traffic sharing by using Simpson 1/3ed rule. Int J Eng Trends Technol 16(4):88–99
22. Shukla D, Verma K, Gangele S (2015) Approximating the probability of traffic sharing by numerical analysis techniques between two operators in a computer network. Am J Comput Sci Inform Technol 3(1):026–039. ISSN: 2349-3917
23. Gangele S, Patil S (2015) Internet traffic distribution analysis in case of multi-operator and multi-market environment of computer network. Int J Adv Netw Appl 130(4):29–36. ISSN: 0975-8887
24. Gangele S, Patil S (2016) Two-call and rest state based internet traffic sharing analysis in two market environment. Int J Eng Sci Technol 6(6):07–17. ISSN: 2250-3498

Anomaly Detection Using Dynamic Sliding Window in Wireless Body Area Networks

G. S. Smrithy, Ramadoss Balakrishnan and Nikita Sivakumar

Abstract Anomaly detection is one of the critical challenges in Wireless Body Area Networks (WBANs). Faulty measurements in applications like health care lead to high false alarm rates in the system which may sometimes even causes danger to human life. The main motivation of this paper is to decrease false alarms thereby increasing the reliability of the system. In this paper, we propose a method for detecting anomalous measurements for improving the reliability of the system. This paper utilizes dynamic sliding window instead of static sliding window and Weighted Moving Average (WMA) for prediction purposes. The propose method compares the difference between predicted value and actual sensor value with a varying threshold. If average of the number of parameters exceed the threshold, true alarm is raised. Finally we evaluate the performance of the proposed model using a publicly available dataset and has been compared with existing approaches. The accuracy of the proposed system is evaluated with statistical metrics.

Keywords Anomaly detection · Dynamic sliding window · Prediction
Weighted moving average · Wireless body area networks

G. S. Smrithy (✉) · R. Balakrishnan
Department of Computer Applications, National Institute of Technology,
Tiruchirappalli, Tiruchirappalli, India
e-mail: smrithygs1990@gmail.com

R. Balakrishnan
e-mail: brama@nitt.edu

N. Sivakumar
Department of CSE, National Institute of Technology, Tiruchirappalli, Tiruchirappalli, India
e-mail: nikita.siva@gmail.com

D. K. Mishra et al. (eds.), *Data Science and Big Data Analytics*,
Lecture Notes on Data Engineering and Communications Technologies 16,
https://doi.org/10.1007/978-981-10-7641-1_8

1 Introduction

Wireless Body Area Network (WBAN) [1] is a recent advancement in real time healthcare systems. WBAN offers medical professionals the ease of continuous monitoring of patients by enabling them to do it remotely. In contrast to the traditional healthcare systems, modern healthcare systems utilizing WBAN can effectively reduce prolonged stay in hospital, betterment the patient treatments by continuous monitoring rather than occasional assessments, affordable treatment expenditure, etc. Figure 1 shows a typical system model for WBAN in a healthcare system. It consists of wearable and implanted sensors, a base station and healthcare professionals. A WBAN consists of wireless sensors which are used to monitor vital human body actions (e.g., motion sensors) and parameters [e.g., body temperature, blood pressure, pulse (heart rate), and breathing rate (respiratory rate)]. These sensors are either embedded (wearable sensors) or implanted (implanted sensors) in the human body. Implanted devices endure from resource constraints such as battery power, storage, etc. On the contrary, wearable devices have fewer resource constraints. The sensors transmit the data to the base station which can be a mobile device (smart phone) having higher computational power, storage capacity, and longer transmission range. The base station analyses the received data (different body parameters) and sends to healthcare professionals that refer to the doctors and nurses or other experts.

The accuracy of diagnosis heavily depends on the reliability of sensor data. Sometimes sensor data becomes unreliable due to reasons such as hardware failure, damaged sensors, flaws, mischievous data injection, etc. These faulty sensor data may lead to large number of false alarms which are not at all acceptable in a critical scenario like healthcare. Thus there is a need to reduce false alarms by identifying such unexpected observations for reliable monitoring system. Haque et al. [2] proposed an algorithm for anomaly detection in wireless sensor networks for health care. The major limitation is they used static size sliding window of historical data for predic-

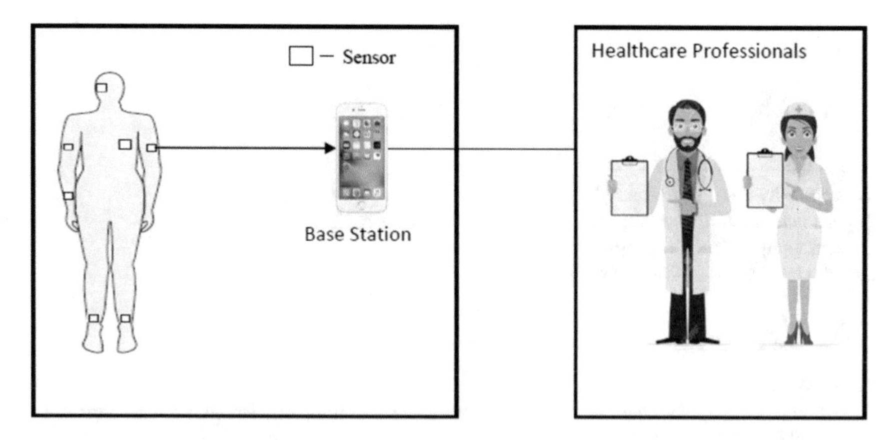

Fig. 1 System model for WBAN in a healthcare system

tion. The proposed work in this paper uses dynamic sliding window to reduce the overhead of considering huge volume of historical data for prediction.

1.1 Major Contributions

The contributions of this paper are highlighted as follows:

1. We propose an algorithm for identifying anomalous sensor measurements in WBAN using dynamic sliding window and WMA for prediction.
2. The performance of the proposed method is compared with existing methods [2] in terms of statistical metrics.

1.2 Paper Organization

The rest of this paper is organized as follows. Section 2 covers dynamic sliding window. Section 3 describes about WMA for prediction purposes. Section 4 details the proposed work. Section 5 provides experimental results and discussion. Section 6 concludes the paper.

2 Dynamic Sliding Window Approach

Prediction models usually make use of large amount of historical data or make use of sliding windows with static size. In [3], the authors proposed a dynamic sliding window for traffic prediction in a cloud computing environment. But no work has ever been reported which makes use of a dynamic sliding window for prediction in a healthcare environment. This paper makes use of dynamic window concept for prediction of physiological parameter value for comparison purposes.

Input:
$\mathbf{CurSW_{\mu}}$: Mean of Current Sliding Window
$\mathbf{PrevSW_{\mu}}$: Mean of Previous Sliding Window
CurSW: Current Sliding Window
$\boldsymbol{\alpha}$: Significance Level
Output:
Successor Sliding Window Size, $\mathbf{SucSW_{size}}$

1 **Procedure** Successor_Sliding_Window_Size
2 **Begin**
3 $\mathbf{SucSW_{size}} = \text{find.size}(\mathbf{CurSW})$
4 $\mathbf{Variation_{avg}} = \left[\frac{\mathbf{CurSW_{\mu}}^2 - \mathbf{PrevSW_{\mu}}^2}{\mathbf{CurSW_{\mu}.PrevSW_{\mu}}}\right]$
5 **If** $\left(\mathbf{Variation_{avg}} > (\mathbf{1} + \boldsymbol{\alpha})\right)$
6 $\mathbf{Size} = \frac{\mathbf{Variance_{max}}}{\mathbf{Variance}}$
7 **If** $\left(\mathbf{CurSW_{\mu}} > \mathbf{PrevSW_{\mu}}\right)$
8 $\mathbf{SucSW_{size}} = \mathbf{SucSW_{size}} + \mathbf{Size}$
9 **Else**
10 $\mathbf{SucSW_{size}} = \mathbf{SucSW_{size}} - \mathbf{Size}$
11 **End If**
12 **End If**
13 **return** $\mathbf{SucSW_{size}}$
14 **End**

The algorithm takes as input mean of the current sliding window, mean of previous sliding window, current sliding window, and a significance level. The algorithm predicts the size of the successive sliding window based on the variance between the predecessor sliding window and current sliding window. A larger variance indicates a large variation from the mean and a smaller variance indicates a value closer to the mean. The variance of the extreme values of a window is considered as the maximum variance in this algorithm [3]. Significance level is Type 1 error which is the probability of rejecting the null hypothesis when it is true. As a preliminary work, we are sticking for a significance value of 0.05. The algorithm starts with finding the size of the current sliding window and is stored in $SucSW_{size}$. In step 4, the algorithm finds the average variation between the current sliding window and the predecessor sliding window. If the variation is greater than the threshold which is $(1 + \alpha)$, the value which should be either added or subtracted with the current sliding window to get the new sliding window size is calculated as shown in step 6. If the mean of the current sliding window is greater than the predecessor sliding window, *Size* is added to $SucSW_{size}$, otherwise *Size* is subtracted to $SucSW_{size}$ to get the new sliding window size of the successor window.

3 Weighted Moving Average (WMA)

A Moving Average (MA) [4] is a statistical technique to analyze data points by creating sequence of averages of any subset of numbers. It is also referred as rolling average, running average, moving mean, and rolling mean. Moving Average methods is simple and low complex compared to other complex techniques such as Autoregressive Integrated Moving Average (ARIMA), Neural Networks etc. which demands large volume of data. The major advantage of using MA techniques is it can be used for short range time series data. Different types of moving average are Simple Moving Average (SMA), Cumulative Average (CMA) and Weighted Moving Average (WMA). A SMA is the unweighted mean of n past data points. A CMA is the running means of an ordered data stream. A WMA [5] assigns weights to data points that decrease in an arithmetic progression such that the most recent data point in the sample window will get the highest weight. Previous studies have used WMA for static size sliding windows. But no work has been reported that uses WMA in dynamic size sliding windows in WBAN scenario. In this paper, we use Weighted Moving Average for prediction purpose. The mathematical expressions for WMA in this context are explained in this section.

The initial computation of WMA for the first window at time t can be expressed as follows:

$$WMA_t = \frac{\sum_{i=1}^{t} W_i D_i}{\sum_{i=1}^{t} W_i}, \quad W_1 < W_2 < \cdots < W_t, \tag{1}$$

where W_i = Weight of ith data point and D_i = Data point at time i.

For calculating WMA across successive values (i.e., the sliding window size can be either expanded or diminished according to the output from the algorithm explained in section above), the following expression can be used:

$$WMA_{t+1} = \frac{\sum_{i=((t+1)-SucSW_{size}+1)}^{t+1} W_i D_i}{\sum_{i=((t+1)-SucSW_{size}+1)}^{t+1} W_i}, \quad W_{((t+1)-SucSW_{size}+1)} < \cdots < W_{(t+1)}, \tag{2}$$

where $SucSW_{size}$ = Successor Sliding Window Size.

4 Proposed Anomaly Detection Algorithm

The main objective of the proposed algorithm is to detect anomalies and to reduce false alarms. In this model we are considering *N* parameters. The algorithm takes as input actual value of *i*th parameter at time *t*, predicted value of *i*th parameter at time *t*, current sliding window of *i*th parameter. We initialize a counter *Pos* that represents the benign parameter count, PF_i and NF_i are set to zero.

Input:
$\boldsymbol{N}$: Number of parameters
$\boldsymbol{AValue_i}$: Actual Value of $\boldsymbol{i^{th}}$ parameter at time $\boldsymbol{t}$
$\boldsymbol{PValue_i}$: Predicted Value of $\boldsymbol{i^{th}}$ parameter at time $\boldsymbol{t}$
$\boldsymbol{W_i}$: current Sliding Window of $\boldsymbol{i^{th}}$ parameter at time $\boldsymbol{t}$, $(i = 1,2,\ldots,N)$
$\boldsymbol{Pos}$: Benign Parameters Count, Initially $\boldsymbol{Pos = 0}$
$\boldsymbol{PF_i}$: Positive flag for benign parameter $(i=1,2,\ldots,N)$, Initially set $\boldsymbol{PF_i} = \boldsymbol{0}$
$\boldsymbol{NF_i}$: Negative flag for abnormal parameter $(i=1,2,\ldots,N)$, Initially set $\boldsymbol{NF_i} = \boldsymbol{0}$
Output:
True sensor alarm or false sensor alarm

1 ***Procedure*** *Proposed_Anomaly_Detection*
2 **Begin**
3 $\boldsymbol{For}(i = 1\ to\ N)$
4 $n = find.size(\boldsymbol{W_i})$
5 $\boldsymbol{For}(j = 1\ to\ n - 1)$
6 $\boldsymbol{TH_i} = find.standarddeviation(\boldsymbol{W_i}[\boldsymbol{j}])$
7 ***End For***
8 $\boldsymbol{diff_i} = |\boldsymbol{AValue_i} - \boldsymbol{PValue_i}|$
9 $\boldsymbol{If}(\boldsymbol{diff_i} \leq \boldsymbol{TH_i})$
10 **Retain** $\boldsymbol{AValue_i}$
11 $\boldsymbol{Pos} = \boldsymbol{Pos} + \boldsymbol{1}$
12 $\boldsymbol{PF_i} = 1$
13 ***Else***
14 $\boldsymbol{NF_i} = 1$
15 ***End For***
16 $\boldsymbol{If}\left(\boldsymbol{Pos} \geq \left(\left\lceil \frac{\sum_{i=1}^{N} N}{N} \right\rceil\right)\right)$
17 *True Sensor Alarm*
18 ***Else***
19 *False Sensor Alarm*
20 **Update** all $\boldsymbol{AValue_i} = \boldsymbol{PValue_i}$ **in** $\boldsymbol{W_i}$ at time $\boldsymbol{t}$ with $\boldsymbol{NF_i} = 1, \forall \boldsymbol{i}, \boldsymbol{i} = \boldsymbol{1}, \boldsymbol{2}, \ldots, \boldsymbol{N}$
21 ***End If***
22 **End**

Initially the algorithm finds the size of sliding window of the ith parameter which is represented as n. Thus a sliding window W_i with size n has n elements in it. Then the algorithm computes the standard deviation of window W_i for $n - 1$ elements which is taken as the threshold value TH_i for the subsequent steps. The absolute difference between actual value and predicted value of ith parameter is represented as $diff_i$. If the absolute difference value is less than or equal to threshold, then actual value is retained in the sliding window and *Pos* is incremented by one and set positive flag PF_i for benign parameter (i = 1, 2, …,) to one. Positive flag for benign parameter (i = 1, 2, …,). Otherwise negative flag NF_i to one. Repeat steps 3–15 for all the N parameters. If the positive counter is greater than or equal to the floor value of mean of total number of parameters N in step 16, then raise true sensor alarm. Otherwise false sensor alarm is generated. Step 20 updates the actual sensor values of all the parameters with negative flag set to 1 with predicted sensor value for future processing.

5 Experimental Results and Discussions

The experiment was conducted on a performed on a desktop machine with an core i7-2600 processor with 3.40 GHz, windows 7 Professional (64-bit) operating system and 8 GB RAM.

5.1 Dataset Organization

For evaluating the proposed work, we have used a publically available dataset from Multiple Intelligent Monitoring in Intensive Care (MIMIC) database (MIMIC DB datasets 221) of Intensive Care Unit patients [6]. MIMIC DB datasets 221 covers logs from 1995-05-18, 08:00 am to 1995-05-19, 07:33 am with five physiological parameters namely Arterial blood Pressure (ABP), Heart Rate (HR), Pulse, Respiration, Oxygen Saturation (SpO2).

5.2 Results

In our experimentation, an anomaly is said to be detected if an alarm is raised in a window containing the anomaly. For experimentation purposes we injected anomalies randomly. For prediction of dynamic window size, we set the significance level $\alpha = 0.05$ for test runs. In contrast to [2], we utilize dynamic sliding window for prediction purposes which helps to reduce the overhead of considering huge volume of historical data for prediction purposes.

Table 1 Comparison of proposed method with existing methods

Metrics	Proposed method	Existing method [2]	SVM [8]	J48 [9]	MD [7]
TPR (%)	100	100	100	100	67
FPR (%)	4.04	5.08	20	30	36
TNR (%)	95.96	94.92	80	70	64
FNR (%)	0	0	0	0	33

Based on experimentation, we are considering a total of 10,449 windows. Initially, the sliding window size for each parameter is fixed as 100. Then the model automatically calculates the dynamic sliding window size based on the variance of current and predecessor sliding window. Out of 10,449 windows, we randomly injected anomalies in 1449 windows. Thus we have 9000 benign windows out of which 8636 have been correctly classified and 1449 anomalous windows out of which all have been correctly classified. The proposed method correctly classified all the anomalous windows, but misclassified 364 benign windows as anomalous. The accuracy results of the proposed algorithm are based on the following statistical metrics.

$$\text{True Positive Rate (TPR)} = \frac{\text{True Positive}}{\text{True Positive} + \text{False Negative}}$$

$$\text{False Positive Rate (FPR)} = \frac{\text{False Positive}}{\text{False Positive} + \text{True Negative}}$$

$$\text{True Negative Rate (TNR)} = \frac{\text{True Negative}}{\text{True Negative} + \text{False Positive}}$$

$$\text{False Negative Rate (FNR)} = \frac{\text{False Negative}}{\text{True Positive} + \text{False Negative}}$$

Table 1 reflects the overall accuracy statistics for the proposed algorithm when compared with existing approach [2], Mahalanobis Distance (MD) [7], Linear SVM [8] and J48 [9].

Figure 2 depicts the performance of the algorithm in terms of specified statistical metrics for *MIMIC DB datasets 221*. Thus it shows the higher accuracy of our proposed method.

6 Conclusions

The proposed method used dynamic sliding window instead of static sliding window which mitigated the overhead of considering huge volume of historical data for prediction purposes. For predicting the sensor value, we used Moving Average (WMA) which is a simple and efficient technique for short range predictions. The proposed

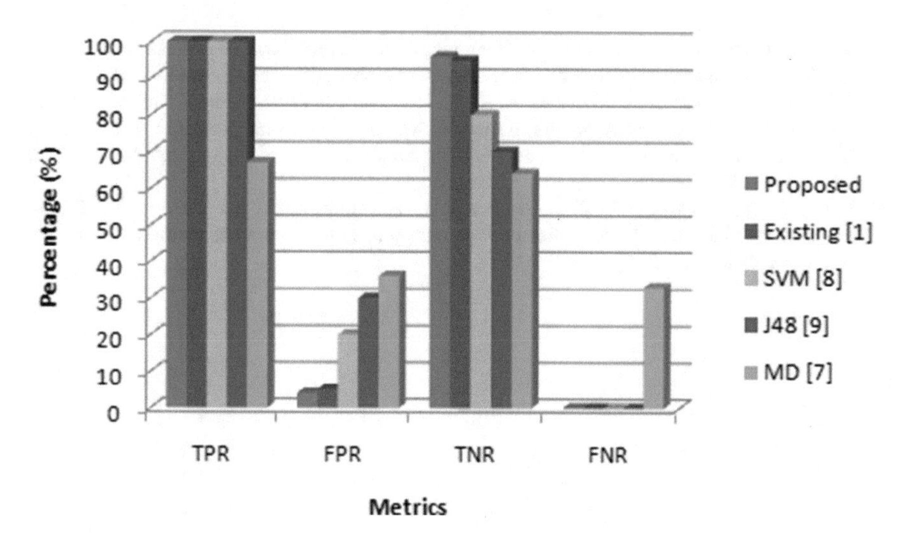

Fig. 2 Performance of the proposed method with existing methods

method was evaluated with MIMIC DB datasets 221. The proposed method achieved 100% TPR (Detection Rate) which is same as existing method [2], but reduces the FPR by 20.47% when compared with existing method [2].

Acknowledgements This research work was supported by Department of Electronics and Information Technology (DeitY), a division of Ministry of Communications and IT, Government of India, under Visvesvaraya Ph.D. scheme for Electronics and IT.

References

1. Li F, Hong J (2016) Efficient certificateless access control for wireless body area networks. IEEE Sens J 16(13):5389–5396
2. Haque SA, Rahman M, Aziz SM (2015) Sensor anomaly detection in wireless sensor networks for healthcare. Sensors 15(4):8764–8786
3. Dalmazo BL, Vilela JP, Curado M (2014) Online traffic prediction in the cloud: a dynamic window approach. In: 2014 international conference on future internet of things and cloud (FiCloud). IEEE, pp 9–14
4. Murphy C, Moving averages tutorial. http://courses.jmsc.hku.hk/jmsc7008spring2012/files/2010/02/MovingAverages.pdf
5. Dash S, A comparative study of moving averages: simple, weighted, and exponential. http://www.tradestation.com/education/labs/analysisconcepts/a-comparative-study-of-moving-averages
6. PhysioNet. http://www.physionet.org/physiobank/database/mimicdb/

7. Salem O, Guerassimov A, Mehaoua A, Marcus A, Furht B (2014) Anomaly detection in medical wireless sensor networks using SVM and linear regression models. Int J E-Health Med Commun
8. Salem O, Guerassimov A, Mehaoua A, Marcus A, Furht B (2013) Sensor fault and patient anomaly detection and classification in medical wireless sensor networks. In: Proceedings of 2013 IEEE international conference on communications (ICC), Budapest, Hungary, 9–13 June 2013, pp 4373–4378
9. Liu F, Cheng X, Chen D (2007) Insider attacker detection in wireless sensor networks. In: Proceedings of 26th IEEE international conference on computer communications, Anchorage, AK, USA, 6–12 May 2007, pp 1937–1945

Effective Healthcare Services by IoT-Based Model of Voluntary Doctors

Bharat B. Prajapati, Satyen M. Parikh and Jignesh M. Patel

Abstract There is dearth of skilled doctors in developing countries like India, various health challenges and high growth of population patient are required to treat in hospital. In this research, we proposed model aimed to design and develop a system, which connect doctors to hospital needs their expertise for treatment of patient. The proposed system allows capture patient healthcare data, store in database and transmit on cloud through various sensors attached to patient bodies. Based on patient data basic, voluntary doctors suggest appropriate treatment and medicine doze based on healthcare data and treatment requirement. It may save life of the patient and further this platform may be helpful to share opinion by analyzing changing capture patient healthcare data.

Keywords Internet of things · Voluntary · Time · Doctor · Sensor

1 Introduction

In last decade human population is increasing day by day. Various health challenges and high growth of population patient are required to treat in hospital. At one side some of the country facing problem of inadequate doctors to treat patient while other side there are some skilled doctors who are voluntary ready to treat the patient. Voluntary skilled doctors want to contribute the human society but due to certain problems, i.e., war, security, traveling, cost, border limitation, etc., they are unable to contribute. In this proposed model, we are proposing system which will provide

B. B. Prajapati (✉) · S. M. Parikh (✉)
AMPICS, Ganpat Vidyanagar, Gozaria Highway, Mehsana 384012, Gujarat, India
e-mail: bbp03@ganpatuniversity.ac.in

S. M. Parikh
e-mail: satyen.parikh@ganpatuniversity.ac.in

J. M. Patel (✉)
DCS, Ganpat Vidyanagar, Gozaria Highway, Mehsana 384012, Gujarat, India
e-mail: jmp03@ganpatuniversity.ac.in

D. K. Mishra et al. (eds.), *Data Science and Big Data Analytics*,
Lecture Notes on Data Engineering and Communications Technologies 16,
https://doi.org/10.1007/978-981-10-7641-1_9

platform to voluntary doctors and health center whom require skilled doctors to treat the patient. Patient profile which include healthcare data, photos and type of treatment require transmitted on cloud through various sensors attached to patient bodies and caretaker. Caretaker person from health center, create the patient pro le with types of treatment require. Doctors suggest appropriate treatment and medicine doze based on patents medical history uploaded by caretaker.

2 Related Work

Proposed research efforts are encouraged to treat the patient from remote place through mobile application, web services, and cloud computing [1–4]. A conceptual framework also proposed for IoT which stores the capture the real-time data and store it on cloud for further process [5]. Automated ECG signals analyze through an IoT-based intelligent nurse for detection of occurrence of Arrhythmia [6]. IoT-based remote application used to take care of the elders who are alone at home. Application gives notification if any health emergency occurs [7–9]. Individual medical data measured through wearable health devices which further send to the medical experts for advice [10]. eBPlatform helps to establish the link between medical experts and patients. eBPlatform also provide common platform in which medical experts give the service and patient get the bene t of medical treatment [11]. Acquisition of mobile health data, i.e., blood sugar level, ECG, blood pressure, asthma, etc., gathered via medical gadgets, wearable and application [12]. e-Health monitor system help to measure dynamic health parameters of patient and generate alert notification after analyzing data [13, 14].

3 System Design and Implementation

The proposed system is designed to get health information of useful to extend medical support by volunteer doctors to distance patients in emergency real time patient health care data make it more efficient. Devices are available, which can gather the data from patient's body and display it. We are extending facility to pass this information to communicate further and process it in desired way. Bedside monitors are devices which continuously reading patient data. Bedside patient monitors measure different human body parameter mention in Table 1. To achieve objective to store, communicate data from bedside monitors or other wearable devices, the following components would be needed.

Table 1 Typical health parameters

Typical health parameters
Arterial Carbon Dioxide Partial Pressure (PaCO2)
Arterial Oxygen Partial Pressure (PaO2)
Arterial Oxygen Saturation (SaO2)
Central Venous Pressure (CVP)
Central Venous Oxygen Saturation (ScvO2)
End Tidal Carbon Dioxide (ETCO2)
Mean Arterial Pressure (MAP)
Mixed Venous Oxygen Saturation (SvO2)
Oxygen Saturation At The Peak Of Blood Pulsation (SpO2)
Positive End Expiratory Pressure (PEEP)
Pulmonary Artery Catheter (PAC)
Systemic Vascular Resistance (SVR)

3.1 Sensor

Sensors are preliminary responsible to capture continuously patients health data. It can be part of wearable devices or bedside monitor system (Fig. 1).

Pulse oximeter is a sensor-based compact medical equipment used to capture an approximation of the amount of oxygen in the lifeblood and pulse rate. Healthcare organizations mostly used it in daily routine as it has characteristic of low cost, light weight, compact, easy to use.

Typical sensors used in bedside monitor measured following parameters (Fig. 2).

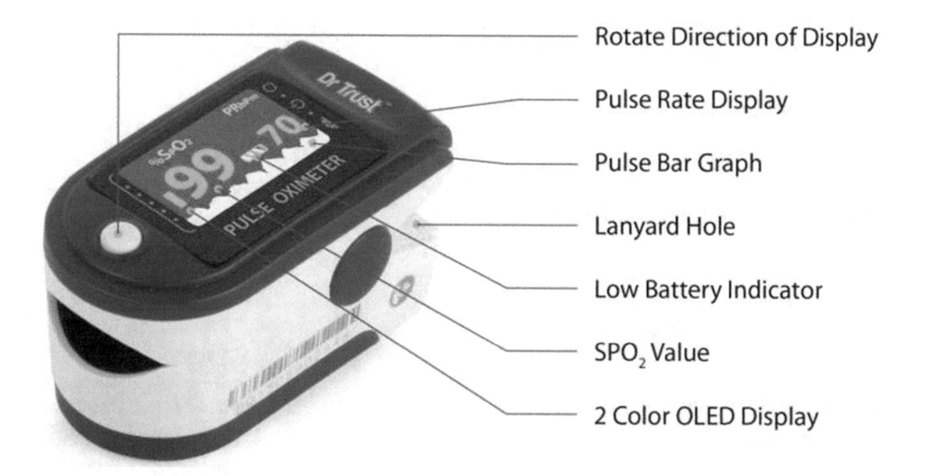

Fig. 1 Pulse oximeter

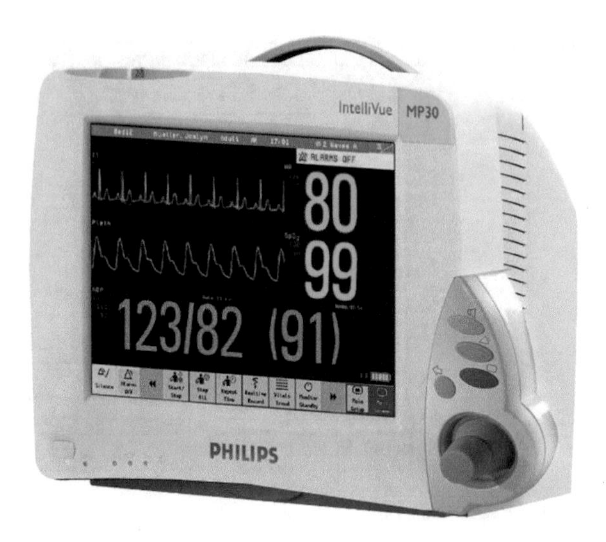

Fig. 2 Bedside patient monitor

Table 2 Standard health parameters (blood pressure)

Blood pressure (levels)	Superior number (systolic mm Hg)		Inferior number (diastolic mm Hg)
Acceptable	<120	&	<80
Pre-hypertension	120–139	or	80–89
(First level) high pressure—tension	140–159	or	90–99
(Second level) high pressure—tension	160 or higher	or	100 or higher

3.2 *Interconnection Networks*

Interconnection networking allows sensor to transmit captured health data to the other systems component like server, etc.

3.3 *Server and Database*

Serves need to manage all the received real-time patient data. Also server has database for standard health parameters. The system should be able to update standard health parameters as per WHO indicators. The following tables give an example of sample of standard health parameters (Tables 2 and 3).

In health care, critical patient are treating in ICU. Due to critical, health doctors needs to continuously monitor the patient. Doctor use bed side monitors for observ-

Table 3 Standard health parameters (heart rate for men and women)

Preferable heart beats (per minute) for male						
Age-interval	18–25	26–35	36–45	46–55	56–65	>65
Sportsperson	49–55	49–55	50–56	50–57	51–56	50–55
Outstanding	56–61	55–61	57–62	58–63	57–61	56–61
Good	62–65	62–65	63–66	65–67	62–67	66–69
Above-average	66–69	66–70	67–70	68–71	68–71	66–69
Below-average	74–81	75–81	76–82	77–83	76–81	74–79
Bad	Beyond 82	Beyond 82	Beyond 83	Beyond 84	Beyond 82	Beyond 80
Preferable heart beats (per minute) for female						
Age-interval	18–25	26–35	36–45	46–55	56–65	>65
Sportsperson	54–60	54–59	54–59	54–60	54–59	54–59
Outstanding	61–65	60–64	60–64	61–65	60–64	60–64
Good	66–69	65–68	65–69	66–69	65–68	65–68
Above-average	70–73	69–72	70–73	70–73	69–73	69–72
Below-average	79–84	77–82	79–84	78–83	78–83	77–84
Bad	Beyond 85	Beyond 83	Beyond 85	Beyond 84	Beyond 84	Beyond 84

ing the patient. Bedside monitors continuously transmitting patient data, i.e., Heart Rate, Blood Pressure, Oxygen Level, Pressure in Brain, etc., to local server for the analyzing purpose. An Intelligent real-time monitor system analyze the human body parameters. Following are the different standards recommended by the American Heart Association for blood pressure

In recent development, we have blood oxygen monitors, which are able to check blood and oxygen level from the human body, so it is easy to connect output of this monitor to IRTBS.

Pressure in the Brain (Intracranial Pressure) The pressure inside the head may rise in patients with head injuries or after a stroke. The kind of brain pressure is known as intracranial pressure, it may block the blood flow in brain. A probe can be inserted in the brain to measure and help doctors provide therapies to reduce it. Value up to 20 mm Hg are acceptable, but the team may decide to use higher values on some patients.

3.4 *Cellular Phone or Personal Digital Assistant (PDA)*

People from the different location, communicate which each other through cellar phone. Communication through mobile phones not only reduces the cost but it also provides faster way of communication between peoples. Doctors can take decision and provide medical treatment to critical patient.

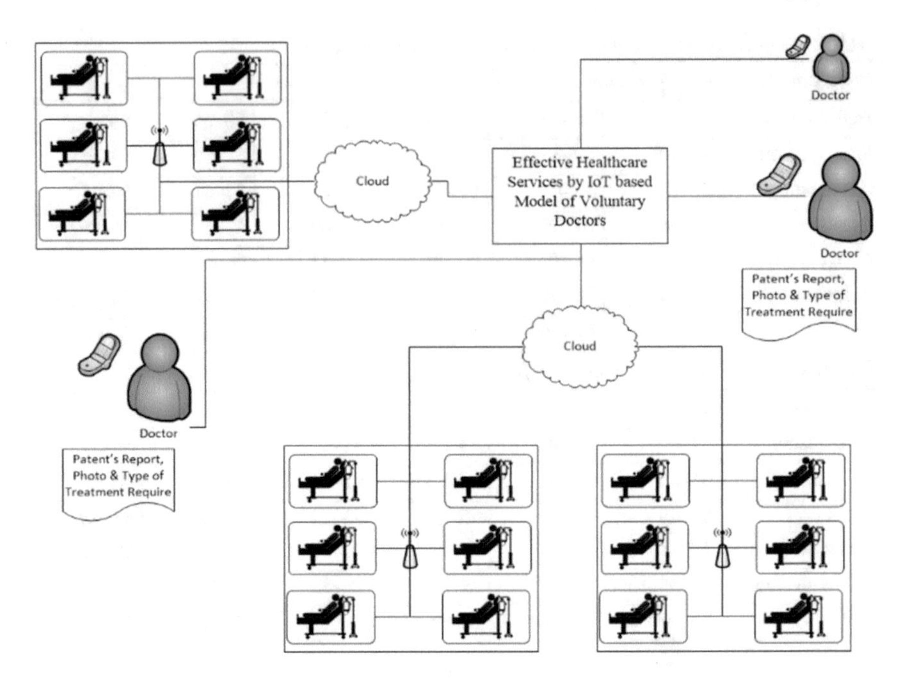

Fig. 3 Effective healthcare services by IoT-based model of voluntary doctors

3.5 Intelligent Software Agent

Intelligent software agent are automated code preliminary responsible to do essential processing on health data captured by the sensors. The processing may have different level. At very basic level the captured data stored in database. The intelligent automated software agents help for following activities.

- Voluntary doctors who want to serve the human society needs to register their self-using smart-phone.
- On other hand, care taker person from various hospitals capture the different health data with the help of medical equipment.
- Care take person upload the patient's profile of the patient.
- Patient profile includes real-time monitoring data—Blood Pressure, Oxygen Level, Pressure of Blood in brain, Heart Beat and Type of treatment required.
- Voluntary Doctors suggest the treatment based on patient's profile via smart-phone (Fig. 3).

3.6 System Architecture

The following diagram represents the architecture of system for typical patient monitoring. This research work aimed to achieve following objectives:

- Capture real time patient healthcare data for storage (future reference) and transmit to doctors (Analyzing and advise).
- Extend medical support by volunteer doctors to distance patients in emergency
- Provide platform to share opinion by analyzing changing capture patient healthcare data between doctors.
- Provide opportunity to doctors to voluntary serving the society betterment.

4 Conclusion and Future Work

This research work will be useful to extend medical support by volunteer doctors to distance patients in emergency and real-time patient healthcare data make it more efficient. Research will also be useful as in India, there are dearth of qualified doctors especially in case of emergency at rural areas. At international level at the time of war, security, traveling, cost, border limitation such project would be useful.

Acknowledgements We are thankful to Dr. Pritesh D. Shah, Mahavir Hospita, Ahmedabad, Dr. Mukul M. Shah, Sterling Hospital and Civil Hospital to extend their support to pursue research work and providing access to their knowledge of various sensor-based healthcare devices. We have received valuable inputs regarding challenges and restriction to implement the system. We are also thankful to below-mentioned researcher in references, who has inspired us by their research work and publications.

References

1. Mohammed J, Lung C-H, Ocneanu A, Thakral A, Jones C, Adler A (2014) Internet of things remote patient monitoring using web services and cloud computing. In: IEEE international conference on internet of things (iThings). IEEE, pp 256–263
2. Ghosh AM, Halder D, Alamgir Hossain SK (2016) Remote health monitoring system through IoT. In: 5th international conference on informatics, electronics and vision (ICIEV), pp 921–926
3. Kumar R, Pallikonda Rajasekaran M (2016) An IoT based patient monitoring system using raspberry Pi. In: International conference on computing technologies and intelligent data, vol 35, no 2, pp 1–4
4. Dhanaliya U, Devani A (2016) Implementation of e-health care system using web services and cloud computing. In: International conference on communication and signal processing—(ICCSP), pp 1034–1036
5. Tyagi S, Agarwal A, Maheshwari P (2016) A conceptual framework for IoT-based healthcare system using cloud computing. In: 6th international conference—cloud system and big data engineering, pp 503–507

6. Nigam KU, Chavan AA, Ghatule SS, Barkade VM (2016) IoT-beat an intelligent nurse for the cardiac patient. In: International conference on communication and signal processing (ICCSP), pp 0976–0982
7. Al-Adhab A, Altmimi H, Alhawashi M, Alabduljabbar H, Harrathi F, ALmubarek H (2006) IoT for remote elderly patient care based on fuzzy logic. In: International symposium on networks, computers and communications (ISNCC), pp 1–5
8. Stutzel MC, Fillipo M, Sztajnberg A, Brittes A, da Motta LB (2016) SMAI mobile system for elderly monitoring. In: IEEE international conference on serious games and applications for health (SeGAH), pp 1–8
9. Blumrosen G, Avisdris N, Kupfer R, Rubinsky B (2011) C-smart: efficient seamless cellular phone based patient monitoring system. In: IEEE international symposium on a world of wireless, mobile and multimedia networks, pp 1–6
10. Lee BM, Ouyang J (2014) Intelligent healthcare service by using collaborations between IoT personal health devices. Int J Bio-Sci Bio-Technol 6(1):155–164
11. Liu Y, Niu J, Yang L, Shu L (2014) eBPlatform an IoT-based system for NCD patients homecare in China. In: IEEE global communications conference, pp 2448–2453
12. Almotiri SH, Khan MA, Alghamdi MA (2016) Mobile health (m-health) system in the context of IoT. In: IEEE 4th international conference on future internet of things and cloud workshops (FiCloudW), pp 39–42
13. Biswas S, Misra S (2015) Designing of a prototype of e-health monitoring system. In: IEEE international conference on research in computational intelligence and communication networks (ICRCICN), pp 267–272
14. Hassanalieragh M, Page A, Soyata T, Sharma G, Aktas M, Mateos G, Kantarci B, Andreescu S (2015) Health monitoring and management using internet-of-things (IoT) sensing with cloud-based processing: opportunities and challenges. In: IEEE international conference on services computing, pp 285–292

Emotional State Recognition with EEG Signals Using Subject Independent Approach

Pallavi Pandey and K. R. Seeja

Abstract EEG signals vary from human to human and hence it is very difficult to create a subject independent emotion recognition system. Even though subject dependent methodologies could achieve good emotion recognition accuracy, the subject-independent approaches are still in infancy. EEG is reliable than facial expression or speech signal to recognize emotions, since it can not be fake. In this paper, a Multilayer Perceptron neural network based subject-independent emotion recognition system is proposed. Performance evaluation of the proposed system, on the benchmark DEAP dataset shows good accuracy compared to the state of the art subject independent methods.

Keywords Electroencephalogram · Affective computing
Multilayer perceptron

1 Introduction

Emotions recognition comes under the field of study like affective computing and brain–computer interface (BCI). The electroencephalogram (EEG) signal represents the electrical activity of the neurons in the brain responding to any mental activity. EEG is the recording of the spontaneous electrical activity of the brain over a specific period of time. Electroencephalography (EEG) is a non-invasive method and therefore it is suitable to collect EEG data to recognize emotional states. The main challenge in developing subject-independent emotion recognition system is that the EEG varies from person to person. Moreover, on the same stimulus, one person may be less reactive than other.

P. Pandey (✉) · K. R. Seeja
Department of Computer Science & Engineering, Indira Gandhi Delhi Technical University for Woman, New Delhi, India
e-mail: jipallavi@gmail.com

K. R. Seeja
e-mail: seeja@igdtuw.ac.in

D. K. Mishra et al. (eds.), *Data Science and Big Data Analytics*,
Lecture Notes on Data Engineering and Communications Technologies 16,
https://doi.org/10.1007/978-981-10-7641-1_10

Emotion recognition system with EEG has several applications. For example, acquiring someone's emotion could assist therapists and psychologists in doing their job. In the field of BCI, the computer may adjust its behavior by observing the mood (based on emotion recognition) of the person. Nowadays, communication between people involves a large set of smileys and BCI with emotion recognition can add emotions automatically between communications. If someone's face is paralyzed or burned and not capable to show his emotion on his face then emotion recognition system can help them for medical purpose. Similarly, a human can communicate with robot effectively if emotions detection capability is there in the robot.

2 Related Work

A good amount of the literature on emotion recognition from EEG signals is available. Authors have used spectral features and features based on time to recognize emotions. Paul et al. [1] in the year 2015 have used Multifractral Detrended Fluctuation Analysis as features and proposed SVM for the classification of positive and negative emotions. They also compared SVM with other methods like LDA, QDA and KNN. Lahane and Sangaiah [2] have used kernel density estimation as EEG feature and used artificial neural network for classification. For data preprocessing they have used Independent component analysis (ICA). Singh et al. [3] have given a review on emotion recognition with EEG. They analyzed how emotions can be quantized and various steps involved in the process of emotion recognition. Multimodal approaches also exists to recognize emotions and to improve classifier accuracy. Researchers have used facial expression with EEG [4] or combined EEG with speech signals for emotion recognition. Abhang et al. [5] have presented review on emotion recognition from EEG and speech signal.

Chen et al. [6] have used a collection of ontological models as EEG feature set to propose an enhanced emotion assessment system. Gupta et al. [7] have proposed graph theoretic features for EEG with relevance vector machine as classifier to characterize emotional ratings. Atkinson and Campos [8] have combined various features and used kernel classifier. They have used valence arousal model for emotion. Bozhkov et al. [9] have developed subject independent methodology for affective computing. They have discriminated positive and negative emotions with 26 subjects. They have taken data from 21 channel and three minimum values and three maximum values are calculated for each channel. The latency (time of occurrence) is also recorded and are used as temporal features. Lan et al. [10] developed a system that monitors the emotion for real time applications and used subject dependent methodology.

Purnamasari et al. [11] have proposed filtered bi-spectrum as feature extraction method and artificial neural network as classifier. They claimed that filtered bi-spectrum is superior than power spectrum. They have performed several experimentations and obtained good results. Best result for arousal was 76% and for valence was 77.58%. Gómez et al. [12] worked on single-channel EEG and have used various types of wavelets with three types of classifiers with accuracy rate between 50 and 87%. Emotions they have classified were Happy, Sad, and neutral. Zhuang et al. [13] have used empirical mode decomposition (EMD) as feature extraction technique. By using EMD various intrinsic mode functions of EEG signals are calculated and used as features. They classified valence and arousal level and achieved classification accuracy of 69% and 72% respectively. Yohanes et al. [14] have used wavelet coefficients as features and classified happy and sad emotions with SVM and ELM. Nakate and Bahirgonde [15] have also extracted EEG features using wavelet transform. They have decomposed the signal up to four levels and have used daubechies wavelet of order two.

3 Proposed Methodology

The process of emotional state recognition from EEG signals consists of several stages and shown in Fig. 1.

In first stage EEG data will be collected. To collect EEG for emotion recognition, subject will watch a stimulus, wearing an electrode cap in the controlled environment, and the EEG signals will be recorded using some software. Here stimuli can be either video or picture to induce an emotion in the subject. Emotion occurs spontaneously by watching stimuli. Then the recorded signals will be preprocessed to remove noises. Since the size of EEG data is very large, feature selection procedures are applied to extract features. Then the extracted features are fed to a classifier for classifying various emotions like happy, sad, anger, etc.

3.1 DEAP Database

DEAP database [16] for emotion recognition is used in the proposed work. It is a multimodal database which contains EEG recordings of 32 subjects. For 22 out of 32 subjects frontal face video is also recorded. Participants have watched one minute long video and rated it on the scale of valence, arousal, liking/disliking and dominance. Emotions are modeled on valence arousal scale given by Russell [17] and rated using self assessment manikin given by Bradley and Lang [18].

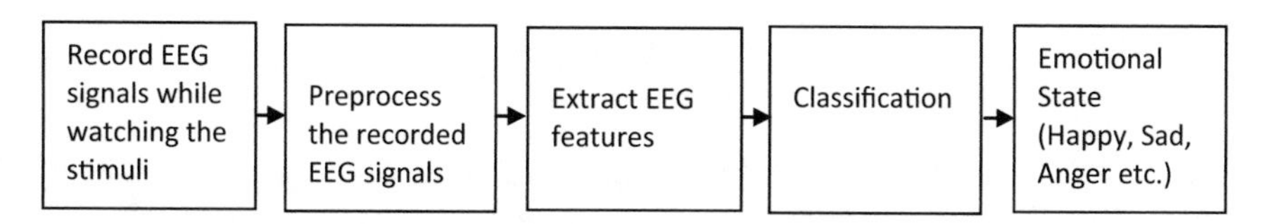

Fig. 1 Process of emotional state recognition

Table 1 EEG frequency bands

EEG bands	Size
Gamma	2027
Beta	1021
Alpha	518
Theta	266
Delta	266

Data is collected using electrode cap with 10–20 international electrode placement system and uses 40 electrodes to record various data among which 32 electrodes are there to record EEG and rest are for other peripheral physiological signals. There are 32.mat files, one for each subject. There are total 40 videos out of which 20 videos data is selected. For each video, this database contains readings for 40 electrodes and each electrode contains 8064 voltage values. That means for one subject data is of the form 40 × 40 × 8064 in which first dimension of the data is for video, second for electrode position and third for EEG. From the selected 20 videos for this experimentation, 10 videos falls in the happy quadrant of valence arousal model and other 10 videos corresponds to sad quadrant of valence arousal model. Authors of the data base have taken forty videos corresponding to four quadrants. They have already preprocessed the data. Original biosemi files of EEG recordings as well as Pre processed signals in the form of .mat, implementable in MATLAB are available [19].

3.2 *Wavelet Coefficients as Features*

Discrete wavelet transform is used to find out different coefficients of the signal which corresponds to different frequency bands of EEG. Wavelet function used was db8. Discrete wavelet transform represents the degree of correlation between the signal under analysis and the wavelet function at different instances of time. For non stationary data, wavelet transform is more suitable because it contains time as well as frequency information both. The Eqs. (1) and (2) are used to obtain coefficients of the original signal $X(n)$. For $j_0 \geq j$ and n from 0 to N – 1 we have,

$$W_{\Phi} = \frac{1}{\sqrt{N}} \sum X(n) \Phi_{j_0,k}(n) \quad (1)$$

$$W_{\Psi} = \frac{1}{\sqrt{N}} \sum X(n) \Psi_{j,k}(n) \quad (2)$$

Here $\Phi(n)$ and $\Psi(n)$ are transformation kernals used to transform the original signal, $X(n)$. The sizes of different frequency bands obtained are given in Table 1.

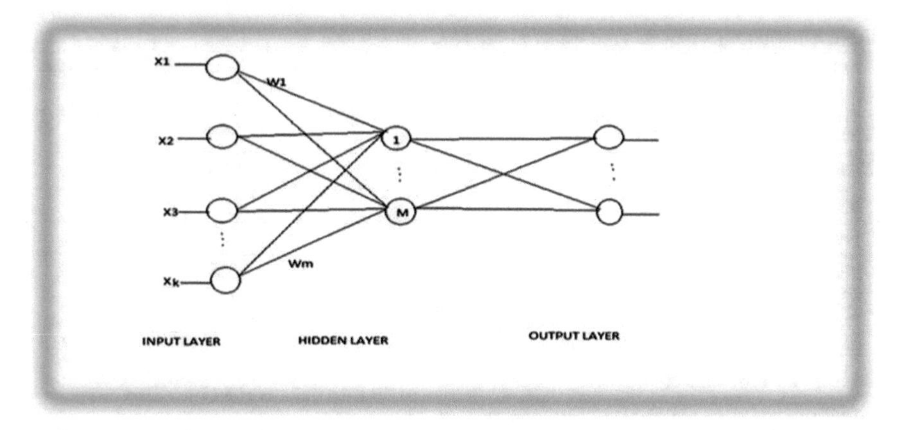

Fig. 2 Architecture of MLP neural network used for the proposed work

3.3 Classification

Multi layer perceptron (MLP) is selected for classification as the data is nonlinear and it is found to be suitable for EEG signals classification from the literature reviewed. The model of MLP used for the proposed work is given in Fig. 2. The input neurons correspond to the features (wavelet coefficients). The output neurons correspond to the emotions.

4 Implementation

This work is implemented in MATLAB R2015a. The proposed work is purely subject (user) independent and test data is completely different from train data. The data of those subjects are used for testing which were not used for training. DEAP database used in this research contains EEG signals of 32 subjects. These signals are already down sampled to 128 Hz. At a time data for all the 32 subjects corresponds to one electrode position are considered. Then using wavelets various frequency bands of EEG are extracted. Wavelet coefficients corresponding to alpha, beta, gamma, theta and delta bands are extracted. Theta band of one electrode for all 32 subjects is taken as input features. Features for 30 subjects were used to train the classifier and the other two subject data are used to test the classifier. For 20 videos and one subject, data is of the form 20×266 where 266 correspond to theta coefficients obtained from one electrode data. For 30 subjects, data will be of size 600×266 (i.e., $30 \times 20 \times 266$) which is used to train the classifier. The size of alpha coefficients was 518. Alpha band is also tested for four electrodes (i.e., Fp1, Fp2, F3 and F4). The training data size with alpha band was 600×518. Test data was of size 40×518 where 518 corresponds to alpha coefficients obtained from single electrode.

Table 2 Accuracy obtained with different electrodes and bands

EEG band	Electrode	Accuracy on training data (%)	Accuracy on test data (%)
Theta	Fp1	64	50
Theta	Fp2	67.33	50
Theta	F3	65	48
Theta	F4	58	**58.5**
Theta	Fp2 + F4	**75**	55
Alpha	F4	**80**	**58**

Table 3 Performance comparison

Article	Database	MLP classifier accuracy (%)	Remark
Zhang et al. [20]	DEAP	57.67 (Four emotions and 12 channels and with all **32 channels** accuracy was 58.75%)	With all 32 channels data training time would be high
Jirayucharoensak et al. [21]	DEAP	Valence 53, Arousal 52	Further mapping is required to evaluate emotions
Proposed	DEAP	**58.5** (Two emotions with **single channel** only)	Training time is less as only single channel is involved

Bayesian regularization is used as training function for neural network. The transfer function at hidden layer is taken as "tansig" and at output layer it was "logsig". The MLP model described in Sect. 3.3 is used. The number of input is 266 if theta band is used and 518 if alpha band is used. The number of neurons at hidden layer was taken as 10 by experimenting on it and varying it from 5 to 20. Good results are obtained at 10. There are two neurons at output layer corresponds to the happy and sad emotions. The results are given in Table 2.

From Table 2, it can be seen that the highest accuracy obtained is 58.5% with electrode F4 (frontal) and Theta band. From the obtained results it can be concluded that theta and alpha band with electrode F4 gives higher accuracy. The result obtained by the proposed method is also compared with that of the other research in this field and is summarized in Table 3.

5 Conclusion

This paper proposes a subject independent emotion recognition system using DWT and MLP. The previous reported works are either subject dependent or gender specific and mainly used self created databases. The proposed method is better than the earlier methods as it has used publically available benchmark database which contains data from 32 subjects from both the genders. Moreover it has employed single channel data to recognize happy and sad emotional states. In future, the work may be extended to more than two emotional states. The other classifier techniques may also be explored and the suitability of these classifiers for emotional state recognition will be examined.

References

1. Paul S, Mazumder A, Ghosh P, Tibarewala DN, Vimalarani G (2015) EEG based emotion recognition system using MFDFA as feature extractor. In: International conference on robotics, automation, control and embedded systems (RACE) IEEE, pp 1–5
2. Lahane P, Sangaiah AK (2015) An approach to EEG based emotion recognition and classification using kernel density estimation. In: International conference on intelligent computing, communication and convergence (ICCC-2015), Odisha, India. Procedia Comput Sci 48:574–581
3. Singh M, Sing M, Gangwar S (2013) Emotion recognition using electroencephalography (EEG): a review. IJITKM 7(1):1–5
4. Soleymani M, Asghari-Esfeden S, Pantic M, Fu Y (2014) Continuous emotion detection using EEG signals and facial expressions. In: IEEE international conference on multimedia and expo (ICME), pp 1–6
5. Abhang P, Rao S, Gawali BW, Rokade P (2011) Emotion recognition using speech and EEG signal—a review. Int J Comput Appl 15(3):0975–8887
6. Chen J, Hu B, Moore P, Zhang X, Ma X (2015) Electroencephalogram based emotion assessment system using ontology and data mining technique. Appl Soft Comput 30:663–674
7. Gupta R, Laghari KR, Falk TH (2016) Relevance vector classifier decision fusion and EEG graph-theoretic features for automatic affective state characterization. Neurocomputing 174:875–884
8. Atkinson J, Campos D (2016) Improving BCI-based emotion recognition by combining EEG feature selection and kernel classifiers. Expert Syst Appl 47:35–41
9. Bozhkov L, Georgieva P, Santos I, Pereira A, Silva C (2015) EEG-based subject independent affective computing models. In: INNS conference on big data. Procedia Comput Sci 53:375–382
10. Lan Z, Sourina O, Wang L, Liu Y (2016) Real-time EEG-based emotion monitoring using stable features. Vis Comput 32(3):347–358
11. Purnamasari PD, Ratna AAP, Kusumoputro B (2017) Development of filtered bispectrum for EEG signal feature extraction in automatic emotion recognition using artificial neural networks. Algorithms 10(2):63
12. Gómez A, Quintero L, López N, Castro J, Villa L, Mejía G (2017) An approach to emotion recognition in single-channel EEG signals using stationary wavelet transform. In: VII Latin American congress on biomedical engineering CLAIB 2016, Bucaramanga, Santander, Colombia. Springer, Singapore, pp 654–657
13. Zhuang N, Zeng Y, Tong L, Zhang C, Zhang H, Yan B (2017) Emotion recognition from EEG signals using multidimensional information in EMD domain. BioMed Res Int
14. Yohanes RE, Ser W, Huang GB (2012) Discrete wavelet transform coefficients for emotion recognition from EEG signals. In: 2012 annual international conference of the engineering in medicine and biology society (EMBC). IEEE, pp 2251–2254
15. Nakate A, Bahirgonde PD (2015) Feature extraction of EEG signal using wavelet transform. Int J Comput Appl 124(2):0975–8887
16. Koelstra S, Muhl C, Soleymani M, Lee J, Yazdani A, Ebrahimi T, Pun T, NIjhilt A, Patras I (2012) DEAP: a database for emotion analysis; using physiological signals. IEEE Trans Affect Comput 3(1):18–31

17. Russell JA (1980) A circumplex model of affect. J Pers Soc Psychol 39(6):1161–1178
18. Bradley MM, Lang PJ (1994) Measuring emotion: the self-assessment manikin and the semantic differential. J Behav Ther Exp Psychiatry 25(1):49–59
19. DEAP database description. https://www.eecs.qmul.ac.uk/mmv/datasets/deap/readme.html
20. Zhang J, Chen M, Zhao S, Hu S, Shi Z, Cao Y (2016) ReliefF-based EEG sensor selection methods for emotion recognition. Sensors 16(10):1558
21. Jirayucharoensak S, Pan-Ngum S, Israsena P (2014) EEG-based emotion recognition using deep learning network with principal component based covariate shift adaption. World Sci J 10 pp. Article ID: 627892

Development of Early Prediction Model for Epileptic Seizures

Anjum Shaikh and Mukta Dhopeshwarkar

Abstract Epilepsy is the neurological disorder of brain electrical system causes the seizure because of that the brain and body behave abnormally (Yadollahpour, Jalilifar, Biomed Pharmacol J 7(1):153–162, 2014) [1]. Epilepsy is the result of recurrent seizure, i.e., if the person has single seizure in their whole lives then that person is not affected by epilepsy but if that person has more than two seizures in their lives then that person is affected by Epilepsy. Near about 0.8–1% of population all over the world is affected by an epilepsy, epilepsy is not able to cure but able to controlled by using anti epileptic medicine or by performing resective surgery then also in 25% epileptic patients no present therapy is used to controlled the epilepsy. Epilepsy is unpredictable in nature so it increases the risk of end dangerous accident when person work with heavy machineries like driving a car, cooking or swimming, again a patient always have fear of next seizure it really affect on their daily lives so to minimize the risk and to improve the quality of life of such patient it is necessary to predict the epilepsy before its onset. In the present study by using 21 patients EEG database which consist of 80 seizure, learn the 336 predictive model using four different classifier, i.e., ANN, KNN, MC-SVM using 1-against-1 approach and MC-SVM using 1-against-all approach and make possible to predict epilepsy 25 min before onset with the maximum average accuracy 98.19% and sensitivity 98.97% and predict 30 min before onset with the average maximum accuracy 98.04% and sensitivity of 98.85%.

A. Shaikh (✉) · M. Dhopeshwarkar
Department of CS & IT, Dr. Babasaheb Ambedkar Marathwada University, Aurangabad, India
e-mail: anjumshaikhcs@gmail.com

M. Dhopeshwarkar
e-mail: mukta_d@rediffmail.com

D. K. Mishra et al. (eds.), *Data Science and Big Data Analytics*,
Lecture Notes on Data Engineering and Communications Technologies 16,
https://doi.org/10.1007/978-981-10-7641-1_11

1 Introduction

The basic unit of the central nervous system is the neuron [2]. Neuron carries electrical pulses and transmits these pulses from one neuron to other neurons in a proper order if this order is disturbed (disorder) then the normal functioning of the brain is disturb and because of that brain and body behave abnormally it results in a seizure. Epilepsy is the result of recurrent seizures, i.e., if the person has a single seizure in their whole lives then that person is not affected by epilepsy but if that person has more than two seizures in their lives then that person is affected by epilepsy. Epilepsy may develop after a particular event such as head injury, brain tumor, and asphyxia that type of epilepsy is called symptomatic epilepsy or it may cause without any event such type of epilepsy is called idiopathic epilepsy. Some of the epilepsy happened to particular age groups. Some suffer from it their whole lives and some are for few years [3]. Before seizure happens number of characteristic clinical symptoms occurs that is increases in oxygen availability changes the rate of blood flow to the brain [4], decreases oxygen level in the blood, increase the heart rate and pulse rate, and increased number of critical interactions among the neurons in the focal region of the brain. Among all these characteristic clinical symptoms the most important symptom is critical interaction among the neuron in the focal region because the main source of epilepsy is the brain. The critical interaction among the neurons in the focal region of the brain is recorded by the device called EEG (Electroencephalogram) then by proper analysis of these signal it is possible to predict epilepsy before its onset and minimize the risk.

2 Methodology

2.1 Database

In the current predictive model we use the database which makes available ASPPR [5] in the form of feature vector of epileptic EEG signal. ASPPR uses the Freiburg EEG database which was previously publicly available but now that database is not publicly available. The Freiburg database contains 24 hours long pre surgical invasive EEG recordings of 21 patients suffering from epilepsy. The patients are varying in gender, age, and seizure location but all are suffer from focal medically intractable epilepsy. Out of 21 patients in 52% patients the location of epilepsy was the neocortical region of the brain where the brain is divided into four different lobe, in 38% patients the location of epilepsy was hippocampus in the temporal lobe, and in 10% patients the location of epilepsy was in both that is neocortical region as well as in hippocampus. The EEG data of these patients were recorded by using invasive grid, strip and depth electrodes using 128 channels and 256 Hz sampling rate [6]. For each of the patients, the two datasets files were prepared one is ictal file and another is interictal, the ictal file containing data with epileptic seizures with 50 min preictal data and the interictal file containing data without seizure activity. For each patient out of 128 electrodes, six electrodes are use to prepare final database that is three from focal region and three non-focal region electrodes are used to prepare dataset [6].

ASPPR [5] used only data from the Ictal file which contain seizure and pre-seizure activity. A total of 204 time series features were extracted for each patient, 34 different features from each of the six electrodes, then label it appropriately as ictal, preictal, interictal and postictal and make publicly available this database in the form of feature vector. We can download it from http://code.google.com/p/asppr.

2.2 Database Description

Database consist of features vector of 204 feature extracted from EEG database of Freiburg database, description of feature is as shown in the following Table 1.

These are 34 distinct features were extracted from each electrodes so from six electrodes total 204 feature were extracted.

2.3 Dataset Preparation and Experimental Work

The four different datasets were prepared for this predictive model two for prediction 25 min before a seizure and another two for prediction 30 min before seizure by using following steps for each of the patients separately:

1. Relabeling the interictal data as preictal for 25 and 30 min
2. Normalized the data standard deviation equal to 1 and mean equal to 0 [7]
3. Selecting more optimal features from the feature vector using SFS (Sequential Forward Selection) feature subset selection algorithm and MI (Mutual Information) feature ranking algorithm.

ASPPR [5] Database consist of feature vector of 204 features extracted from 6 EEG channels out of these three from focal region and three from non-focal region 34 distinct features value from each of the electrodes were recorded the electrical activity of brain every after 5 s and continuing till the end of ictal file from the Freiburg EEG Database and then label each data instances by using appropriate label as ictal which shows seizure activity, preictal [8, 9] shows time period just 300 s before the seizure, postictal is period 300 s after seizure and interictal is period between two seizure excluding preictal and postictal period [10].

Preictal state of EEG signal is very important for seizure prediction medically it is proved that the Seizure activity starts several hrs before the seizure. The Accumulated energy increases 50 min before seizure [11]. The dataset used in this predictive model in which the 300 interictal instances is relabeling as preictal for learning predictive model 25 min prediction and relabeling 360 interictal instances as the preictal instance for learning predictive model 30 min prediction for each of the

Table 1 Description of feature vector

Concept	Feature
Signal energy	(1) Accumulated energy (2) Energy level (3) Energy variation [short term energy (STE)] (4) Energy variation [long term energy (LTE)]
Wavelet transform	(1) Energy STE 1 (0–12.5 Hz) (2) Energy STE 2 (12.5–25 Hz) (3) Energy STE 3 (25–50 Hz) (4) Energy STE 4 (50–100 Hz) (5) Energy LTE 1 (0–12.5 Hz) (6) Energy LTE 2 (12.5–25 Hz) (7) Energy STE 3 (25–50 Hz) (8) Energy LTE 4 (50–100 Hz)
Nonlinear system dynamics	(1) Correlation dimension (2) Max Lyapunov exponent
Standard statistical measures	(1) Mean of STE (2) Skewness of STE (3) Kurtosis of STE (4) Mean of LTE (5) Skewnss of LTE (6) Kurtosis of LTE
Spectral band power (SBP)	(1) Spectral band power STE (0.5–4 Hz) (2) Spectral band power STE (4–8 Hz) (3) Spectral band power STE (8–13 Hz) (4) Spectral band power STE (13–30 Hz) (5) Spectral band power STE (30–48 Hz) (6) Spectral band power LTE (0.5–4 Hz) (7) Spectral band power LTE (4–8 Hz) (8) Spectral band power LTE (8–13 Hz) (9) Spectral band power LTE (13–30 Hz) (10) Spectral band power LTE (30–48 Hz)
Spectral edge frequency (SEF)	(1) Spectral edge frequency STE (2) Spectral edge frequency LTE (3) Median frequency STE (4) Median frequency LTE

patients separately. After the relabeling data were normalized by mean equal to zero and standard deviation equal to one then optimal feature set is selected by using SFS (Sequential Forward Selection) algorithm [12, 13]. SFS start optimal feature selection with the vacant set and then sequentially add the feature into the vacant set by analyzing the performance of the classifier if the performance of classifier increases after adding the new feature into set then that feature is permanently added into subset otherwise that feature was discarded [14]. By using this iterative process SFS finally gives the subset of optimally selected features that are different in size and type for the different patient. By using this method on EEG database of 21 different patients separately finally 2 different datasets were prepared for prediction before 25 min and prediction before 30 min before learning the predictive model. The another feature ranking method is also used to select the feature set to compare the result with ASPPR algorithm which was used the ReliefF [15, 16] feature ranking algorithm to rank the features and used top rank 14 features to learn the predictive model for this predictive model MI (Mutual Information) feature ranking algorithm [17–19] was used to rank the features and after ranking the same number of top rank features were selected into subset which was selected by SFS algorithm like for patient 2 SFS gave the subset of seven features so for patient 2 by using MI algorithm top rank seven features are selected into the subset in these way by applying MI feature ranking algorithm separately on EEG database of 21 patient two datasets were prepared one for prediction before 25 min and another for prediction before 30 min before learning the predictive model. These finally prepared 4 datasets are then used to learn the predictive model.

2.4 Learning Predictive Model

In the present study total 336 predictive model learned by using 4 different learning algorithm that are ANN, KNN, MC-SVM using 1-against-1 approach and MC-SVM using 1-against-all approach by using four different prepared datasets that two dataset was prepared by using SFS feature subset selection algorithm for prediction before 25 and 30 min before its onset and another two different datasets were prepared by using MI feature ranking algorithm for prediction before 25 and 30 min before its onset

2.4.1 Learning Predictive Model Using ANN

The present predictive model learned by using Artificial Neural Network (ANN) [20] pattern recognition tool. It uses two-layer feed forward neural network [21] to learn the predictive model by randomly dividing the dataset for 70% training and 30% testing and train the model using Bayesian regulation training function and performance is evaluated by using mean square error (MSE) performance function. By using this classifier 21 predictive models were learning for prediction 25 min before onset Fig. 1 shows the performance of the classifier, confusion matrix and the network used by the neural network for patient 2.

By using ANN classifier total 84 predictive model learned for four prepared data set, i.e., 21 predictive model for each data set and classify data instances into four different classes that are interictal, preictal, ictal, and postictal and evaluate the accuracy measures for correct classification by using formula Accuracy = ((TPc1 + TPc2 + TPc3 + TPc4)/total No. of instances) * 100) where c1, c2, c3, and c4 are the notation used for class label interictal, preictal, ictal and postictal respectively and TP is the number of data instances that were correctly classified as respective class. In a similar manner the sensitivity, specificity also measures in terms

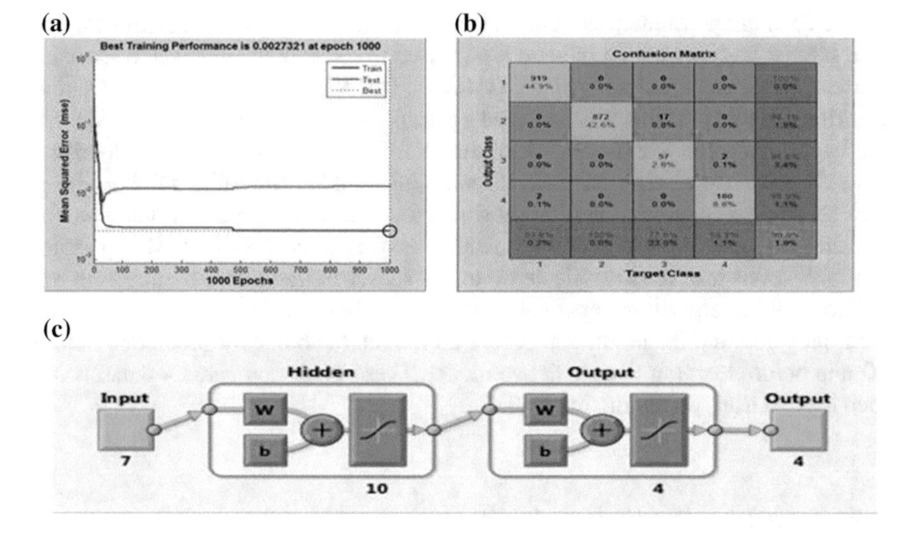

Fig. 1 Predictive model learn for patient 2 using ANN pattern recognition tool, **a** shows performance of classifier, **b** confusion matrix generated by classifier with 99% accuracy for prediction before 30 min, **c** network diagram of ANN

Table 2 Result of prediction using ANN classifier

Sr. no.	Time of prediction	Dataset prepared by algorithm	Accuracy (%)	Sensitivity (%)	Specificity (%)	S1-score (%)
1	Prediction 25 min before on set	SSF	97.96	98.67	99.16	98.84
		MI	87.12	82.19	82.23	81.11
2	Prediction 30 min before on set	SSF	97.77	97.93	98.29	98.10
		MI	82.91	78.99	72.61	74.47

of correct classification for the c2 (preictal) class also measures harmonic mean of sensitivity and specificity, i.e., S1-score for each of patient separately as follows.

$$\text{Sensitivity} = (\text{TP}_{c2})/(\text{TP}_{c2} + \text{FNc}_2) \tag{1}$$

$$\text{Specificity} = ((\text{TP}_{C1} + \text{TP}_{C3} + \text{TP}_{C4}) - (\text{FN}_{C1} + \text{FN}_{C3} + \text{FN}_{C4}))/(\text{TP}_{C1} + \text{TP}_{C3} + \text{TP}_{C4})) * 100 \tag{2}$$

$$\text{S1-score} = 2((\text{Sensitivity} * \text{Specificity})/\text{Sensitivity} + \text{Specificity}), \tag{3}$$

where TP is the number of data instances that were correctly classified as a respective class and FP is the number of data instances that were incorrectly classified as all other classes. By using these evaluations measures present predictive model noted the average Accuracy, Sensitivity, Specificity and S1-score for four prepared datasets using SSF and MI is as shown in following Table 2.

Table 3 Results for epileptic seizure prediction 25 min before seizure using KNN for dataset prepared using SFS algorithm

Data	Accuracy (%)	Sensitivity (%)	Specificity (%)	S1-score (%)
Data 1	98.21	98.89	97.33	98.10
Data 2	98.44	99.18	97.57	98.36
Data 3	97.88	99.40	95.76	97.54
Data 4	98.97	99.56	98.11	98.82
Data 5	98.74	99.50	97.68	98.58
Data 6	98.78	99.30	98.27	98.78
Data 7	98.20	98.66	97.88	98.26
Data 8	96.42	98.69	94.12	96.35
Data 9	98.89	99.45	97.89	98.66
Data 10	97.87	98.79	97.21	97.99
Data 11	98.83	98.71	98.95	98.82
Data 12	99.16	99.15	99.16	99.15
Data 13	98.10	99.36	96.37	97.84
Data 14	98.65	99.43	98.00	98.70
Data 15	98.28	99.39	96.99	98.17
Data 16	94.64	96.04	93.37	94.68
Data 17	98.45	99.00	97.96	98.47
Data 18	98.03	98.74	97.20	97.96
Data 19	98.83	99.30	98.02	98.65
Data 20	97.95	98.69	97.35	98.01
Data 21	98.59	99.13	98.12	98.62
Average	**98.19**	**98.97**	**97.30**	**98.12**

2.4.2 Learning Predictive Model by KNN

In the present predictive model the second experiment is performed by using the KNN [22, 23] classifier and learn 84 predictive models to classify data instances into 4 different classes using 4 different prepared dataset, i.e., 21 predictive model for each dataset using the distance metric as Euclidean distance and Consensus rule, while using the KNN classifier care should be taken to select the value of k that is number of nearest neighbor because if we choose small value of k that means there is more noise and higher effect on the result and if choose the large value of k then it is computationally expensive [24, 25], by considering this fact here the value of $k=4$ is used and recorded the evaluation measures Accuracy, Sensitivity, Specificity and S1-score for each of the patient separately as well as average measure of all were also recorded in following Tables 3 and 4.

Table 4 Results for epileptic seizure prediction 30 min before seizure using KNN for dataset prepared using SFS algorithm

Data	Accuracy (%)	Sensitivity (%)	Specificity (%)	SI-score (%)
Data 1	97.99	98.96	97.03	97.98
Data 2	98.24	98.70	97.87	98.28
Data 3	97.77	99.26	96.13	97.66
Data 4	98.89	99.41	98.27	98.83
Data 5	98.21	99.03	97.36	98.18
Data 6	97.85	99.11	96.88	97.98
Data 7	97.90	98.28	97.69	97.98
Data 8	97.17	98.52	96.08	97.28
Data 9	99.03	99.51	98.31	98.90
Data 10	97.74	98.50	97.31	97.90
Data 11	97.95	98.98	97.16	98.06
Data 12	98.57	98.84	98.30	98.56
Data 13	98.02	99.17	96.71	97.92
Data 14	98.50	98.95	98.22	98.58
Data 15	98.35	99.22	97.58	98.39
Data 16	98.31	99.25	97.62	98.42
Data 17	98.27	98.68	98.00	98.33
Data 18	98.03	98.84	97.26	98.04
Data 19	96.08	97.94	93.44	95.63
Data 20	97.48	97.97	97.21	97.58
Data 21	98.42	98.70	98.25	98.47
Average	**98.04**	**98.85**	**97.27**	**98.05**

2.4.3 Learning Predictive Model by MC-SVM One-Against-All Approach

The dataset that we have used for predictive model is multiclass data so simple binary SVM classifier is not work with dataset. So extended approach of SVM is MC-SVM one-against-all approach [26, 27] is used for predictive model for classification using LIBSVM software library function in MATLAB [28, 29]. This classifier make n number of binary classifier for n classes data that is for ith class it create a binary pair ith class verses rest of all classes. In this pair the feature space which belongs to ith class is assign label 1 and for all other instances which belongs to other classes assign class label zero, so in this way it makes n binary pair and for every binary pair it classify data. For each binary pair find the probability of prediction of data instances belonging to that class. Final output of this classifier is with the class who are having maximum probability of prediction of data instances belonging to that class. By using this approach 84 different predictive models were learned for 4 different datasets, i.e., 21 different predictive model for each data set and noted the Average evaluation

Table 5 Result of epileptic seizure prediction using MC-SVM 1-against-all approach

Sr. no.	Time of prediction	Dataset prepared by algorithm	Accuracy (%)	Sensitivity (%)	Specificity (%)	S1-score (%)
1	Prediction 25 min before on set	SSF	85.13	87.84	89.98	88.67
		MI	89.03	90.48	84.98	86.90
2	Prediction 30 min before on set	SSF	85.22	90.40	90.48	90.28
		MI	87.42	91.30	79.43	83.48

Table 6 Result of epileptic seizure prediction using MC-SVM 1-against-1 approach

Sr. no.	Time of prediction	Dataset prepared by algorithm	Accuracy (%)	Sensitivity (%)	Specificity (%)	S1-score (%)
1	Prediction 25 min before on set	SSF	92.67	94.15	96.27	95.13
		MI	94.46	91.24	93.73	90.19
2	Prediction 30 min before on set	SSF	91.47	92.20	92.62	92.40
		MI	93.04	92.12	92.43	91.08

measures accuracy, sensitivity, specificity and S1-score for prediction 25 and 30 min before as shown in Table 5.

2.4.4 Learning Predictive Model by MC-SVM One-Against-One Approach

The next experiment has been performed to build the predictive model by using MC-SVM using one-against-one approach. The one-against-one [30, 31] strategy uses the concept of binary SVM classifier by creating a binary pair for each of different class label and apply soft margin SVM. This approach constructs C(C – 1)/2 binary classifiers where C is the number of classes. Database that were used in this study has 4 classes so it make 6 pair of binary classifiers, i.e., (C1, C2), (C1, C3), (C1, C4), (C2, C3), (C2, C4), and (C3, C4). The collection of these binary pair classifiers are trained using the voting strategy, the predicted class is that class which receives the maximum votes [32]. The 84 different predictive models were learn for 4 different datasets, i.e., 21 different predictive model for each data set and noted the Average evaluation measures accuracy, sensitivity, specificity and S1-score for prediction 25 and 30 min before as shown in Table 6.

3 Results

We analyze the performance of the predictive model by performing the comparative analysis of 16 different experiments, eight experiments for prediction 25 min before onset and eight experiments for prediction 30 min onset. Each of this experiment is performed on 21 patients finally prepared four different datasets, two datasets are prepared by using SFS (Sequential Forward Selection) feature subset selection algorithm one for prediction 25 min before onset and another for prediction 30 min before onset. The another two datasets were prepared using MI (Mutual Information) feature ranking algorithm one for prediction 25 min before onset and another for prediction 30 min before onset. By using this four finally prepared dataset total 336 predictive models were learn using four different classifiers ANN. KNN, MC-SVM using 1-against-all approach and MC-SVM using 1-against-1 approach for each of the datasets of 21 patient and reported the average evaluation measures Accuracy, Sensitivity, Specificity and S1-score for each classifier. Following Table 7 shows evaluation measure for all four classifiers for finally prepared dataset using SFS feature subset selection algorithm and Fig. 2 shows its graphical representation.

Table 7 summarized the average result of eight experiments, four for epileptic seizure prediction 25 min before onset, and four for prediction 30 min before onset. The following Table 8 shows the evaluation measure for all four classifiers for finally prepared dataset using MI feature ranking algorithm and Fig. 3 shows its graphical representation.

Table 8 summarized the average result of 8 Experiments, four for epileptic seizure prediction 25 min before onset and four for prediction 30 min before onset.

Table 7 Result of epileptic seizure prediction subset selection using SFS

Sr. no	Classifier	Result of prediction 25 min before seizure				Result of prediction 30 min before seizure			
		Accuracy (%)	Sensitivity (%)	Specificity (%)	sl-score (%)	Accuracy (%)	Sensitivity (%)	Specificity (%)	sl-score (%)
1	ANN	97.96	98.67	99.16	98.84	97.77	97.93	98.29	98.10
2	KNN	98.19	98.97	97.30	98.12	98.04	98.85	97.27	98.05
3	MC-SVM 1-against-1 approach	92.67	94.15	96.27	95.13	91.47	92.20	92.62	92.40
4	MC-SVM 1-against-all approach	85.13	87.84	89.98	88.67	85.22	90.40	90.48	90.28

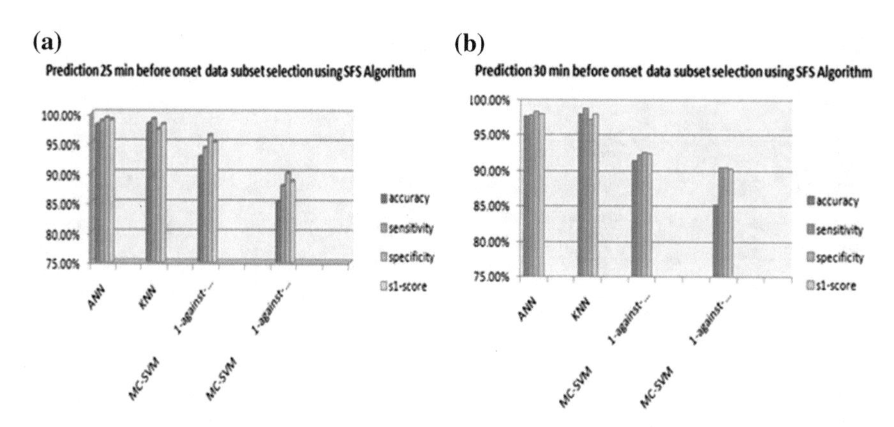

Fig. 2 **a** Graphical result of prediction 25 min before onset data subset selection using SFS algorithm, **b** graphical result of prediction 30 min before onset data subset selection using SFS algorithm

Table 8 Result of epileptic seizure prediction subset selection using MI

Sr. no	Classifier	Result of prediction 25 min before seizure				Result of prediction 30 min before seizure			
		Accuracy (%)	Sensitivity (%)	Specificity (%)	sl-score (%)	Accuracy (%)	Sensitivity (%)	Specificity (%)	sl-score (%)
1	ANN	87.12	82.19	82.23	81.11	82.91	78.66	72.61	74.47
2	KNN	78.46	82.71	74.08	77.98	82.07	84.52	79.80	81.99
3	MC-SVM 1-against-1 approach	94.46	91.24	93.73	90.19	93.04	92.12	92.43	91.08
4	MC-SVM 1-against-all approach	89.03	90.48	84.98	86.90	87.42	91.30	79.43	83.48

4 Conclusion

The second row of Table 7 is highlighted with yellow color indicates that the KNN classifier gives highest average accuracy 98.16% and sensitivity 98.97% for prediction 25 min before onset and average accuracy 98.04% and sensitivity 98.85% for prediction before 30 min before onset for 21 patients finally prepared dataset using SFS algorithm. Whereas the third row of Table 8 highlighted with yellow color indicates that 1-against-1 classifier gives highest average accuracy 94.26% and sensitivity 91.24% among the four classifiers for prediction 25 min before onset and average accuracy 93.04% and sensitivity 92.12% for prediction 30 min before onset for 21 patients finally prepared dataset using MI algorithm. In the present study MI ranking algorithm is used to compare the result with the ASSPR algorithm which was used Relieff ranking algorithm to select the top rank features and MC-SVM 1-against-1

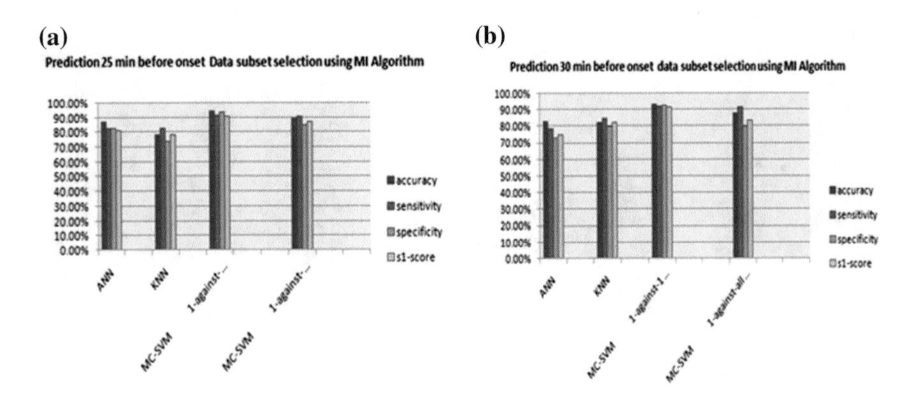

Fig. 3 **a** Graphical result of prediction 25 min before onset data subset selection using MI algorithm, **b** graphical result of prediction 30 min before onset data subset selection using MI algorithm

classifier and able to report S1-score 94.2% for prediction 20 min before onset, the present predictive model reported average S1-score 90.19% for prediction 25 min before onset and 91.08% for prediction 30 min before onset for MI dataset and same classifier. From overall analysis of this study concludes that SFS algorithm gives more optimal feature subset than ranking algorithm which varies in size and type of features and the KNN classifier gives highest average evaluation measures among all other classifiers, i.e., accuracy 98.16%, sensitivity 98.97%, specificity 97.30% and S1-score 98.12% for 25 min prediction and accuracy 98.04%, sensitivity 98.85%, specificity 97.27% and S1-score 98.05% for 30 min prediction.

References

1. Yadollahpour A, Jalilifar M (2014) Seizure prediction methods: a review of the current predicting techniques. Biomed Pharmacol J 7(1):153–162
2. Fullick A (2011) Edexcel IGCSE biology revision guide. Pearson Education, p 40. ISBN: 9780435046767
3. http://www.who.int/mental_health/media/en/639.pdf
4. Adelson PD, Nemoto E, Scheuer M, Painter M, Morgan J et al (1999) Noninvasive continuous monitoring of cerebral oxygenation periictally using near-infrared spectroscopy: a preliminary report. Epilepsia 40:1484–1489. https://doi.org/10.1111/j.1528-1157.1999.tb02030.x
5. Moghim N, Corne DW (2014) Predicting epileptic seizures in advance. PLoS ONE 9(6):e99334. https://doi.org/10.1371/journal.pone.0099334
6. Epilepsy.uni-freiburg.de (2007) EEG database—seizure prediction project
7. https://in.mathworks.com/matlabcentral/answers/216489-why-we-need-to-normalize-the-Data-what-is-normalize-data?requestedDomain=www.mathworks.com

8. De Clercq W, Lemmerling P, Van Huffel S, Van Paesschen W (2003) Anticipation of epileptic seizures from standard EEG recordings. The Lancet 361:971–971. https://doi.org/10.1016/s0140-6736(03)12780-8
9. Martinerie J, Adam C, Le Van Quyen M, Baulac M, Clemenceau S et al (1998) Epileptic seizures can be anticipated by non-linear analysis. Nat Med 4:1173–1176. https://doi.org/10.1038/2667
10. Costa RP, Oliveira P, Rodrigues G, Leitão B, Dourado A (2008) Epileptic seizure classification using neural networks with 14 features, pp 281–288
11. Litt B, Esteller R, Echauz J, D'Alessandro M, Shor R et al (2001) Epileptic seizures may begin hours in advance of clinical onset: a report of five patients. Neuron 30:51–64. https://doi.org/10.1016/s0896-6273(01)00262-8
12. Khalil M, Al Hage J, Khalil K (2015) Feature selection algorithm used to classify faults in turbine bearings. Int J Comput Sci Appl 4(1 April 2015) 12324-7037/15/01 001-08 https://doi.org/10.12783/ijcsa.2015.0401.01
13. Ladha L, Deepa T (2011) Feature selection methods and algorithms. Int J Adv Trends Comput Sci Eng 3(5):1787–1797. ISSN: 0975-3397
14. Rathore SS, Gupta A (2014) A comparative study of feature-ranking and feature-subset selection technique for improved fault prediction. In: Conference Paper, · Feb 2014. https://doi.org/10.1145/2590748.2590755
15. Kononenko I, Šimec E, Robnik-Šikonja M (1997) Overcoming the myopia of inductive learning algorithms with RELIEFF. Appl Intell 7(1):39–55
16. Robnik-Šikonja M, Kononenko I (2003) Theoretical and empirical analysis of ReliefF and RReliefF. Mach Learn 53(1–2):23–69
17. Yin C, Feng L, Ma L, Yin Z, Wang J (2015) A feature selection algorithm based on Hoeffding inequality and mutual information. Int J Signal Process Image Process Pattern Recognit 8(11):433–444. http://dx.doi.org/10.14257/ijsip.2015.8.11.39
18. Fleuret F (2004) Fast binary feature selection with conditional mutual information. Mach Learn Res 5:1531–1555
19. Chandrashekar Girish, Sahin Ferat (2014) A survey on feature selection methods. Comput Electr Eng 40:16–28
20. Haykin S (1999) Neural networks a comprehensive foundation, 2nd edn. Prentice Hall Inc., Upper Saddle River, NJ, USA
21. Hornik K, Stinchcombe M, White H (1989) Multilayer feedforward networks are universal approximators. Neural Netw 2:359–366
22. Cover TM, Hart PE (1967) Nearest neighbor pattern classification. IEEE Trans Inf Theory 13:21–27
23. Devroye L (1981) On the asymptotic probability of error in nonparametric discrimination. Ann Statist 9(1320):1327
24. Gou J, Du L, Zhang Y, Xiong T (2012) A new distance-weighted k-nearest neighbor classifier. J Inf Comput Sci 9(6):1429–1436
25. Gil-Garcia R, Pons-Porrata A (2006) A new nearest neighbor rule for text categorization. Lecture notes in computer science, vol 4225. Springer, New York, pp 814–823
26. Knerr S, Personnaz L, Dreyfus G (1990) Single-layer learning revisited: a stepwise procedure for building and training a neural network. Springer, Berlin, Heidelberg, pp 41–50. https://doi.org/10.1007/978-3-642-76153-9_5
27. Krebel UHG (1999) Pairwise classification and support vector machines. MIT Press. 14 pp
28. Chang C-C, Lin C-J (2011) LIBSVM: a library for support vector machines. ACM Trans Intell Syst Technol (TIST) 2:27
29. LIBSVM (2013) LIBSVM—A library for support vector machines. http://www.csie.ntu.edu.tw/~cjlin/libsvm/. Accessed 18 May 2014

30. Oladunni OO, Trafalis TB (2006) A pair wise reduced kernel-based multi-classification Tikhonov regularization machine. In: Proceedings of the international joint conference on neural networks (IJCNN'06), Vancouver, BC, Canada, July 2006, on CD-ROM. IEEE Press, pp 130–137
31. Chamasemani FF, Singh YP (2011) Multi-class support vector machine (SVM) classifiers—an application in hypothyroid detection and classification. In: The 2011 sixth international conference on bio-inspired computing, pp 351–356. https://doi.org/10.1109/bic-ta.2011.51
32. Milgram J, Cheriet M, Sabourin R (2006) "One against one" or "one against all": which one is better for handwriting recognition with SVMs? Guy Lorette. In: Tenth international workshop on frontiers in handwriting recognition, Oct 2006, La Baule (France), Suvisoft

Research Issue in Data Anonymization in Electronic Health Service: A Survey

Amanuel Gebrehiwot and Ambika V. Pawar

Abstract At today time, the rapid change of technology is changing the day-to-day activity of human being. Healthcare data and practice also made use of these technologies; they change its way to handle the data. The electronic health Service (EHS) is increasingly collecting large amount of sensitive data of the patient that is used by the patient, doctors and others data analysts. When we are using EHS we should concern to security and privacy of the medical data, because of medical data is too sensitive due to their personal nature. Especially privacy is critical for the sensitive data when we give for medical data analysis or medical research purpose, first we should do sanitization or anonymized of the data before releasing it. Data anonymization is the removing or hiding of personal identifier information like name, id, and SSN from the health datasets and to not to be identified by the recipient of the data. To anonymize the data we are using different models and techniques of anonymization. This paper is survey on data anonymization in Electronic Health Service (EHS).

Keywords Data anonymization · Electronic health service (EHS) · Privacy Sensitive data · Anonymization models

1 Introduction

Health organizations collect the increasingly large amount of sensitive medical data and used as a source of relevant research, data holders often anonymized or remove explicit or public identifiers, such as names and unique security numbers. Data Anonymization is critical to hiding the medical datasets that is by removing per-

A. Gebrehiwot (✉) · A. V. Pawar
CSE Department, SIT, Symbiosis International University, Pune, India
e-mail: amanuel.gebrehiwot@sitpune.edu.in

A. V. Pawar
e-mail: ambikap@sitpune.edu.in

D. K. Mishra et al. (eds.), *Data Science and Big Data Analytics*,
Lecture Notes on Data Engineering and Communications Technologies 16,
https://doi.org/10.1007/978-981-10-7641-1_12

sonal identifiable information. By using the data anonymity, we increase the degree of privacy in the sensitive medical data [1]. When we use anonymization the character of sensitive medical data about the patients to be covered up. It even accepts that sensitive data should be retained for analysis. According to [2] before we anonymize the data we should classify the attributes into *explicit identifier* or *a public identifier*, *quasi identifier*, *sensitive identifiers,* and *nonsensitive identifier*. *Explicit identifiers* are directly or uniquely identifying the owner of the data that is like name, id and SSN, *quasi identifier* is an identifier that directly identified when the data is combined with publicly available data. *Sensitive identifier* is an identifier that contains the sensitive data of the patient like a disease. *Nonsensitive identifier* has No effect on the privacy of the data even if when we are revealed to the public.

1.1 Basic Privacy Models of Data Anonymization

A. K-Anonymization

In order to preserve privacy by using Anonymization method aims at making the individual record be indistinguishable among a group of records by using the K-Anonymization model it has been proposed by the Sweeney [3]. The objective of k-anonymization is to make every tuple in identity-related attributes of a published table identical to at least (k-1) other tuples. Identity-related attributes are those which potentially identify individuals in a table. And it uses two major of techniques, namely generalization and suppression [4, 5, 6].

However, while k-anonymity protects the privacy of the data, it does not provide sufficient protection against attributing disclosure it can be easily attacked by the intruders using homogeneity and background knowledge attack.

B. l-Diversity

L-Diversity is a group based anonymization model that assists to preserve the privacy of data through reducing the granularity of a data representation using generalization and suppression. It is an extension of K-anonymity [7]. It proposed to overcome the problem of homogeneity attack and background knowledge attacks.

C. t-closeness

t-Closeness is another group-based privacy model that extends the l-diversity model. It treats the values of an attribute distinctly, and considers the distribution of data values of the attribute to preserve the privacy [8] (Table 1).

1.2 Anonymization Techniques

Data anonymization is removing personal identifiable information from the given data sets. But to anonymize the data it uses different anonymization techniques on

Table 1 Comparison of anonymization models

S. no	Type of model	Advantage	Possible attack	Disadvantage
1.	K-Anonymity [3]	• Easily method to implement	• Homogeneity attack, background knowledge attack	• It fails to prevent homogeneity and knowledge background attack • It's not protect for the massive amount of data
2.	l-diversity [7]	• Reduce the dataset into a summary form. Sensitive attribute has at most the same frequency • Prevent from homogeneity attack and background knowledge attack	• Skewness, similarity attack	• It depends upon the range of sensitive attributes. For the l-diversity should be l different value of sensitive attributes
3.	t-closeness [8]	• It solves the attribute disclosure Vulnerabilities inherent to l-diversity: skewness and Similarity attacks	• EMD distance	• It limits the amount of useful information that is released by the threshold

Table 2 Sample medical data

Name	Age	Gender	Zip-code	Diseases
John	45	M	213201	Cancer
Miki	26	M	225001	Flu
Kahn	47	M	213112	Flu
Ellen	34	F	213667	Heart diseases
Ronald	32	M	225987	Cancer
Rihanna	36	F	225123	Cancer
Ciara	29	F	213645	Flu
Johnson	31	M	225976	Viral infection

the data set. For example, as illustrated in Table 2, the sample medical data contain name, age, gender, zip-code, and diseases of the patient.

As illustrated from the above Table 3 before applying the anonymization techniques they remove the public identifiers from the medical data then we can apply different anonymization techniques.

Table 3 Public identifier removed data

Age	Gender	Zip-code	Diseases
45	M	213201	Cancer
26	M	225001	Flu
47	M	213112	Flu
34	F	213667	Heart diseases
32	M	225987	Cancer
36	F	225123	Cancer
29	F	213645	Flu
31	M	225976	Viral infection

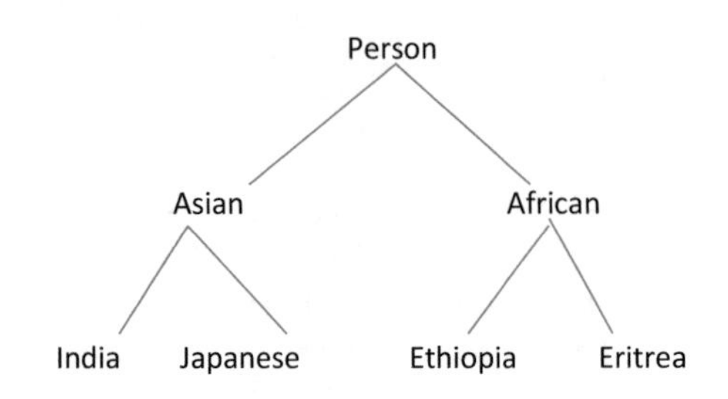

Fig. 1 Taxonomy for attribute generalization

Table 4 Published data by generalization

Age	Gender	Zip-code	Diseases
44–49	Person	213201	Cancer
25–30	M	225001	Flu
44–49	Person	213112	Flu
31–36	Person	213667	Heart diseases
31–36	M	225987	Cancer
31–36	F	225123	Cancer
25–30	Person	213645	Flu
31–36	M	225976	Viral infection

A. **Generalization**: this technique, it replaces the values of the data with a more general value or replaced the individual value by a border category that is to hide the details of attributes, to making the quasi identifier less identifying [4, 9] (Fig. 1).

If the value is numeric, it may be changed to a range of values [5]. And if the value is categorical data changed with most general value, for example from the above table the Gender statues M or F change into a more general value like the person (Table 4).

A. **Suppression**: means removing an entire tuple or attribute value from the table then Replaces tuple or attribute values with special symbols such as asterisk "*" that means some or all value of the attribute is replaced by the "*" [4, 9, 5]. The suppression form of the medical data displayed as follows (Table 5).

Table 5 Published data by suppression

Age	Gender	Zip-code	Diseases
*	*	213201	Cancer
*	*	225001	Flu
*	*	213112	Flu
*	*	213667	Heart diseases
*	*	225987	Cancer
*	*	225123	Cancer
*	*	213645	Flu
*	*	225976	Viral infection

Table 6 Published data by slicing

Age	Gender	Zip-code	Diseases
45	M	213201	Cancer
26	M	225001	Flu
47	M	213112	Flu
34	F	213667	Heart diseases
32	M	225987	Cancer
36	F	225123	Cancer
29	F	213645	Flu
31	M	225976	Viral infection

But by using suppression the quality of the data extremely reduces because it completely removes the value of the attribute.

B. **Slicing**: it performs its task based on the partitioning of the data. That is either by vertically or horizontally [10]. Vertical partitioning is applied by classifying attribute values into columns manner on the correlations between the attributes. Each column contains a subset of attributes that are highly correlated. Horizontal partitioning is applied by classifying tuples into buckets. Finally, within each bucket, values in each column are randomly permutated to break the linking between different columns [11, 12] (Table 6).

C. **Bucketization**: it divides the whole records into a group of partitions, then it assigns a value for the partitioned content. Bucketization is similar to generalization, but the only difference is generalization is containing more general value and Bucketization is containing the actual value of tuple [13, 12] (Table 7).

To summarize the anonymization techniques and models in medical data (Table 8).

Then the content of the paper is in Sect. 2 reviews the challenges in data anonymization and also in Sect. 3 it discuss about the methods and existing work for the privacy of Electronic Health Service (EHS) for privacy using data anonymization, then Sect. 4 reviews the open research problems and finally at the Sect. 5 we conclude the survey paper.

Table 7 Published by Bucketization

A. Quasi identifier table				B. Sensitive table	
Age	Gender	Zip-code	GID	GID	Diseases
45	M	213201	1	1	Cancer
26	M	225001	1	1	Flu
47	M	213112	2	2	Flu
34	F	213667	2	2	Heart diseases
32	M	225987	3	3	Cancer
36	F	225123	3	3	Cancer
29	F	213645	4	4	Flu
31	M	225976	4	4	Viral infection

Table 8 Anonymization model and techniques

Anonymization	Type
Model	1. K-Anonymization 2. l-diversity 3. t-closeness
Techniques	1. Generalization 2. Suppression 3. Slicing 4. Bucketization

2 Challenges to Privacy in Data Anonymization

The goal data anonymization medical data is to ensure that to share confidential medical data between the electronic health service (EHS) and data analyzer. But at the time of anonymized the data it faces some challenges that to lose the privacy of the patient. Some of the common challenges in data anonymization are like information loss, homogeneity attack and background knowledge attack.

I. **Homogeneity attack**
While k-Anonymity protects against identity disclosure, it does not provide sufficient protection against attributing disclosure, because of that it exposes into intruders attack. While the anonymized groups are not identical to each other and the value may be similar to each other, at this time the intruder it is easy to conclude the sensitive information based on the sensitive values [14].

II. **Background knowledge attack**
Another attack of k-anonymization is a background knowledge attack that is attacked by the general knowledge of the intruder. For example Cancer is a higher incidence in the USA, and Malaria is extremely incidence in sub-Sharan countries, then based on this knowledge the intruder can easily attack the sensitive data of the patient [14].

III. **Composition attack**
An individual privacy attack is caused by the combination of independent data sets, the intruder uses the intersection of the datasets. For example, a patient may visit two different hospitals for the same disease, then their records anonymized differently and published by the hospitals. Then the combination of these two hospitals may contain the sensitive value in both datasets, then it can be easily attacked by the intruders [15].

IV. **Probabilistic inference attack**
When the sensitive attribute value is repeated in frequently number, then it is easy to guess the value by the intruder. For example, if we have an anonymized table, then in the sensitive attribute when some value is repeated frequently it is easy to guess the sensitive value by the intruder [14].

V. **Similarity attack**
Similarity attack has happened if the table containing similar values of record and sensitive value in the same table, then it can be attacked by the intruder [14].

3 Existing Solution and Open Issue for Privacy of Data Anonymization

In recent years, many data anonymization solutions or techniques have proposed. The most popular k-anonymization for the first time was proposed by Sweeney [4] and it uses generalization and suppression techniques. Balusamy and Muthusundari [16] proposed a new privacy specialization method that is used by generalization to provide secure data generalization to preserve the sensitive data and provide the security against scalability in an efficient way with the help of MapReduce.

Li et al. [17] introduced two k-anonymity algorithms to create an accurate anonymization table that is to maximize the classification capability that is by combining global attribute generalization and local value suppression and the proposed model have better and faster classification than the benchmark utility aware data anonymization method. Soria-Comas et al. [18] described the t-closeness that is an extension of k-anonymization, ε-differential privacy and k-anonymization are two privacy models. And it discusses about privacy preserving based on Bucketization used to achieve the t-closeness that is related to the ε-differential privacy and it solve the assign disclosure issue.

Domingo-Ferrer et al. described about nominal attributes and anonymized by knowledge base numerical mapping for the attributes, then it computes semantically and mathematically coherent mean, variance, and covariance of the attributes [19]. Rose et al. it [20] discussed about how to securely transfer the medical data from one to another especially the sensitive data by storing the data in the cloud because of its large scalability of the data it uses MapReduce Algorithm. At the cloud it encrypted the data by using different encryption algorithms. Parmar and Shah [21]

Table 9 Review of previous works

S. no	Research title	Publication/year	Methodology	Advantage	Limitation
1	A hybrid approach to prevent composition attacks for independent data releases [22]	Elsevier/2016	It uses hybrid algorithm	It prevents attack causes by independent datasets	It depends on the knowledge of data publisher
2	A clustering approach for anonymizing distributed data Streams [23]	IEEE/2015	It uses CASTLE algorithm	It uses for anonymized distributed data streams and minimized its information loss	It experimented in a small number of datasets
3	Privacy preserving big data publishing [24]	ACM/2015	It uses MapReduce to partition the data	It uses MapReduce to preserve privacy of massive amount data	Generalization doesn't handle massive amount of data
4	Data Anonymization through generalization using MapReduce on cloud [25]	IEEE/2015	It uses MapReduce to partition the data into chunks, then it uses Generalization	By partition the data into chunks it solves the problem of scalability	Generalization it may attack by intruders
5	Slicing: a new approach for privacy preserving data publishing [26]	IEEE/2012	It uses slicing technique	It has better data utility compare to generalization and its handle high dimensional data	It provides less data utility compare to duplicate attribute

give a review of the data anonymization methods like K-Anonymization, l-Diversity, t-Closeness and its review of data mining anonymization approaches and Algorithms in K-Anonymization.

Dubli and Yadav [13] have explored about various data anonymization techniques like generalization, suppression, Bucketization, and slicing and also it proposed new slicing technique that gives high level of security data even when applied to large datasets and it minimizes the risk of identity and membership disclosure (Table 9).

4 Open Research Problem

Opportunities

The paper [24] discusses about privacy preserving for the huge amount of dataset by using MapReduce and to experiment the result, it uses K-Anonymity and l-diversity, but generalization or K-anonymity is not supported for the huge amount of data. And it can be easily attacked by the homogeneity attack and background knowledge attack [4, 5, 9, 14]. The paper [27] uses generalized clustering algorithm for achieving k-Anonymization that is by dividing the dataset into clusters, but the algorithm is applied in small amounts of datasets and also it does not check the information loss by applying in different anonymization models.

Possible Approach
Electronic Health service highly needed privacy because it contains sensitive data of the patients. The paper [14] explains about the data anonymization and possible attacks in the anonymization and it proposed a clustering-based algorithm to prevent the attack in the anonymization. Then the newly proposed clustering algorithm is better than the previous algorithm in terms of accuracy and execution time.

5 Conclusion

This paper gives the overview of the Electronic health service privacy when given to another data recipient for different purposes like for data analysis or data mining purpose, then before releasing the data it should be anonymized or removing personal identifiable information. And it gives a summary of data anonymization techniques, models and its possible attacks and also it discussed about the advantage and limitation of data anonymization methods. Finally, it gives an overview of the previous works that used to prevent privacy and securely transfer data among different recipients.

References

1. Popeea T, Constantinescu A, Rughinis R (2013) Providing data anonymity for a secure database infrastructure. In: Roedunet international conference (RoEduNet), 11 Feb 2013. IEEE, pp 1–6
2. Presswala F, Thakkar A, Bhatt N (2015) Survey on anonymization in privacy preserving data mining
3. Sweeney L (2002) K-anonymity: a model for protecting prvacy. Int J Uncertai Fuzziness Knowl Based Syst 10(5):557–570
4. Sweeney L (2002) Achieving k-anonymity privacy protection using generalization and suppression. Int J Uncertai Fuzziness Knowl Based Syst 10(05):571–588
5. Fung B, Wang K, Chen R, Yu PS (2010) Privacy-preserving data publishing: a survey of recent developments. ACM Comput Surv (CSUR) 42(4):14
6. LeFevre K, DeWitt DJ, Ramakrishnan R (2005) Incognito: efficient full-domain k-anonymity. In: Proceedings of the 2005 ACM SIGMOD international conference on management of data. ACM, pp 49–60
7. Machanavajjhala A, Kifer D, Gehrke J, Venkitasubramaniam M (2007) L-diversity: privacy beyond k-anonymity. ACM Trans Knowl Discov Data (TKDD) 1(1):3
8. Li N, Li T, Venkatasubramanian S (2007) t-closeness: privacy beyond k-anonymity and l-diversity. In: 2007 IEEE 23rd international conference on data engineering, ICDE 2007. IEEE, pp 106–115
9. Rodiya K, Gill P (2015) A review on anonymization techniques for privacy preserving data publishing
10. Li T, Li N, Zhang J, Molloy I (2012) Slicing: a new approach for privacy preserving data publishing. IEEE Trans Knowl Data Eng 24(3):561–574

11. Sreevani P, Niranjan P, Shireesha P (2014) A novel data anonymization technique for privacy preservation of data publishing. Int J Eng Sci Res Technol
12. Patil SA, Banubakod DA (2015) Comparative analysis of privacy preserving techniques in distributed database. Int J Sci Res (IJSR) 4(1)
13. Dubli D, Yadav DK (2017) Secure techniques of data anonymization for privacy preservation. Int J 8(5)
14. Nayahi JJV, Kavitha V (2017) Privacy and utility preserving data clustering for data anonymization and distribution on Hadoop. Future Gener Comput Syst 74:393–408
15. Li J, Baig MM, Sattar AS, Ding X, Liu J, Vincent MW (2016) A hybrid approach to prevent composition attacks for independent data releases. Inf Sci 367:324–336
16. Balusamy M, Muthusundari S (2014) Data anonymization through generalization using map reduce on cloud. In: 2014 international conference on computer communication and systems. IEEE, pp 039–042
17. Li J, Liu J, Baig M, Wong RCW (2011) Information based data anonymization for classification utility. Data Knowl Eng 70(12):1030–1045
18. Soria-Comas J, Domingo-Ferrert J (2013) Differential privacy via t-closeness in data publishing. In: 2013 eleventh annual international conference on privacy, security and trust (PST). IEEE, pp 27–35
19. Domingo-Ferrer J, Sánchez D, Rufian-Torrell G (2013) Anonymization of nominal data based on semantic marginality. Inf Sci 242:35–48
20. Rose PS, Visumathi J, Haripriya H (2016) Research paper on privacy preservation by data anonymization in public cloud for hospital management on big data. Int J Adv Comput Technol (IJACT)
21. Parmar K, Shah V (2016) A review on data anonymization in privacy preserving data mining. Int J Adv Res Computd Commun Eng 5(2)
22. Li J, Baig MM, Sattar AS, Ding X, Liu J, Vincent MW (2016) A hybrid approach to prevent composition attacks for independent data releases. Inf Sci 367:324–336
23. Mohamed MA, Nagi MH, Ghanem SM (2016) A clustering approach for anonymizing distributed data streams. In: 2016 11th international conference on computer engineering and systems (ICCES). IEEE, pp. 9–16
24. Zakerzadeh H, Aggarwal CC, Barker K (2015) Privacy-preserving big data publishing. In: Proceedings of the 27th international conference on scientific and statistical database management. ACM, p 26
25. Balusamy M, Muthusundari S (2014) Data anonymization through generalization using map reduce on cloud. In: 2014 international conference on computer communication and systems. IEEE, pp 039–042
26. Li T, Li N, Zhang J, Molloy I (2012) Slicing: a new approach for privacy preserving data publishing. IEEE Trans Knowl Data Eng 24(3):561–574
27. Xu X, Numao M (2015) An efficient generalized clustering method for achieving k-anonymization. In: 2015 third international symposium on computing and networking (CANDAR). IEEE, pp 499–502

Prediction of Cervical Cancer Based on the Life Style, Habits, and Diseases Using Regression Analysis Framework

K. Vinoth Raja and M. Thangamani Murugesan

Abstract Cervical cancer is the most common disease in the woman nowadays. Even though its panic diseases, we can control and prevent it by finding the symptoms of growing cancer. It is the disease formed in the genital area of the woman and later it spreads to all the parts of the body and makes the system collapse to stop functioning of the organs. Condylomatosis, wrong sexual practices and hormonal contraceptives are one of the major primary factors for getting the cervical cancer very easily via Human Papilloma Virus. The secondary factors for causing the cervical cancer is smoking and alcoholic usage. Along with these factors molluscum contagiosum, HIV and Hepatitis B also make the humans to get affected by the cervical cancer very easily. All these factors are to be considered for analysing the patient whether they got affected by the cervical cancer. Regression Analysis model framework is used for comparing the various factors to determine the diseases vastly.

Keywords Condylomatosis · HPV (human papilloma virus) · Hormonal contraceptives · Regression analysis · Molluscum contagiosum

1 Introduction

Cervical cancer is the second most leading cause of cancer death in women worldwide. Cervical cancer is the malignant tumor which is produced in the lower part of the uterus. It occurs when the cells in the cervix are growing abnormally. The disease has been identified by some of the conditions and symptoms of the STD (Sexually Transmitted Diseases) diseases like *condylomatosis* such as *cervical condylomatosis, vaginal condylomatosis, vulvo-perineal condylomatosis* and *syphilis*. The main

K. Vinoth Raja (✉)
Anna University, Chennai, Tamil Nadu, India
e-mail: oniv4uever@gmail.com

M. Thangamani Murugesan
Kongu Engineering College, Perundurai, Erode, Tamil Nadu, India
e-mail: manithangamani2@gmail.com

D. K. Mishra et al. (eds.), *Data Science and Big Data Analytics*,
Lecture Notes on Data Engineering and Communications Technologies 16,
https://doi.org/10.1007/978-981-10-7641-1_13

cause for the cervical cancer is the individual lifestyle such as wrong sexual practices, number of pregnancies, drugs usage and some pregnancy prevention steps such as intrauterine device. IUD (Intrauterine Device) is sometimes allergetic to the users and makes the possibility of high risk of getting affected by the HPV very easily by their lifestyle. All these factors are considered for the prediction of the cervical cancer growth in the genital area of the woman. Even though there are many factors available for the risk of getting affected by the HPV (Human Papilloma Virus) [1], they are all unique from each other. So it is quite tough for us to consider all the factors simultaneously to predict the affection of the disease. In US, various steps have been taken to vaccinate the people to avoid the morbidity and mortality caused by the HPV. It plays the major role in the economical and social level in US. Daley et al. [2] proposed the various methods to control the HPV and its related illness to the both gender. Male also get affected by the HPV and though he may not get the cervical cancer, but they are all suffered by their resulting illness.

Normally the cervical cancer is affected in the age ranges from age 15–60 years. So the age is to be considered as the primary factors for analysis and the prediction of the diseases. This mainly depends upon the lifestyle lead by the individual. *Oropharyngeal squamous cell carcinoma* (OPSCC) is increased the risk factors of Human papillomavirus (HPV). *Oropharyngeal subsite* includes tonsil, base of tongue, soft palate and posterior wall. Haeggblom et al. [3] calculated that *oropharyngeal subsite* calculated as 56%, 40%, 12% and 19% for tonsil, base of tongue, soft palate and posterior wall respectively. The prevalence of HPV was significantly higher in "*lymphoepithelial*" sites of the *oropharynx,* i.e., tonsil and base of tongue, and lower in "*non-lymphoepithelial*" sites of the *oropharynx*, i.e, soft palate and *oropharyngeal*. Panici et al. [4] measured that 83% of oral condyloma are not visible to the naked eyes. 45% of people are affected by the human papilloma virus were found out by taking the samples from the oral scrapings. Genital human papillomavirus can establish a local infection in the oral cavity and human papillomavirus oral lesions in patients with *genital condyloma*. These are happened by the non-safety relationship with the various sexual partners. The screening of cervical cancer is done in visually by the Papanicolaou test (Pap smear), visual inspection of the cervix with acetic acid (VIA) or Lugol's iodine (VILI), cervicography and HPV testing. VIA is very high in sensitivity and specificity compared to the other tests, which is more cost efficient and applicable to the rural and urban area. Li et al. measured that 50% in sensitivity and 97.7% in specificity [5].

Condylomatosis is inspected by the skins of the genital part of the patience after the operations. Even though there are no reported cases of primary *neovaginal carcinoma* associated with extensive *neovaginal condylomatosis*. *McIndoe vaginoplasty* patients are susceptible to complications related to HPV infection [6]. Cirrhosis is the risk factor established by alcohol. The hepatic sensitivity of the alcohol leads to the cirrhosis. Alcohol can also lead to a specific inflammatory state in the liver, alcoholic hepatitis (AH). There is an increased risk for fibrosis progression and development of hepatocellular carcinoma specifically for smoking [7] In smokers who also have hepatitis C or hepatitis B virus infection the risk is increased. Smokers who have alcohol in large quantity may have a risk of almost 10 times higher than people who don't smoke or drink. HPV DNA was detected in 3.0% (2/67) of lung cancer cases was found in tissue samples examined and there is no E6/E7 mRNA of five high-risk HPV types was found in it [8]. Lung cancer-related HPV infection rate is fluctuated in between 10 and 80%, depending on the various research methods and geographical factors. 22.4% of the lung cancer patients are suffered with the HPV infection.

Most intrauterine device wearers has increased the risk of pelvic inflammatory disease persists for only a few months after insertion. They are suffered by the premenopausal symptoms and bleeding [9]. The copper T gets corrosion by the uterine fluid. Zhang et al. [10] found that the pH of the fluid played an important role. They have measured that the corrosion rate at pH 6.3 was several times higher than that at pH 8.0. Both cathodic reduction of surface film and chemical analysis of solution were also measured by them.

The devices used in the years as follows: he TCu380A and the TCu220 for 12 years, the Multiload Cu-375 for 10 years, the frameless GyneFix R (330 mm^2) for 9 years, the levonorgestrel intrauterine system 52 mg (Mirena) for 7 years and the Multiload Cu-250 for 4 years [11]. The Dalkon Shield appears produces a higher risk of pelvic inflammatory disease than the other devices. The selection of IUD plays the important role in the pelvic inflammatory disease [12]. Syphilis is generally sexually transmitted, it can also be spread by contact with open lesions, blood transfusions, needle sharing, or congenitally. It is caused by infection with Treponema pallidum. Risk of transmission is highest in the first 2–3 years of infection. Without the proper treatment of syphilis can lead to irreversible neurological or cardiovascular complications [13].

Pelvic inflammatory disease (PID) is a cause of morbidity in young women and it is secondary to a sexually transmitted infection. Untreated PID can result in infertility, ectopic pregnancy or chronic pelvic pain in up to 40% of women. Once PID gets affected, reproductive tract may be less able to fight a new infection because of scar tissue from past PID. When *Chlamydia* is come with PID, it causes more irritation and pelvic organ damage that is worse than the first time [14, 15]. *Gonorrhea and chlamydia* are the two sexually transmitted diseases and are the most common causes of PID. There is greater risk for the person who are sexually active and younger than 25, have more than one sex partner, and Douche, it can cause vaginal and pelvic infections [16]. Anal cancer is a rare disease with an incidence of 1–3% of all tumors of the digestive system [17]. Anal canal squamous intraepithelial lesions (SILs) caused by certain human papilloma viruses (HPVs) which cause anal carcinogenesis [18]. Situ hybridization is used by Hwang et al. [19] to identify the HPV types.

Castel et al. [20] evaluated NA for 14 carcinogenic HPV genotypes on a sample of liquid cytology specimens ($n = 531$). Most of the *cervical intraepithelial neoplasia* (CIN3) cases testing positive for carcinogenic HPV E6/E7 mRNA. They have tested positive for carcinogenic HPV E6/E7 mRNA than for carcinogenic HPV DNA, especially in women with <CIN1 ($P < 0.0001$). Wang et al. [21] detected 32 HPV genotypes based on the reverse blot hybridization assay (REBA) for the detection of oncogenic HPV infection according to cytological diagnosis. HPV 16 was the most common HPV genotype among women with high- grade lesions. The women with age of 40–49 with confirmed lesions are having E6/E7 mRNA expression in them.

Human papillomaviruses (HPV) is a double-stranded DNA virus, which is causing many *mucocutaneous* diseases, benign or malignant, ranging from common warts to malignancies in the area of aerodigestive tract and the anogenital sphere. Eide [22] proposed the two types of HPV identification methods namely signal amplification methods (hybridization techniques in liquid phase) and target amplification methods (the techniques of gene amplification or Polymerase Chain Reaction [PCR]). Genotyping techniques are used for the quantitative detection of viral DNA of HPV genotype and so monitor changes in viral load over time. It uses the amplification method followed by the hybridization methods for identification of the genotypes. Cytology and HPV DNA testing combination is used to evaluate the 11 possible cervical screening strategies. The outcome of this test is sensitivity for detection of cervical intraepithelial neoplasia grade 3 or at colposcopy for women with a persistent type-specific HPV infection [23]. Somlev et al. [24] proposed the separate univariate logistic regression was used to evaluate the association between β- and γ-HPV and HIV status (positive versus negative). They have found out that anal prevalence β- and γ-HPV detection was significantly higher in HIV+ than in HIV− men.

DNA virus has over 200 serotypes, of which 18 are classified as oncogenic. Among this 18, 12 of these are high-risk HPV prototypes. Almost it is taken 20 years for the progressive development of invasive cancer. Nayak [25] instructed to do pap smear test and HPV DNA test for early detection of the disease. All the other habits such as drinking, smoking, STDs and promiscuity are increasing the multiparty of the cervical cancer. Inverted papilloma is a rare *sinonasal tumor*, which is very different by three characteristics namely a relatively strong potential for local destruction, high rate of recurrence, and a risk of carcinomatous evolution. 40% of this disease is diagnosed with HPV. The only solution for this disease is operation [26]. In European population, there is increased in OPSCC incidence and it is mainly attributed to HPV, and the HPV status significantly affected prognosis [27]. Cigarette smoking is increased the risk of cervical cancer and its immediate precursor, cervical intraepithelial neoplasia grade 3 (CIN3), among women infected with oncogenic human papilloma virus (HPV). Multiplex real-time PCR is used to measure the viral genome copies per nanogram of cellular PCR. Regression models were used to assess the relationship between cigarette smoking and baseline viral load. Among non-smokers, baseline HPV16 DNA load was related to race, parity, and referral Pap [28]. A total of 1,007 patients were recruited among women, and grouped them. Statistically the difference between the groups is considered as significant at $P<0.05$. Risk of CIN 2–3 or cervical carcinoma cervical is increased 1.642 times among smoking patients than the non smoking patients [29]. Smoking is increasing the HPV infection prevalence and the risk is also increased with more smoking per day.

2 Materials and Methods

2.1 Data Collection

Data has been collected from the 800 woman about their age, personal life style and habits, who is taking medication and having the symptoms of cervical cancer. The information is included such as number of sexual partners, number of pregnancies obtained, their smoking habits and the quantity of smoking is also recorded. The patients are also suffered by hormonal contraceptives and the duration of hormonal contraceptives also observed. Many women have worn IUD (Intraurine Devices) for safety intercourse and the years of wearing are also recorded. Some ladies affected by the STDs (Sexually Transmitted Diseases) such as various types of *Condylomatosis, Syphilis, Genital herps*, and AIDS.

Hormonal Contraceptives—Hormones is the messenger for impacting the function of the organ and the cells. The lower secretion and higher secretion of the hormones may leads to the infection of the person. Each hormone plays the important role in the function of the body such as reproduction, respiration, etc. In the case of the cervical cancer, the estrogen plays the important role, which is secreted in the area of the uterine. It influences physiological processes in various tissues/systems including the female reproductive tract, breast, colon, brain, bone, cardiovascular, and immune systems. The nuclear receptors ERα and ERβ are responsible for these influences in the tissues of the body. It creates the changes in the epithelium of the cervix. The changes of the cervical epithelium lead the easiest way of HPV to attack via the cervical area. It shows in Fig. 1.

The least risky type represents only mild dysplasia, or abnormal cell growth. CIN1 is confined to the basal one-third of the epithelium. CIN2 is moderate dysplasia confined to the basal two-third of the epithelium. CIN3 is the severe dysplasia that spans more than two-third of the epithelium, and may involve the full thickness. This lesion may sometimes also be referred to as cervical carcinoma in situ.

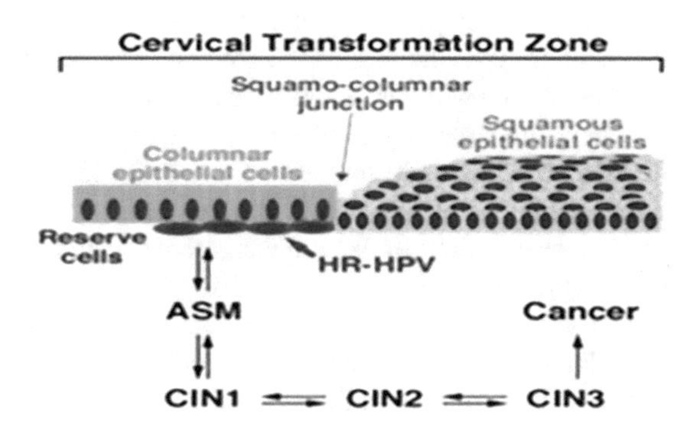

Fig. 1 Progression of cervical cancer

2.2 Methods

In our problem, we have various independent variables which are given as the results of cervical cancer. The regression analysis framework is used to analyze the data and predict the possibilities of affected by the cervical cancer. The prediction of the cervical cancer is considered as the dependent variable Y and the number of sexual partners is considered as the variable x_1, the number of years of smoking is x_2, IUD wearing years count is considered as x_3, the number of harmonal contraceptive suffering year is considered as x_4, and the number of STDs affected is calculated as x_5. The multiple linear regression models are used to calculate the possibilities of having cervical cancer, because there are numerous independent variables which is partially related with each other. The multiple regressions is calculated using Eq. (1)

$$Y = \beta_0 + \beta_1 X_1 + \cdots + \beta_n X_n + \epsilon \tag{1}$$

where ϵ is the error rate. According to Eq. (1) whenever the value of X_1 is increased, the value of β_1 will leads to increase the value of Y [30]. Every independent variables value is raised; the coefficient estimation β_n is varied directly proportional to them. The two ways of finding the multiple linear regression models is by scattering plot and the correlation between the independent variable has to determine for finding the values of multiple regression model. By this method we have to find the correlation between the independent variables such as number of sexual partners, number of years smoking, number of IUD wearing years, number of years suffered by the hormonal contraceptives and number of STDs affected the patients. All the predictors' values is the collection of different life style habits and the various diseases symptoms are closely related for the risk of the cervical cancer.

In the case of parameters such as IUD wore women, smoking women and STDs affected are having their sub parameters such as number of years smoking, number of cigarette smoke per day, number of years IUD wore, types of STDs affected. For calculating the main parameters such as IUD, Smoking and STD, we have to determine the values by considering their relevant sub-parameters of the parameters. Let us consider the case of smoking; there are three parameters available to give the details about the smoking of the women. The sub-parameters of the smoking are whether she is having the habits of smoking, how many years they are smoking, and how many cigarette they smoke in a day. Here the cancer risk factor is depends upon the number of cigarette smoke by them in the long years. Because some may have started the habits of smoking long ago, but the cigarette per day is very less and in some cases they may smoke more cigarettes from long year ago. In some cases they may start recently with heavy quantity.

The cervical cancer risk factor is depended upon the large quantity consumption in the long years. So we have to determine the values for the number of cigarettes smoke by them in the life time to find the probability of getting the cervical cancer by this habit. By considering the number of cigarettes and number of years, we have to find the frequency of smoking of each woman. Depends upon the frequency of smoking the value of possibility is found out by using the binomial distribution, by which we can easily find out the probability of getting affected or not [30]. The binomial distribution is shown in Eq. (2)

$$P(X = x) = nC_x p^x q^{n-x} \tag{2}$$

Here n is the number of trial, x is getting affected by cancer and $n - x$ is not getting affected by the cancer. p is the probability of getting cancer and q is the probability of getting cancer. By this way we can find out the probabilities of getting the cervical cancer by considering the number of STDs affected the woman and the number of years having the hormonal contraceptives.

3 Results and Discussion

3.1 Results

The sample of the collected data from the 800 women is shown in Table 1. In Table we have shown the summary of the parameters which is having some sub parameters with them such as smoking, IUDs. The mean age of the woman is calculated as 26. The minimum age of the woman attended in the data collection is 13 and the maximum age is 84. The average sexual partners having by the women are calculated as 0.26, their minimum and maximum partner's ranges from 0 to 9. The average of first intercourse is calculated as 16 years, the range of their intercourse age is from 0 times to 32 years. The number of pregnancies attempted is received from 0 to 6 times. The smokes habit is ranges from 0 to 37 years. The average smoking years of the women is calculated as 1.2 years. The women wearing IUD is from 0 to 19 years. The summary of the data collection is shown in Fig. 2. In our datasets the ladies are suffered by the hormonal contraceptives up to 30 years. They have affected maximum of four STD diseases and minimum of disease free, i.e., 0. The proportional of the ladies affected by the cervical cancer is calculated in the average of 20.98 and the minimum chance for the risk of the cervical cancer is 0 and the maximum of 64. The summary of the parameters in the data collection is shown in Table 1.

We have considered the ladies suffered by the hormonal contraceptives to find the risk of having the cervical cancer. Their proportional is directly proportional to each other which are shown in Fig. 2. Likewise we considered the case is the number of sexual partner's role in having the risk of having cervical cancer. The comparison of the cervical risk is shown in Fig. 3. It gives us the clear effect of having more sexual partners in the chance of getting cervical cancer by their illegal relationships. Likewise we can deal the conditions with the various types of STDs to get the possibility of getting cervical cancers by the related diseases. The variations of the predicted variable values is calculated which is used to determine the average level of the values in the data collection.

Table 1 Data collection parameters

	Age	Number of sexual partners	First sexual intercourse	Number of pregnancies	Smokes	IUD	STDs	Hormonal contraceptives	Affected
1	18	4	15	1	0.000000	0.00	0	0.000000	20.00000
2	15	1	14	1	0.000000	0.00	0	0.000000	16.00000
3	34	1	0	1	0.000000	0.00	0	0.000000	2.00000
4	52	5	16	4	37.000000	0.00	0	3.000000	65.00000
5	46	3	21	4	0.000000	0.00	0	15.000000	43.00000
6	42	3	23	2	0.000000	0.00	0	0.000000	28.00000
7	51	3	17	6	34.000000	7.00	0	0.000000	67.00000
8	26	1	26	3	0.000000	7.00	0	2.000000	39.00000
9	45	1	20	5	0.000000	0.00	0	0.000000	26.00000
10	44	3	15	0	1.266973	0.00	0	0.000000	19.26697
11	44	3	26	0	0.000000	0.00	0	2.000000	31.00000
12	27	1	17	0	0.000000	0.00	0	8.000000	26.00000
13	45	4	14	0	0.000000	5.00	0	10.000000	33.00000
14	44	2	25	0	0.000000	0.00	0	5.000000	32.00000
15	43	2	18	0	0.000000	8.00	0	0.000000	28.00000
16	40	3	18	0	0.000000	0.00	0	15.000000	36.00000
17	41	4	21	0	0.000000	0.00	0	0.250000	25.25000
18	43	3	15	0	0.000000	0.00	0	3.000000	21.00000
19	42	2	20	0	0.000000	6.00	2	7.000000	37.00000

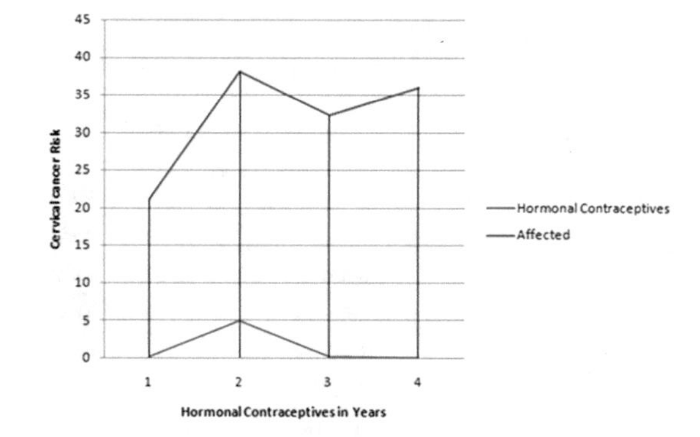

Fig. 2 Cervical cancer versus hormonal contraceptives

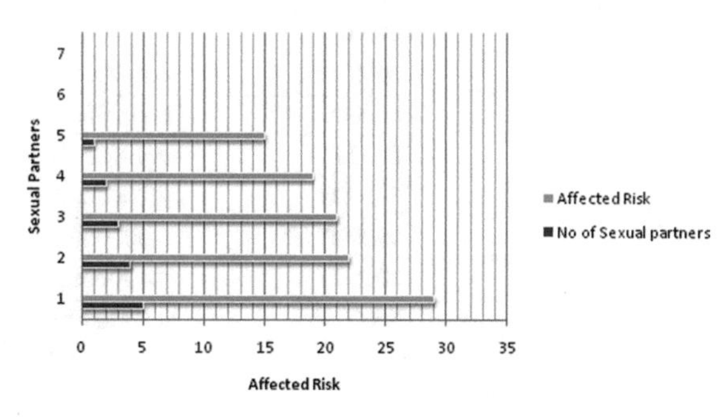

Fig. 3 Number of sexual partners versus cervical cancer affected risk

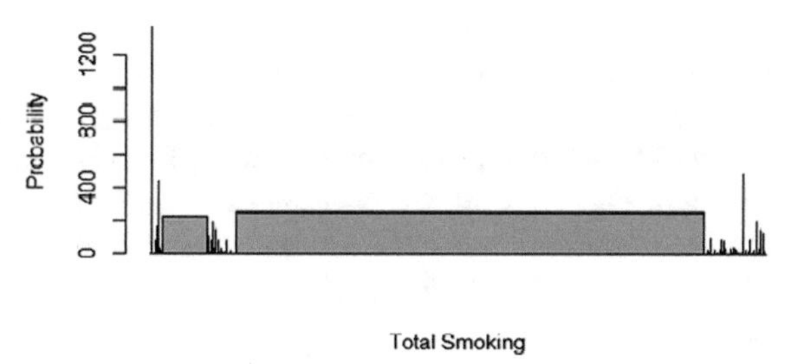

Fig. 4 Smoking versus cervical cancer affected risk

We have considered the case of smoking ladies who are all getting the possibility of cervical cancer easily. It depends upon the number of years and the number of cigarette they are smoking per day. The smoking is also one of the leading cancer producing factors. It kindles the cells to enhance the growth nature of the cancer. The risk factors of the cervical cancer to the women who are smoking is shown in Fig. 4. The problems caused by the smoking, hormonal contraceptives and illegal sexual practices is explained in Sect. 1.

In Table 1, we have shown the main attributes which is considered for the determination of the risk factors of the cervical cancer. Among the above attributes, smoking, IUDs, and STDs are having their own sub-attributes. By consolidating the sub-attributes only we have shown the consolidated attributes in above. Each attributes have their own unique sub attributes. In smoking attributes, we get the

Table 2 Smoking attributes

	Smokes years	Smokes packs year	Total smoking
41	0.000000	0.0000000	0.0000000
42	0.0000000	0.0000000	0.0000000
43	1.266973	2.4000000	3.0407350
44	0.000000	0.0000000	0.0000000
45	12.000000	6.0000000	72.0000000
45	0.000000	0.0000000	0.0000000
47	0.000000	0.0000000	0.0000000
48	0.000000	0.0000000	0.0000000
49	0.000000	0.0000000	0.0000000
50	0.000000	0.0000000	0.0000000
51	0.000000	0.0000000	0.0000000
52	0.000000	0.0000000	0.0000000
51	0.000000	0.0000000	0.0000000
54	18.000000	9.0000000	162.0000000
55	0.000000	0.0000000	0.0000000
56	0.000000	0.0000000	0.0000000
57	0.000000	0.0000000	0.0000000

information such as having smoking habit or not, number of years, number of packs per day. The frequency of the patient smoking habit is measured by considering the above parameters. It is shown in Table 2. Among the 800 values we have shown the sample dataset in Table 2. The risk factor for 859 patients by the smoking habit is shown in Fig. 4. In Table 2 we have shown the number of years they are smoking, number of packs they smoke in the year and number of packs smoke by them in their life time to find the possibility of the risk factors increasing in the cervical cancer by this habit.

3.2 *Discussions*

Variables Selection—To predict the values of the risk of cervical cancer using the multiple regression models we have to scatter the plot to identify the variables which is significant since we are using eight independent variables to predict the response values. In order to predict the values we have to correlate the relationship between the independent variables such as age, STDs, IUDs, and Hormonal Contraceptives. The scattering plot of the parameters is shown in Fig. 3. The plots show as the significance among the parameters with respect to the predictive values. The correlations values of age with the number of sexual partners, first sexual intercourse, number of pregnancies, smokes, IUD, STDs, Hormonal contraceptives is calculated as 0.26534, 0.3001, 0.1944, 0.2182, 0.2154, −0.0013, and 0.2988 respectively. In the correlations the STDs is negligible which very low value is. All the others are considered as the significant variables with respect to the age of the women. The correlation with respect to the number of sexual partners, the values are 0.265345189, 0.017164979, 0.49991575, 0.09589243, 0.057627735, −0.024414539, and 0.056307928. Among these values, age and the number of pregnancies is considered the significant values. Likewise the correlation value of the first sexual intercourse has only one significant value is the age. In the case of number of pregnancies attend by the women has the correlation with the age, number of sexual partners and smokes are the significant values to consider while we predict the values of them.

The correlation matrix is represented in Table 3. It represents the correlation between all the parameters which is to be considered significant parameters to retrieve the features of the parameters in the data collection. After find the significant parameters related to the independent variables. The value of the response variable has to be predicted by using the Eq. (1). The multiple regression model is shown as summary in Table 4. The value of β_0 is determined as 3.878e−11 to predict the values of the risk of the cervical cancer. Likewise the coefficient values of age, number of sexual partners, STDs and hormonal contraceptives is estimated as 7.081e−13, 8.230e−10, 9.037e−10 and 8.150e−09 respectively. The standard error value ϵ is estimated at 2.70×10^{-11}. The scatter plot of the data collections to select the significant data is shown in Fig. 5.

Smoking—The quantity of smoke which inhales by the person only determines the risk factors of the cancer. The quantity may vary upon the smoking habits of the individual character. The person may have the large quantity in the large duration by having small quantity in their subsequent years. Some may have large quantity of cigarettes in their short duration of habits. The average years of smoking is found out as 1.2 years but the maximum value is found out as 37 years and the minimum year is 0 year, i.e., non-smoking habit which is shown in Table 5. The total cigarettes had by the patients are 6031 cigarettes in their life time for the 859 patients. The average cigarette consumption is calculated as 7.02 cigarettes per woman in their life time. The standard deviation of the smoking habits is calculated as 56.857. The variance of the smoking per year is determined as 4.8565. The binomial calculation of the cervical cancer gets affected by the smoking habit is shown in Fig. 6 for 859 patients.

Hormonal Contraceptives—As we have discussed about the hormones in the Sect. 2. Hormones play the vital role for the normal function of the organs in the body. The people who are all suffered by the contraceptives of the hormone may get affected by the abnormal activities of the glands which are responsible for the activities of the human body. In our case we have to find whether the person is get affected by the hormonal contraceptive. If so we have to know how many years they are suffered by the problem. As per the duration the range of the infection and the problem is also increased. The median value of the hormonal contraceptive is 0.250 and the mean range is calculated as 1.972 and the maximum of they are affected in 30 years. Since the hormonal contraceptive is directly proportional to the risk factor ratio. The risk factor is varied upon the number of years get affected by this

Table 3 Correlation of the parameters in the data collection

	Age	Number of sexual partners	First sexual intercourse	Number of pregnancies
Age	1.000000000	0.26534519	0.300110402	0.19440383
Number of sexual partners	0.265345189	1.00000000	0.017164979	0.49991575
First sexual intercourse	0.300110402	0.01716498	1.000000000	0.04732011
Number of pregnancies	0.194403834	0.49991575	0.047320111	1.00000000
Smokes	0.218260785	0.09589243	−0.055023457	0.17436241
IUD	0.215426549	0.05762774	−0.004362686	0.04710288
STDs	−0.001329562	−0.02441454	0.027036931	−0.02926987
Hormonal contraceptives	0.298891901	0.05630793	0.043468506	0.01403478
Affected	0.518886429	0.27017842	0.450005043	0.27634918
SAffected	0.518886429	0.27017842	0.450005043	0.27634918

	Smokes	IUD	STDs	Hormonal contraceptives	Affected	SAffected
Age	0.21826079	0.215426549	−0.001329562	0.298891901	0.5188864	0.5188864
Number of sexual partners	0.09589243	0.057627735	0.056307928	0.056307928	0.2701784	0.2701784
First sexual intercourse	−0.05502346	−0.004362686	0.027036931	0.043468506	0.4500050	0.4500050
Number of pregnancies	0.17436241	0.047102883	−0.029269873	0.01403477	0.27634982	0.2763492
Smokes	1.000000000	0.038060777	0.088605055	0.052435566	0.6212318	0.6212318
IUD	0.03806078	1.000000000	0.021285943	0.0177955444	0.2998043	0.2998043
STDs	0.08860505	0.021285943	1.000000000	0.002236074	0.140714	0.1401714
Hormonal contraceptives	0.05243557	0.017955444	0.002236074	1.000000000	0.5750137	0.5750137
Affected	0.62123177	0.299804339	0.140171360	0.575013715	1.000000000	1.000000000

problem. The standard error rate is calculated as 1.401e−12. The risk factor analysis of hormonal contraceptive is shown in Fig. 3.

Number of Sexual Partners—The chance of getting infection is very high to the people who are all practicing more sexual partners in their life. The person who is having more sexual partners is becoming transceiver of the infectious diseases such as cervical cancer via HPV and other STDs. The disease is spread very easily by this wrong sexual practice. In our case a women is having maximum of nine sexual partners. Most of the woman is having average of one partner for their sexual life. The disease ratio depends upon the natural habitat of the maximum sexual partners. The standard error rate is found out as 6.252e−12. The error rate is very negligible so their sexual practice plays the important role in the ratio of cervical cancer or HPV infection. In Fig. 4, we have shown the cervical cancer versus number of sexual

Table 4 Coefficient estimation for the prediction of response values

Coefficients:

	Estimate std.	Error	t value	Pr (>|t|)
Intercept	3.878e−11	2.710e−11	1.431e+00	0.153
Age	7.081e−13	6.781e−13	1.590e−01	0.873
Number of sexual partners	7.081e−13	6.252e−12	1.599e+11	0.856
First sexual intercourse	1.000e+00	1.604e−12	6.235e+11	<2e−16***
Number of pregnancies	1.000e+00	1.099e−11	9.101e+10	<2e−16***
Smokes	1.000e+00	1.238e−12	8.076e+11	<2e−16***
IUD	1.000e+00	2.716e−12	3.681e+11	<2e−16***
STDs	9.037e−10	9.101e−12	1.099e+11	0.965
Hormonal contraceptives	8.150e−09	1.401e−12	7.140e+11	1.567

*** The values are relatively very low and its equal to 0 because these attributes does not affect the organ and the values of the cancer is not increased

Table 5 Summary of parameters in the data collection

Age	Number of sexual partners	First sexual intercourse	Number of pregnancies	Smokes
Min.: 13.00	Min.: 0.0000	Min.: 0.00	Min.: 0.00000	Min.: 0.000
1st Qu.: 20.00	1st Qu.: 0.0000	1st Qu.: 15.00	1st Qu.: 0.00000	1st Qu.: 0.000
Median: 25.00	Median: 0.0000	Median: 17.00	Median: 0.00000	Median: 0.000
Mean: 26.82	Mean: 0.2634	Mean: 16.86	Mean: 0.08858	Mean: 1.201
3rd Qu.: 32.00	3rd Qu.: 0.0000	3rd Qu.: 18.00	3rd Qu.: 0.00000	3rd Qu.: 0.000
Max.: 84.00	Max.: 9.0000	Max.: 32.00	Max.: 6.00000	Max.: 37.000
IUD	STDs	Hormonal contraceptives	Affected	SAffected
Min.: 0.0000	Min.: 0.000	Min.: 0.000	Min.: 0.00	Min.: 0.00
1st Qu.: 0.0000	1st Qu.: 0.000	1st Qu.: 0.000	1st Qu.: 16.18	1st Qu.: 16.18
Median: 0.0000	Median: 0.000	Median: 0.250	Median: 19.00	Median: 19.00
Mean: 0.4446	Mean: 0.155	Mean: 1.972	Mean: 20.98	Mean: 20.98
3rd Qu.: 0.0000	3rd Qu.: 0.000	3rd Qu.: 2.000	3rd Qu.: 24.00	3rd Qu.: 24.00
Max.: 19.0000	Max.: 4.000	Max.: 30.000	Max.: 67.00	Max.: 67.000

Table 6 Relationship values of STDs

	STDs	STDs number	STDs condylomatosis
STDs	1.0000000	0.9383819	0.62402572
STDs number	0.9383819	1.0000000	0.85562762
STDs condylomatosis	0.6240257	0.8556276	1.00000000
STDs cervical condylomatosis	NA	NA	NA
STDs vaginal condylomatosis	NA	NA	NA
STDs vulvo-perineal condylomatosis	0.6240257	0.8556276	1.00000000
STDs syphilis	0.3892495	0.2449422	−0.02914820
STDs pelvic inflammatory disease	0.2739983	0.1724183	−0.02051784
STDs genital herpes	NA	NA	NA
STDs molluscum contagiosum	NA	NA	NA
STDs AIDS	NA	NA	NA
STDs HIV	0.3892495	0.2449422	−0.02914820
STDs Hepatitis B	NA	NA	NA
STDs HPV	0.3892495	0.2449422	−0.02914820
STDs number of diagnosis	0.8006408	0.7265579	0.44366294

	STDs vulvo-perineal condylomatosis	STDs syphilis
STDs	0.62402572	0.38924947
STDs number	0.85562762	0.24494219
STDs condylomatosis	1.00000000	−0.02914820
STDs cervical condylomatosis	NA	NA
STDs vaginal condylomatosis	NA	NA
STDs vulvo-perineal condylomatosis	1.00000000	−0.2914820
STDs syphilis	−0.02914820	1.00000000
STDs pelvic inflammatory disease	−0.02051784	−0.01279844
STDs genital herpes	NA	NA
STDs molluscum contagiosum	NA	NA
STDs AIDS	NA	NA
STDs HIV	−0.02914820	−0.01818182
STDs Hepatitis B	NA	NA
STDs HPV	−0.02914820	−0.01818182
STDs number of diagnosis	0.44366294	0.22438728

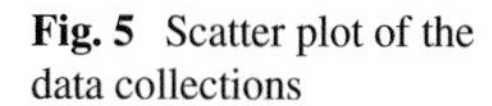
Fig. 5 Scatter plot of the data collections

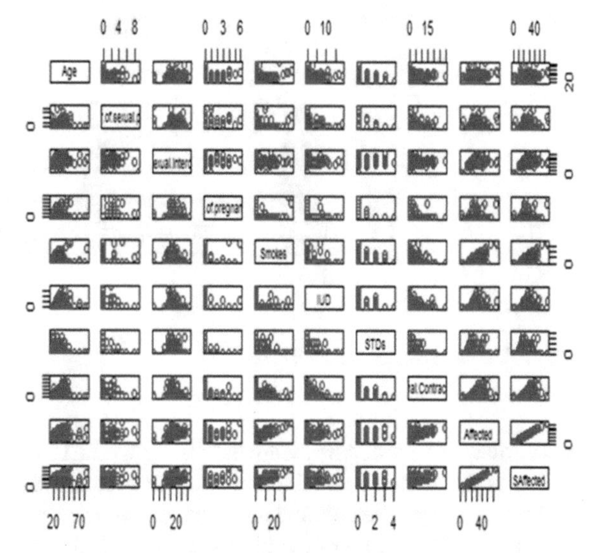


```
Console E:/desktop/Academics/R/Cervicalcancer/
[439] 0.0009765625 0.0009765625 0.0009765625 0.0009765625 0.0009765625 0.0009765625
[445] 0.0009765625 0.0009765625 0.0000000000 0.0009765625 0.0009765625 0.0009765625
[451] 0.0009765625 0.0009765625 0.0009765625 0.0009765625 0.0009765625 0.0009765625
[457] 0.0009765625 0.0000000000 0.0009765625 0.0009765625 0.0009765625 0.0009765625
[463] 0.0009765625 0.0009765625 0.0009765625 0.0009765625 0.0009765625 0.0000000000
[469] 0.0009765625 0.0009765625 0.0009765625 0.0009765625 0.0009765625 0.0009765625
[475] 0.0009765625 0.0009765625 0.0009765625 0.0009765625 0.0000000000 0.0009765625
[481] 0.0009765625 0.0009765625 0.0009765625 0.0009765625 0.0009765625 0.0009765625
[487] 0.0009765625 0.0009765625 0.0009765625 0.0009765625 0.0009765625 0.0009765625
[493] 0.0009765625 0.0009765625 0.0009765625 0.0009765625 0.0009765625 0.0000000000
[499] 0.0009765625 0.0009765625 0.0009765625 0.0009765625 0.0009765625 0.0009765625
[505] 0.0009765625 0.0009765625 0.0000000000 0.0009765625 0.0000000000 0.0009765625
[511] 0.0009765625 0.0009765625 0.0000000000 0.0009765625 0.0009765625 0.0009765625
[517] 0.0009765625 0.0009765625 0.0000000000 0.0009765625 0.0009765625 0.0009765625
[523] 0.0000000000 0.0009765625 0.0009765625 0.0009765625 0.0009765625 0.0009765625
[529] 0.0009765625 0.0009765625 0.0009765625 0.0009765625 0.0009765625 0.0009765625
[535] 0.0009765625 0.0009765625 0.0009765625 0.0439453125 0.0009765625 0.0000000000
[541] 0.0009765625 0.0009765625 0.0009765625 0.0009765625 0.0009765625 0.0009765625
[547] 0.0009765625 0.0009765625 0.0009765625 0.0009765625 0.0000000000 0.0009765625
[553] 0.0009765625 0.0009765625 0.0009765625 0.0009765625 0.0009765625 0.0009765625
[559] 0.0009765625 0.0009765625 0.0009765625 0.0009765625 0.0009765625 0.0000000000
[565] 0.0009765625 0.0009765625 0.0009765625 0.0009765625 0.0009765625 0.0009765625
[571] 0.0009765625 0.0009765625 0.0439453125 0.0000000000 0.0009765625 0.0009765625
[577] 0.0009765625 0.0009765625 0.0009765625 0.0000000000 0.0009765625 0.0009765625
[583] 0.0009765625 0.0009765625 0.0009765625 0.0009765625 0.0000000000 0.0000000000
[589] 0.0009765625 0.0009765625 0.0009765625 0.0009765625 0.0000000000 0.0009765625
[595] 0.0009765625 0.0009765625 0.0009765625 0.0009765625 0.0009765625 0.0000000000
```

Fig. 6 Binomial distribution of smoking patient getting cervical cancer

partner. The probability distribution is used to determine the risk factor caused by the wrong practice of sex.

STDs—Condylomatosis, Cervical Condylomatosis, vaginal condylomatosis, perineal condylomatosis, syphilis, pelvic inflammatory diseases, genital herpes, molluscum contagiosum, AIDS, HIV, Hepatitis B, and HPV are considered in our experiments to determine the ratio of the STDs. These diseases boosts the risk of the cervical cancer very rapidly. Even though there is medicine available for these diseases separately. Once they affected the person, it increases the risk of the cervical cancer by affecting the organs. In our data a woman has maximum of four STDs and the average of 0.155. Here we have the various diseases under the name of the STD. We have to analyze about each types which affected the person at high rate. The values

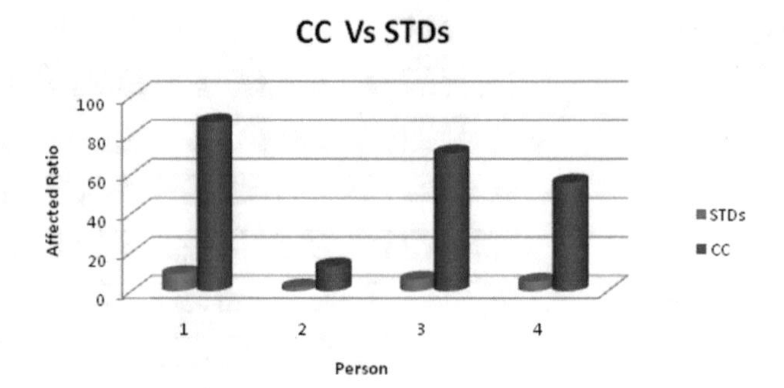

Fig. 7 Cervical cancer versus STDs

of the diseases are determined which is shown in Table 6. The condylomatosis probability is calculated as 0.6240 and the vulvo-perennial condylomatosis is calculated at 0.624 whereas the cervical and vaginal condylomatosis is not affected the people very easily. The value of syphilis and pelvic inflammatory disease is calculated as 0.389 and 0.273 respectively. HIV and HPV is also affected at the rate of 0.3892. The diseases such as molluscum contagiosum, genital herpes, AIDS and Hepatitis B is not affected the women. The probability of getting affected is high at the rate of 0.93838 when the number of STDs is high.

The variables in STDs are independent variables which is not dependent on each other. Totally seven diseases we have to consider in this case, which is affected the human very easily. Even we can analyze these diseases which are very susceptible for the cervical cancer by using the regression analysis. Here we have taken the significant variable to predict the values of the STDs which increased the risk of the cervical cancer by their rapid attack characteristics. We have taken the total of seven diseases in the STDs column in Table 6. The comparison of the cervical cancer with the number of STDs is shown in Fig. 7.

The significant values are having high coefficients values which is affected the values of the risk of cervical cancer response variable Y. The P values of age, number of sexual partners, First sexual intercourse is calculated as 0.153, 0.873, and 0.856, respectively. We can easily estimate the response variable values which are the target of us to find it by applying the coefficient values to the Eq. (1), we can get the affected probability values. The residual standard error for the prediction is estimated as 1.402e−10, F value is calculated as 2.691e+23 and the maximum residual value is estimated as 9.867e−10. The p-value is also less than 2.2e−16. The values of STD and hormonal contraceptives are 0.965 and 1.567. By the calculation of all these residual values, coefficient values, and predicted values; we can easily predict the values of the response variable Y which is the risk factors of the cervical cancer.

4 Conclusion and Future Direction

The cervical cancer possibility is predicted very easily by using the multiple regression models. By this model we can easily find out the significant parameters such as number of sexual partners and hormonal contraceptives required to predict the response variable values such as the prediction of cervical cancer affected women from their life style habits and the diseases affected by them. In our paper we have considered the women who are having practices of having more number of sexual partners and the hormonal contraceptives of long term. Both the cases are directly proportional to the risk factors of cervical cancer.

In future we can use the logistic regression for the types of STDs affected with the risk factors of the cervical cancer in the women. We can also use machine learning algorithm to train the system by giving the sample input data. Moreover we can use the probability distribution function to decide the woman gets affected by the cervical cancer with these same parameters. Likewise we can find the risk factors of cervical cancer with the help of estimating smoke, IUD, other panic diseases such as Hepatitis and the medical diagnosis report.

Here we have considered only the values about the diseases, habits and life styles for prediction of the cervical cancer. We can diagnose the cancer by using the testing report such as HPV test, CIN, Hinselmann, Schiller, Cytology, and Biopsy. In our paper we have shown the method of prediction using the regression analysis. Like this we can consider the other test report values in the regression analysis to diagnose the cervical cancer easily.

References

1. Smith L (2016) Human Papillomavirus (HPV): causes, symptoms and treatments. Elseiver
2. Daley EM, Vamos CA, Thompson EL, Zimet GD, Rosberger Z, Merrell L, Kline NS (2017) The feminization of HPV: how science, politics, economics and gender norms shaped U.S. HPV vaccine implementation. Papillomavirus Res (Elseiver)
3. Haeggblom L, Ramqvist T, Tommasino M, Dalianis T, Näsman A (2017), Time to change perspectives on HPV in oropharyngeal cancer. A systematic review of HPV prevalence per oropharyngeal sub-site the last 3 years. Elseiver
4. Panici PB, Scambia G, Perrone L, Battaglia F, Cattani P, Rabitti C, Dettori G, Capelli A, Sedlis A, Mancuso S (1992) Oral condyloma lesions in patients with extensive genital human papilloma virus infection. Am J Obstet Gynecol 167(2): 451–458
5. Li M, Nyabigambo A, Navvuga P, Nuwamanya E, Nuwasiima A, Kaganda P, Asiimwe FT, Vodicka E, Mugisha NM, Mukose A, Kwesiga DK, Lubinga SJ, Garrison Jr LP, Babigumira JB (2017) Acceptability of cervical cancer screening using visual inspection among women attending a childhood immunization clinic in Uganda. Papillomavirus Res (Elseiver)
6. Schirmer DA, Gordon AN, Roberts CP (2016) Neovaginal condylomatosis and carcinoma following mcindoevaginoplasty. Fertil Steril (Elseiver) 106(3 Suppl): 129
7. Hagström H (2017) Alcohol, smoking and the liver disease patient. Best Pract Res Clin Gastroenterol
8. Argyri E, Tsimplaki E, Marketos C, Politis G, Panotopoulou E (2016) Investigating the role of human papillomavirus in lung cancer. Papillomavirus Res

9. Sivin I (2007) Utility and drawbacks of continuous use of a copper T IUD for 20 years. Contraception 75(6 Suppl): S705
10. Zhang C, Yang B (1996) The corrosion behaviour of copper in simulated uterine fluid. Corros Sci 38: 635–641
11. Wu JP, Pickle S (2014) Extended use of the intrauterine device: a literature review and recommendations for clinical practice. Contraception
12. Grimes DA (1987) Intrauterine devices and pelvic inflammatory disease: recent developments. Contraception 36(1): 97–109
13. Hook EW (2017) Syphilis: an ancient disease in a modern era. The Lancet 389(10078): 1550–1557
14. Ross JDC (2014) Pelvic inflammatory disease. Medicine 42(6): 333–337
15. Brunham RC, Gottlieb SL, Paavonen J (2015) Pelvic inflammatory disease. N Engl J Med 372:2039–2048
16. Mitchell C, Prabhu M (2013) Pelvic inflammatory disease: current concepts in pathogenesis, diagnosis and treatment. Published online 2013 Oct 31
17. Clavero O, McCloskey J, Molinab VM, Quirósa B, Bravoa I, de Sanjoséa S, Boscha X, Pimenoff VN (2016) Squamous intraepithelial lesions of the anal squamocolumnar junction: histopathological classification and HPV genotyping. Papillomavirus Res (Elseiver)
18. Mary Poynten I, Tabrizi SN, Jin F, Templeton DJ, Machalek DA, Cornall A, Phillips S, Fairley CK, Garland SM, Law C, Carr A, Hillman RJ, Grulich AE (2017) Vaccine-preventable anal human papillomavirus in Australian gay and bisexual men. Papillomavirus Res, SPANC Study Team
19. Hwang C-F, Huang C-C, Chien C-Y, Huang S-C, Yanga C-H, Sua C-Y (2012) Human papillomavirus infection in oral papillary and verrucous lesions is a prognostic indicator of malignant transformation. Cancer Epidemiol 36(2): e122–e127
20. Castle PE, Dockter J, Giachetti C, Garcia FA, McCormick MK, Mitchell AL, Holladay EB, Kolk DP (2007) A cross-sectional study of a prototype carcinogenic human papillomavirus E6/E7 messenger RNA assay for detection of cervical precancer and cancer. Clin Cancer Res 13(9): 2599–2605
21. Wang HY, Lee D, Park S, Kim G, Kim S, Han L, Yubo R, Li Y, Park KH, Lee H (2015) Diagnostic Performance of HPV E6/E7 mRNA and HPV DNA assays for the detection and screening of oncogenic human papillomavirus infection among woman with cervical lesions in China. Asian Pac J Cancer Prev 16(17): 7633–7640
22. Eide ML, Debaque H (2012), HPV detection methods and genotyping techniques in screening for cervical cancer. Ann Pathol 32(6): e15–e23, 401–409. https://doi.org/10.1016/j.annpat.2012.09.231. Epub Nov 22, 2012
23. Naucler P, Ryd W, Törnberg S, Strand A, Wadell G, Elfgren K, Rådberg T, Strander B, Forslund O, Hansson BG, Hagmar B, Johansson B, Rylander E, Dillner J (2009) Efficacy of HPV DNA testing with cytology triage and/or repeat HPV DNA testing in primary cervical cancer screening. J Natl Cancer Inst.101(2): 88–99. https://doi.org/10.1093/jnci/djn444. Epub Jan 13, 2009
24. Smelov V, Hanisch R, McKay-Chopin S, Sokolova O, Eklund C, Komyakov B, Gheit T, Tommasino M (2017) Prevalence of cutaneous beta and gamma human papilloma viruses in the anal canal of men who have sex with women. Papillomavirus Res 3: 66–72
25. Nayak S (2015) Human papilloma virus and its relation to cervical cancer prevention strategies. Pediatric Infect Dis 7(1): 27–32
26. Lisan Q, Laccourreye O, Bonfils P (2016) Sinonasal inverted papilloma: from diagnosis to treatment. Eur Ann Otorhinolaryngol Head Neck Dis 133(5): 337–341
27. Tinhofer I, Jöhrens K, Keilholz U, Kaufmann A, Lehmann A, Weichert W, Stenzinger A, Stromberger C, Klinghammer K, Dommerich S, Stölzel K, Hofmann VM, Hildebrandt B, Moser L, Ervens J, Böttcher A, Albers A, Raguse JD (2015) Contribution of human papilloma virus to the incidence of squamous cell carcinoma of the head and neck in a European population with high smoking prevalence. Eur J Cancer 51(4): 514–521

28. Xi LF, Koutsky LA, Castle PE, Edelstein ZR, Meyers C, Ho J, Schiffman M (2009) Relationship between cigarette smoking and human papilloma virus types 16 and 18 DNA load. Cancer Epidemiol Biomark Prev 18(12)
29. Mzarico E, Gómez-Roig MD, Guirado L, Lorente N, Gonzalez-Bosquet E (2015), Relationship between smoking, HPV infection, and risk of Cervical cancer, Eur J Gynaecol Oncol. 2015; 36(6):677-80
30. Devore JL (2011) Probability and statistics for engineering and the sciences, 8th edn. Brooks/Cole Cengage Learning

Novel Outlier Detection by Integration of Clustering and Classification

Sarita Tripathy and Laxman Sahoo

Abstract A unique method of outlier detection consisting of integration of clustering and classification is proposed here. Basically the algorithm is divided into two parts the first phase consists of application of the classical DBSCAN algorithm to the data set which is followed by the second phase which consists of application of decision tree classification algorithm. The analysis on the algorithm states that the accuracy of unwanted data detection is high in the proposed method.

Keywords DBSCAN · Decision · Tree · Clustering · Outlier detection
Classification

1 Introduction

Outlier analysis is an essential step in data mining. Investigating outliers [1] can result in the discovery of very useful data. In wide range of applications where unwanted data detection is used such as, detection of unauthorized access in credit cards, in the field of medicine, in the area of image analysis, detection of intrusion, etc. In both academic field and in industry a number of approaches of noise detection have been suggested and designed. The methods which are most common are [2]: statistical approach, method based on clustering, method based on distance measure and method based on density of a dataset. The assumption made in statistical method is that there is consistency of data with a distribution model of probability and the data which does not follow this model is considered as an outlier. The distribution

---

S. Tripathy (✉) · L. Sahoo
School of Computer Engineering, KIIT University, Bhubaneswar, Odisha, India
e-mail: sarita.tripathyfcs@kiit.ac.in

L. Sahoo
e-mail: lsahoofcs@kiit.ac.in

D. K. Mishra et al. (eds.), *Data Science and Big Data Analytics*,
Lecture Notes on Data Engineering and Communications Technologies 16,
https://doi.org/10.1007/978-981-10-7641-1_14

model is highly required to determine the category to which a given data belongs. For a given data set with high dimension it is not possible to have prior knowledge of the distribution of data which is in practice. In the case of method based on clustering a particular clustering technique [3, 4] is used to recognize the unwanted data. In this approach the connection between the clusters and data is found out, the data that belongs to any cluster with considerable number of data is considered as normal data otherwise, The small clusters or isolated points are taken the outliers. DBSCAN is one of the classical approach for detection of outlier which is based on density. The technique used for clustering is responsible for the effectiveness of the clustering based method. In process where distance measure is taken [5] the data object is taken to be an outlier if the distance of a data object from the neighboring z object highly deviates from the distance of the most of the data objects from the neighboring data objects. For dealing with the mentioned problem above, a method based on density was proposed. It consists of Local Outlier Factor (LOF) and variants of LOF. The elements relative density with respect to the neighbors was taken into consideration and based on it, the LOF determines the extent to which an element is considered to be an outlier. These methods are advantageous in identifying the global and local outliers. The assumptions on the probability distribution is not done. They are highly affected by the parameters and its determination is done based on the knowledge of the domain by the user. These are highly computationally expensive methods as they have to do calculations of the distance of each of the element from all the other elements which are closest to the object considered. A new method was proposed, i.e., Novel Local Outlier Detection (NLOD) algorithm which looks at data set locally. Hence it is clear that it supports high difference in densities. The outliers in each group is determined by the help of chebyshev inequality peak reachability. The organization is: Second section gives overview of works done till date of various outlier detection methods. Third section describes about the DBSCAN algorithm. In fourth section the classification algorithm is described, section five consists of the algorithm which is proposed in this paper, sixth section the experimental analysis is presented finally seventh section draws the conclusions and future work.

2 Related Work

In the year 1994 [6] Ng and Han developed an efficient medoid based clustering algorithm CLARANS, it is an optimization of k-means which involves calculation of mean of the points in the clusters, a medoid which is the point or object itself is chosen for every cluster. If the neighbor is a more appropriate choice medoid then it is chosen, otherwise a local optima needs to be discovered.

In the year 1995 Ester et al. [7] proposed Density Based Spatial Clustering of applications, this method clusters the objects those have high density, and the other objects those are isolated are taken as noisy data.

In the year 1996 Zhang et al. [8] developed BIRCH algorithm which works efficiently for large databases, it is balanced iterative reducing and clustering using hierarchies.

In the year 1998 Guha, Rastogi and Shim developed CURE [9] which uses representatives with clustering this acts as an efficient algorithm for very large database, this algorithm is highly robust to outliers and non-spherical clusters are also identified by it.

Followed by this in the year 1999 Knorr and Ng [5] proposed a method which is a distance-based outlier detection method. Identification of exceptional outliers is one of the objective of this method.

In the year 2000 [10] a method was developed by a Local Outlier Factor (LOF) for each object in the data set, this factor represents the outlierness degree. It is also known as local because it is only limited to the nearest neighbor. It is considered to be more efficient than distance based method as computation of LOF value of an object.

In the year 2005 [11–13] Anguiulli and Pizzuti proposed a method which can find out the unwanted points by considering the whole neighboring points.

In the year 2009 [14] Zhang et al. proposed Local Density based outlier detection method.

3 DBSCAN

DBSCAN is a clustering method [15] which considers density as the important criteria for classification.

The two criterion's that it considers are

(1) Density reachability
(2) Density connectivity

Density reachability
A point "m" is density reachable from another point "n" if "m" is within a distance known as epsilon from the other point and there are enough number of points surrounding the second point within that distance.

Density Connectivity
Points m and n are said to be density connected if there is a third point p which is density reachable by both m and n.

3.1 Advantages

(1) The prior knowledge of number of clusters is not required.
(2) Robust to noisy data and has the ability to find outliers in a data set in the process of clustering.

(3) The various shaped clusters and various sized clusters can be identified.

3.2 Disadvantages

(1) Failure is encountered when there are clusters of different densities in the data set.
(2) It also fails when applied to neck type data set.

4 Decision Tree

It is a well-known classification technique used in various areas. The basic aim in this method is to develop a structure which is able to determine the target variable which is the result of various input variables. The edges are created for each child for all the valid input variable. The leaf denotes the target data which is created by considering values of input variable by a path which goes from root to leaf. It is one of the simplest representation for classifying examples. It is also known as classification tree in which each node apart from the leaf node is labeled with an input feature. The probability distribution over classes are created by considering each leaf node as a class.

The appropriate definition of decision tree [16] is the field of data mining is given by combining mathematical and computational techniques

$$(\mathrm{D}, \mathrm{Q}) = (\mathrm{d}_1, \mathrm{d}_2, \mathrm{d}_3, \ldots, \mathrm{d}_\mathrm{n}, \mathrm{Q})$$

Y is known as the dependent variable that is nothing but target which is to be understood, classified and generalized, for this task vector D is used which is a set of inputs d1, d2, d3, etc.

4.1 Advantages

1. Understanding and interpreting is simple.
2. This method can handle both categorical and numerical data.
3. The data interpretation required is very little.
4. Decision tree uses white box model.
5. The validation of the model can be done by using statistical tests.
6. It can handle large data set more easily.
7. Unlike other approaches this method mirrors easily the decision-making by humans.

4.2 Disadvantages

1. This method do not tend to be more accurate.
2. Any small deviation in training data can result in large change which will affect the final result.

5 Proposed Algorithm DBSDCT

Input: *The original data sets DS, eps is the radius,mps is the minimum number of objects in neighborhood..*

Output: n_o *consists of set of outliers obtained from the dataset DS.*

Step1: *Randomly select a point from the dataset and check for the count of neighbouring points.If the count is more than mps, then take it as a core point, or else, mark it as an outlier.*

Step2: *if the point is a core object, create a cluster with radius eps and the core point. Then add the objects in the cluster into a list container, and check the objects recursively. If the object in the container is a core object, classify it to the same class as the point and add its neighborhood points into the container. Else, mark and delete it from the list container.*

Step3: *The steps 2 and 3 are repeated until all the objects in the dataset DS are processed. Outlier points which are marked are included in* n_0.

Step4: *The m clusters obtained from step 3 are marked as class 1, 2, 3,... m.*

Step5: *Take an appropriate criteria for creation of the decision tree*

Step6: *train and build the decision tree on* DS_t.

Step7: *The noise points set* n_1 *is taken to be the estset.*

Step8: *The decision trees is applied on the test set.*

Step9: *Set of outliers* n_0 *is obtained after the application of the decision tree.*

DBSDCT is analyzed to be an efficient algorithm for large databases.

Table 1 DBSCAN with DBSCDCT for two different Eps and MinPts

Algorithm	Eps	MinPts	Number of noise points	Accuracy
DBSCAN	0.4	4	25	0.8024
DBSDCT	0.4	4	10	0.8400
DBSCAN	0.6	6	11	0.8700
DBSDCT	0.6	6	4	0.9322

6 Result Analysis

An experimental analysis is performed by us for measuring the efficiency of DBSDCT algorithm. The experiment is performed first by taking only the DBSCAN clustering and then by combining it with decision tree. The algorithm is being implemented on the Iris data set which consists of 150 observations which is combination of three species, which is categorized based on four features.

For decision tree classification the data set is divided into two parts they are training and testing parts, in which 75% is taken to be the training set and the rest 25% is taken to be the testing set. The initial step is to use DBSCAN on the data set, after which we will get a set of noise points, which are taken to be the test set in the second stage the algorithm, the rest of the points are not noise points and are grouped into different clusters, each cluster is taken to be a class, with which the data set will be trained.

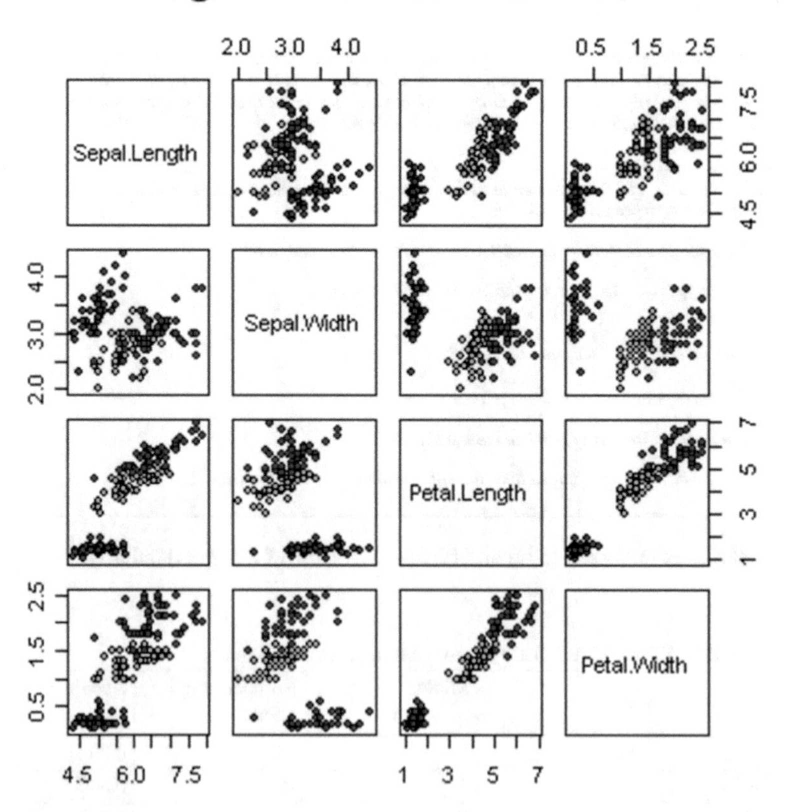

Fig. 1 Features of iris data set

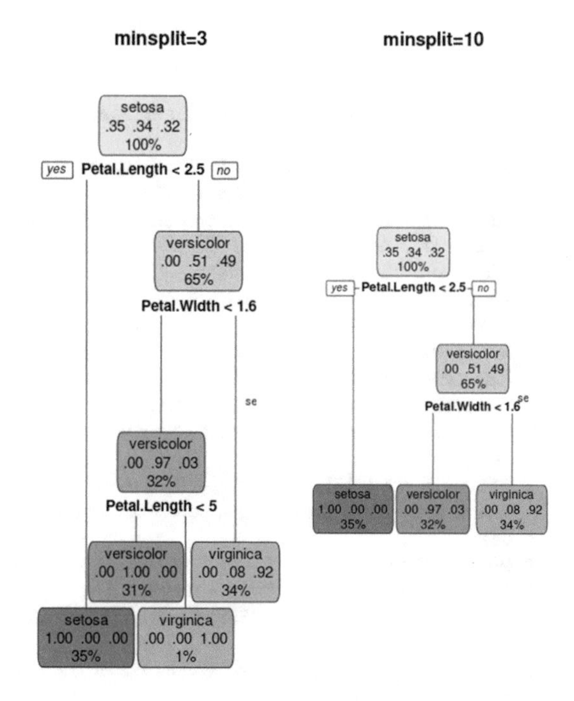

Fig. 2 Structure after application of decision tree on the data set

The efficiency of DBSDCT algorithm can be judged by the result as shown (Table 1, Figs. 1 and 2).

7 Conclusion and Future Work

The new method based on DBSCAN and decision tree classification is proposed in this paper which is more efficient as it minimizes the error of DBSCAN algorithm, as evident from the experimental output the algorithm improves the accuracy of the clustering results. The parameter those are required to be selected are (i) minimum points (ii) Epsilon (iii) the type of decision criteria which depends on the data set used. Manually we can adjust the parameters according to the application area. The accuracy can further be increased by introducing techniques such as bagging and random forest. Future work will consist of finding out a method for automatic selection of the optimal parameters and introduction of the above methods for further increasing the accuracy.

References

1. Tan PN, Steinbach M, Kumar V (2005) Introduction to data mining, 1st edn. Pearson Addison Wesley, Boston, MA
2. Li Y, Nitinawarat S, Veeravalli V (2013) Universal outlier detection. In: Information theory and applications workshop (ITA), pp 528–532
3. Ahmed M, Mahmood AN (2013) A novel approach for outlier detection and clustering improvement. In: Proceeding of the 2013 IEEE 8th conference on industrial electronics and application, ICIEA, pp 577–582
4. Waang HL, Li WB, Sun BY (2013) Support vector clustering for outlier detection. In: Information technology applications in industry, computer engineering and materials science. Advance materials research, pp 493–496
5. Knorr EM, Ng KT (1998) Algorithm for mining distance-based outliers in large datasets. In: Proceedings of the 24th international conference very large data bases, VLDB, pp 392–403
6. Ng RT, Han J (1994) Efficient and effective clustering methods for spatial data mining, pp 144–155
7. Ester M, Kriegel H-P, Xu X (1995) A database interface for clustering in large spatial database. In: Proceedings of 1st international conference on knowledge discovery and data mining (KDD-95)
8. Zhang T, Ramakrishnan R, Living M (1996) Birch: an efficient data clustering method for very large databases. SIGMOD Rec 25(2):103–114
9. Guha S, Rastogi R, shim K (1998) CURE: an efficient clustering algorithm for top databases. SIGMOD Rec 27(2):73–84
10. Breunig MM, Kriegel H-P, Ng RT, Sander J (2000) LOF: identifying density-based local outlier. SIGMOD Rec 29(20):93–104
11. Angiulli F, Basta S, Pizzuti C (2006) Distance-based detection and prediction of outliers. IEEE Trans Knowl Data Eng 18:145–160
12. Angiulli F, Pizzuti C (2002) Outlier mining in large high dimensional spaces. In: PKDD 02: proceedings of the 6th European conference on principles of data mining and knowledge discovery, pp 15–26
13. Angiulli F, Pizzuti C (2005) Outlier mining in large high dimensional data sets. IEEE Trans Knowl Data Eng 17:203–215
14. Zhang K, Hutter M, Jin H (2009) A new local distance-based outlier detection approach for scattered real-world data. In: PAKDD 09: proceeding of 13th Pacific-Asia conference on advances in knowledge discovery and data mining, pp 813–822
15. Ester M, Kriegel H-P, Sander J, Xu X, Simoudis E, Han J, Fayyad UM (eds) (1996) A density-based algorithm for discovering clusters in large spatial databases with noise. In: Proceedings of the second international conference on knowledge discovery and data mining (KDD-96). AAAI Press, pp 226–231
16. Rokach L, Maimon O (2008) Data mining with decision trees: theory and applications. World Scientific Pub Co Inc. ISBN 978-9812771711

A Review of Wireless Charging Nodes in Wireless Sensor Networks

Shiva Prakash and **Vikas Saroj**

Abstract Sensing of data and computation properties of sensors has created a new group of these type devices and to the using more of these types of sensor devices which are established in wireless sensor networks (WSNs). Wireless sensor networks customized sensor nodes which are situated in open areas as well as public places, with a large number that makes some problems for the scholars and designer, for designing the wireless network. There are problems such as security, routing of data and processing of bulk amount of data, etc., and life time of sensor nodes due to limited battery power, charging and replacement of batteries are sometimes not possible. This paper describes the concepts of wireless charging of sensors nodes with energy-efficient manner in WSN. We explore the concepts of wireless charging node in WSN with detail literature review and comparison of well-known works. It helps to new scholar to get decision in existing techniques and more explore about energy transfer to the sensor nodes in wireless sensor network.

Keywords Wireless charging · Charging vehicle · Magnetic resonance coupling Power management and data gathering

1 Introduction

Wireless sensor networks (WSNs) are one of the emerging topics of research nowadays. WSNs are considered as the group of nodes or large number of sensors nodes

S. Prakash (✉) · V. Saroj (✉)
Department of Computer Science & Engineering, M.M.M. University of Technology, Gorakhpur, India
e-mail: shiva_pkec@yahoo.com; shiva.plko@gmail.com

V. Saroj
e-mail: sarojvikascs093@gmail.com

D. K. Mishra et al. (eds.), *Data Science and Big Data Analytics*,
Lecture Notes on Data Engineering and Communications Technologies 16,
https://doi.org/10.1007/978-981-10-7641-1_15

which are deployed in anywhere in environment, i.e., public place and open area, that also creates many problems for network designer and scholars. It takes many numbers of nodes so complexity always appears. These types of network could be a part of such systems, i.e., environmental monitoring, home automation, detection of chemical and biological attacks, surveillances, and many others control system.

The sensor and actor nodes first sense, then collect the data from the environment where are they stabilised and after that perform appropriate action on data base. Short life expectancy of the sensors battery and the likelihood of having harmed nodes during arrangement in WSN, large numbers of sensors are normal in any WSNs applications. Natural disaster event observing used to keep the disaster events. Natural disaster monitoring used to prevent the natural disasters, like floods. Forest fire detection act when fire has started in forest, a network of Sensor Nodes can be installed. With the help of such type of wireless sensor networks we can take early action to protect forest [1]. Another some general applications, i.e., sensing humidity, the current characteristics of object, lightning condition and the absence or presence of certain kinds of objects. With the assistance of such sort of remote sensor systems we can make early to secure forest land [2].

Several categories of the energy related to the layers and other components of energy-efficient sensor nodes in WSN. Above we can see that Dynamic Voltage Scaling (DVS) in the physical layer category dynamic adjustment of the clock speed with supply voltage and another thing instantaneous workload. MAC layer of the network in this chapter reduced the idle listening period and also avoid the collision therefore collision directly call for retransmission of data and ideal state consumed some power for lifetime maximization of the network sensor node in wireless sensor network. After that, another management techniques category in this way minimized the size of routing information table by using real time routing protocols and important thing of this network layer clustering for energy efficient and data gathering techniques. The entire component, i.e., Radio, Batteries, Sensors, and Microcontroller play important role and takes its own important in WSN (Fig. 1).

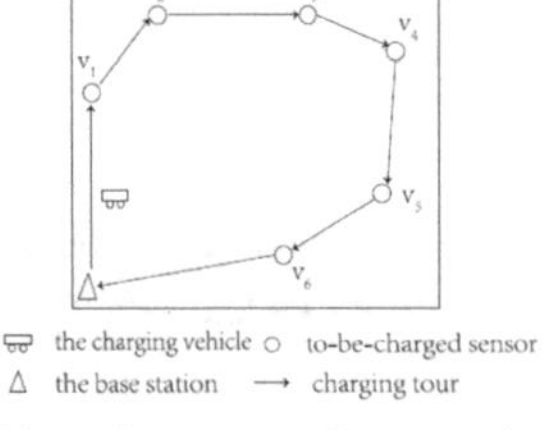

(a) employ just one charging vehicle to charge sensors

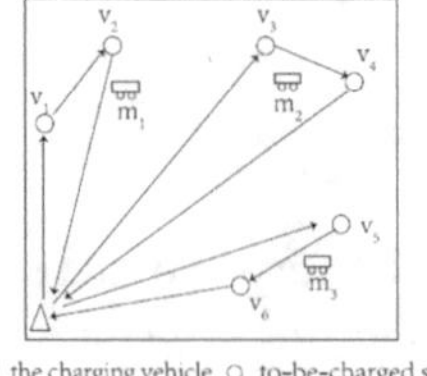

(b) dispatch multiple charging vehicles to charge sensors

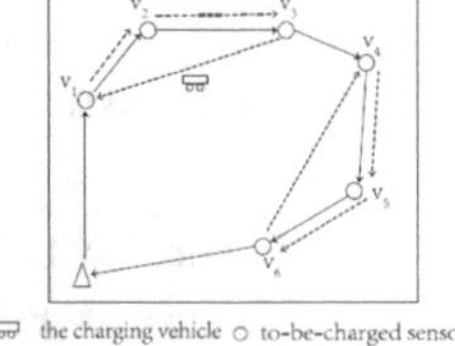

(c) deploy one charging vehicle carrying multiple low-cost removable chargers to replenish sensors

Fig. 1 Various methods to charging the sensor nodes in WSN with the help of mobile charging vehicle

Energy one of the major limitation in the long life of the sensor nodes in WSN. Constrained energy wirelessly supplies limits the lifetime of a sensor node in wireless network, which has retained a large challenge in sensor network design specially when the deployment done of network for long term monitoring. This paper defines the types of wireless sensor networks and the likely solutions for how can we reduced the major problem of energy in network. The main technique uses in this paper wirelessly transfer the energy or power to the sensor nodes which deployed in the network also described of many other solutions. Magnetic resonant based energy transfer is widely regarded as a technology [3]. Energy should be transfers efficiently from a source to a receiver destination coil without any physical contact via electromagnetic field of nonradioactive. So wirelessly power transfer technology particularly attractive due to its wirelessly environment. We discussed however various techniques use for charging the sensor nodes via the travelling path or circuit with the help of wireless charging vehicle.

2 Research Challenges in WSNs

In WSNs there are various research challenges which are a great concern in today's world and cannot be ignored easily. WSNs area units presently obtaining are thought noteworthy because of their unknown and unlimited potential. Though its initial section of the event within the period of such WSNs systems and lots of investigative challenges exists. These challenges have time to time as the researchers feel motivated.

The key challenges (problems) in WSNs are as following:

- Power management
- Data Gathering
- Security and Privacy
- Real-world Protocols
- Brief Literature Survey.

3 Brief Literature Survey

Wireless energy transfer is a technology [4] to prolong the lifetime of wireless sensor networks, by employing charging vehicles to transfer energy to lifetime-critical sensors. Studies on sensor charging assumed that one or multiple charging vehicles being deployed. Such an assumption may take its limitation for a real sensor network. On one hand, it usually is insufficient to employ just one vehicle to charge many sensors in a large-scale sensor network due to the limited charging capacity of the vehicle or energy expirations of some sensors prior to the arrival of the charging vehicle. In this paper, we propose a novel charging model that a charging vehicle can carry multiple low-cost removable chargers and each charger is powered by a portable high-volume battery. When there are energy-critical sensors to be charged, the vehicle can carry the chargers to charge multiple sensors simultaneously, by placing one portable charger near one sensor.

In this paper, author designated wireless energy transfer technology in multi-node and study whether this technology is a scalable to address power problem in a wireless sensor network. These are considered a wireless charging vehicle (WCV) travelling inside a network and charged the sensors node without any physical connection. Charging range of the wireless charging vehicle, they propose a cellular construction that panels into together hexagonal cells. They follow a formal framework by organized adjusting travelling path, charging time, and flow routing. By employing discretization and one new Reformulation Linearization Technique (RLT), they produce a provable approximately to optimal solution for some anticipated level of correctness. Through mathematical results, they prove that that solution can really address the charging efficiency scalable problem in wireless sensor networks [5]. Author [6] proposes a routing method which shows energy efficiency for network. Those techniques choose cluster head with maximum remaining energy in each communication period of transmission and the shortest route from the cluster heads to base station. Because, the major circumstance that sensors run out of energy rapidly has been a problem and some energy-efficient manner of routing protocols. They have been proposed to resolve this problem and reserve the durability of the network. These are the few reasons why routing methods in wireless sensor network emphasis specially on the accomplishment of energy conservation. The most recent journals have shown huge number of protocols designed to optimized energy consumption in wireless sensor nodes in wireless sensor networks.

Author [7] has proven to clustering method for an energy efficient that raises the network lifespan by dropping the power utilization and delivers the necessary scalability. To achieve high scalability and increased energy efficiency and to enhance the network life time the researchers have highly adopted the scheme of forming clusters, i.e., grouping the sensor nodes in large-scale wireless sensor network environments. Basically, a clustering scheme selects a set of sensor nodes that can act as a backbone of other network nodes that connect the nodes to the base station. The type of nodes discussed here are called cluster heads and the remaining nodes of the network are referred to as member nodes. In these approaches [8] stated not only consideration of energy but also balances load in wireless sensor networks. Initially based on weight metric the cluster heads selected in a WSN after that cluster establishment placed. Local clustered mechanism reduced the cost of communication and computation by selected the cluster. Similarly, new technique is exposed for data transmission. A selection of cluster head technique called the Cluster Chain Weight Metrics approach (CCWM) should be considered this attractive service limitation for improving performance of the whole network. In a nodes clustered technique, created method one of the main apprehensions are selecting of suitable clusters head in sensor network and also consideration of balanced clusters. Authors propose a novel energy-efficient protocol for single-hop based on clustering method [9], for heterogeneous wireless sensor networks. In the given protocol, the selection of Cluster Heads used channel state information. Because efficient energy is one of the most significant proposal for wireless sensor networks (WSNs). For improved this, clustering is typically used to extend the lifetime of sensors node in WSNs. It is exposed through simulated in MATLAB this is the anticipated protocol has 1.62–1.89 times given better stability over well-known protocol including DEEC, LEACH and SEP.

Which Protocols deliberated in this paper suggest a considerable enhancement over old-style clustering behaviour [10]. Each protocol takes its own advantages and disadvantages which makes them suitable for specific applications. By the survey detail the current clustering algorithms is discussed and presented energy constrained for WSNs. The energy of the sensor node essential is to be used in an additional security and efficient technique. A brief comparison of protocols has been shorted in table. CEBCRA protocol gives the perfection over old protocol in network lifespan. The author also suggested clustering protocols can be made further effective in time duration by considering metrics related to time constraints and Quality of service, security requirements to be combined in them to protect data from another end. In this work [11], author gives a practically work called wireless energy charged WSN (WINCH), for sensors battery maintenance, it includes recharging sensors battery by using mobile robots or vehicle. This basis integrates process of routing in which the heads of clusters are selected optimally, as in the less energy adaptive clustering centralized protocol and the robots arrive the sites normally based on requirement and place them in optimal positions with respected to the cluster heads Wireless sensor networks too effective and dynamic model for some applications, with health monitoring and tracing systems. However, the transmit of energy to sensor nodes plays a main role in the applications to success them as well as the designing and deployment of networks. Batteries drive now mostly wireless sensor networks, increasing operational and maintenance costs so must be replaced regularly. Author [12], disused over the issues of optimum load determination in a wireless sensor network in energy transfer link among single source and many receivers has been mentioned and solved it. In certain, analytical methods prove that in the situation where the receivers are separated, the optimum loads are decently resistive for not only power but also efficiency maximization, the values corresponding to energy maximization are constantly better than one's necessary to the efficiency maximization. These result belonging an optimal range of the loads which can be derived: corresponding the solution maximizing efficiency of the lower bound of this range, whereas solution maximizing power transfer on the upper bound corresponds to it.

A main contribution of the author [13] is the outline of a power harvesting receiver in a many access network with impact model. Author defines problem as the average rate of information to maximization issues, successfully interpreted at the receiver end. They propose a dynamic structure, which creates an optimal solution. Mathematical results approve the efficiency of the planned scheme. For future work, they will deliberate a combined control of senders and receiver's nodes or ends for the aim of enhancement of the network performance through energy consumption limitations. Another way is to consider power splitting in energy harvesting, whereas the receiver could make not only information decoding but also energy transmitting by the one wireless transmission. A novel routing protocol [14] that is appropriate for low bit rate and low energy in networks. The protocols idea is very simple by using the lowest energy path always is not essentially best for the longstanding strength of the network. Therefore, using a simple method to send traffic via differ routes that helps to using the sensor nodes resources more fairly. Using probabilistic forwarding to sending traffic to the different routes which means provides a simple way to use several paths without taking complication. Both factors are most important for evaluating networks and can be mentioned in isolation. An example of this, the first condition in the network that does not sending (forwarding) packets at every satisfied. Yet, survivability is still an introductory concept and wants to be worked on more and understood well before using it for judging network protocols.

MANETs are rapidly gaining in wireless communications. There are also increased threats and safety topics in a wireless environment by rapidly increasing of applications, security and privacy is a dynamic structure block. Unsecure wireless networks are basically unusable. Generally, there are three types of safety issues on MANET with three different classifications were discussed. In this paper, the authors have introduced a short review of the most common threats and attacks. The discussion on attacks in the scope of this type of classifications aligning within its counter-measures. Mitigation methods of threats were also discussed. MANETs characteristics [15] makes it marked for attacks type this is not possible on other network of wired and (without any physical contacts) wireless networks. For example, makes sensor nodes vulnerable by having constrained battery lifespan for battery draining attacks. A confidence-based system in which the kind and variety of security methods applied rely on the level of confidence for making good direction in study would be the creation. While many research done in this way, especially in terms of routing protocols, it gives fresh area thru large potential.

In this paper, author [16] suggests a numerical model for an efficient energy routing algorithm. The proposed routing method based on the resolve the optimal number of hops to divides the route from the source to the destination or sink. Author [17] presented a LR technique to maximize lifespan by defining serious parameters which are controlled the adaptive hop-by-hop transferring. Their results significantly progressive in lifetime related with the three known algorithms. The Wireless Ad hoc Networks do not have gateway; every node can act as the gateway. Author tried to simplify the description, architecture and the behaviour of MANET, the main issues of creating the MANET. While various works have been done to resolve the issues or problem, they show in paper that it is much difficult to resolve these problems by which makes the Mobile Ad hoc Networks unfairly. Goldsmith [18] have described recent advances in ad hoc wireless network protocols and across different protocol layers the strong interaction that occurs. Whereas there is still many work to be completed on enhanced performance on link, protocols for application, and network the relations across these dissimilar protocol layers offer new cross layer designing that achieved these interdependencies. For the energy restrictions, cross layer designed mostly important, and then energy across the complete protocol stack must be diminished.

4 Comparative Study of Well-Known Wireless Charging Techniques for Sensor Nodes in WSN

This section presents techniques of wireless charging sensor nodes in WSN. For the comparative study of various techniques and their advantages, disadvantages of well-known authors work for wireless charging node in wireless sensor network as shown in Table 1.

Table 1 Comparison of various techniques

Author/Year of publications	Technique used	Advantages	Limitations
Islam et al. [10]	New trends in clustering for energy efficient in WSNs	Network life time has improved over conventional protocol	Improved energy efficiency and effective in prolonging the network lifetime
Liguang Xie [5]	Used multi-node wireless energy charging	Wireless energy transfer technology to charge the batteries of sensor nodes	Based on charging range of the wireless charging vehicle (WCV)
Mahajan et al. [8]	Energy balanced QoS based cluster head selection strategy	Reduces energy consumption and balances load by selecting the cluster head nodes	Technique is proposed to limit the node degree
Nuray and Daraghma [9]	CHs are selected on the according to nodes residual energy and the greatest channel	Enhanced period of stability of WSN to the other LEACH, DEEC, and SEP protocols	Overhead is increases instead of improving nodes energy and overall network lifetime
Shankar and Jaisank [6]	Selects cluster head with highest residual energy	Transmission energy takes shortest distance	Prediction of transmission energy within shortest distance to the base station may be right
Giuseppina Monti [12]	The energy maximization of multiple receivers by optimized development of wireless power transfer	Optimum range of the loads can be derived the lower and upper bound corresponds to the solution maximizing power transfer	Compensating spontaneous elements in series to the loads
Baroudi [11]	Wireless charging nodes in WSN via robotic vehicle	Solves the problem of energy transfer to node placed at difficult area via robot positioning	There is no way to ensure highest harvesting of power by wireless sensor nodes
Yunus Sarikaya and Ozgur Ercetin [13]	Self-sufficient receiver with wireless energy	Energy harvesting receiver in a multi-access channel with collision model	Battery recharged only from the energy harvested from incoming RF signals
Hasan et al. [16]	Transfer in a multi-access network. lifetime maximization	Significantly improve the lifetime with the well-known algorithms	cut-off for the optimal number of hops to partition the path from the source to the sink
Zou et al. [4]	Deploying one mobile vehicle with multiple removable chargers	Shortest charging trajectory while ensuring that the longest dead duration of sensors is minimized	If the residual lifetime of each sensor can be ignored, devise a novel heuristic algorithm

Comparative study as shown in Table 1 for some well-known wireless energy transfer techniques which are directly or indirectly related to the wirelessly charging of sensor nodes in WSN. One of the authors Md. Faruqul Islam introduced new trend in clustering and provides some improves over traditional protocols in form of enhancement of the network life time. Author Liguang Xie gives wireless energy transfer technology by using multi-nodes to provides some energy to the sensor nodes for maximized the lifetime of the deployed nodes in the network and the transfer of power based on charging range of the wireless charging vehicle (WCV).

In 2014 Shilpa Mahajan has presented the energy consumption and balances load by selecting the cluster head nodes but there are remarks point such as technique is proposed to limit the node degree in this way based cluster head selection strategy. These authors introduced its own techniques to produced outstanding performance of the sensor devices which may be after some time disconnected due to the lack of energy and these sensor devices could not perform tasks without energy. One of the author Achyut Shankar of the researcher who works on to the highest residual energy by selecting cluster head amongst the thousands of sensors node which connected in wireless network to take its performance. Not always all the factors of researches take advancement but also some disadvantages or limitation may be occurred therefore in this research topic basic concept prediction of transmission energy via a shortest distance to the base station it takes both advantage and disadvantage because prediction not always right some time may or not. Concluding all that authors have expressed techniques and clamed some improvement over others works but everyone has some limitations also.

5 Gaps in Study

The knowledge gap is that which needs to be filled by new research either because we know little or nothing. Research gap is a problem which has not been addressed so far in a field. In this literature review, there are many authors discussed on its own method to how reduced the complexity and get the better performance evaluation. Context is not the research gap but one that will surely indicate or lead you to the problem. Limitations of continuous energy supply. Generally, batteries are used to power the network nodes and usually they are not being recharged or replaced. In such environment, the network is considered as terminated end if energy minimum. Efficient power consumption is essential in any protocol for developed this type of networks. Many authors give various techniques and method which directly or indirectly related to the energy-aware and energy efficient in WSN. Energy lifetime improved by various technique such as energy-efficient routing protocol, clustering base protocol, wireless charging vehicle, etc. In the future might be wireless takes one of the best techniques to charging without any physical connectivity.

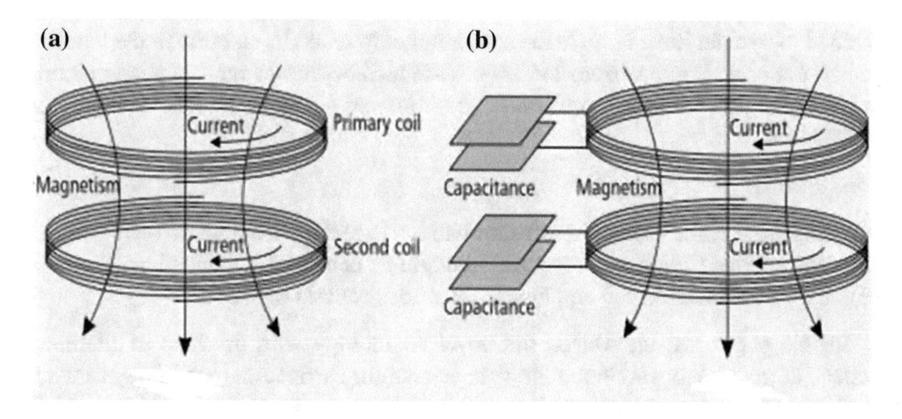

Fig. 2 Magnetic resonant coupling

6 Problem Formulation

Problem formulation is one of the important term in the research paper because it plays the major role. Without problem formulated we cannot simulate any relevant data in the simulator for getting proper results. In this paper, many models work for properly charging the sensor node in the deployed network. Wireless energy transfer, retained the ability to transfer electrical energy from one node to another node without any plugs or wires, in the other hand has been proposed to use in place of power transfer method to traditional power harvesting methods. Wireless energy transfer technologies can be categorized into radioactive radio frequency and nonradioactive coupling based charging. The former consists of three techniques, i.e.:

(1) Inductive coupling,
(2) Capacitive coupling and
(3) Magnetic resonance coupling (Fig. 2).

These can be further sorted based on to directive Radio frequency power beam forming and non-directive RF power transfer to the other devices. It is essential to fill the power level of battery some specific time interval for prolong the lifetime of wireless sensor networks. How can we formulated the charger or energy transfer operational activity in modeling form (mathematical) to maximize important properties and minimize disadvantage factors such as the charging efficiency and the energy balance and many model factors?

In literature review paper [5] author takes the one wireless charging vehicle to charge the multi-nodes simultaneously in wireless sensor networks, the charging range decides on the basis of vehicles capability to charging range. Vehicle start from the service station to charging the nodes travels into the centre of the entire cell in Wireless sensor network. Range of the wireless charging vehicle is same as the cell sides length. And another wireless charging related work is done by the author [4] of this paper. In both paper author introduced the magnetic resonance coupling method to transfer the energy to the sensor nodes to retain its stability in the wireless sensor network. Some various techniques are also described such as single mobile vehicle charging the entire deployed node in sensor networks with advantages and drawbacks.

Limitations:

(1) There are some assumptions as limitations in real environment.
(2) The service station and base station might be deployed at same place.
(3) How could decide cell and how many nodes contain by a cell?

In this paper, we introduced the novel technology with the help of literature papers, at previously single mobile vehicle charging arrived in technology to wirelessly charge the sensor node and some various techniques with one mobile vehicle with multi- or many removable chargers. Now we describe two mobile vehicles for charging the wireless sensors wirelessly with the help of clustering mechanism illustrate the charging trajectory.

7 Objectives

Replacement of the batteries of sensor nodes is not effective and sometimes it is not possible to replace the batteries of sensor node and provide continuous power supply to nodes of the sensors is also may not be. Hence to enhance the energy of every sensor node after the deployment in the network. Because if once sensor node deployed in environment they cannot be capable to move away to network without break the network. So, energy of the sensor node is one of the major issue in this way, then main objective of this paper to improved or enhanced the stored energy of the chargeable battery of the sensor nodes with respect to the life maximization of the sensor node to perform their task without any energy related problem.

There are main objectives in this study including investigating behaviours of low-power wireless communication with respect to varied transmission energy power and proposing a new scheme on sensor nodes capable of retaining its stability in network without any energy issues and the corresponding results were used for analysis.

8 Methodology for Enhancement of Existing Work

As describe the limitations after detailed literature review, we identified the problems in traveling of WCV (Wireless Charging Vehicle) on salesman path but this path is not decided based on energy awareness. Therefore, it may possible that when WCV reached to node for charging them before that node may not be live mean that network disconnection occurs. If we follow the energy-aware WCV traveling path then we will be possibly able to avoid such type of situations. Developing a frame work which is capable to choose direction of the movement means that WCV will move first in which direction where nodes energy level is less as compared to other direction node. Also, we will divide the network in two parts and each part will use separate WCV for each part. So, this method probably will enhance the performance of the wireless sensor network. We will be providing verification of this concept that enhancement of performance as well as reducing the complexity of wireless sensor network in my upcoming research works.

9 Conclusion and Future Work

In this paper, we briefly explore the ideas of energy transfer to sensor nodes in the wireless sensor network. We described recent advances in wireless energy transfer technology to charge the batteries of sensor nodes in a WSN. It also concludes based on various heuristic or review experiences that show the limitations of existing works. We have provided comparative study of well-known wireless charging techniques to charge sensor nodes in WSN. In our proposed approach sensor network is divided in two clusters and each cluster have separate charging vehicle, the travelling path of charging vehicle will be decided on the concept of energy awareness and the direction from starting point of wireless charging vehicle will also decide which is based on nodes energy. Proposed method will have improved the performance in terms of reducing the charging time and complexity. We will provide proof of our proposed approach in my upcoming research work via writing algorithms and their implementation and verification of them by simulation using simulation tools.

References

1. Kumar Y (Fulara PEC University of Technology) (2015) Some aspects of wireless sensor networks. IJANS 5(1):15–24
2. Akyildiz IF, Su W, Sankarasubramaniam Y, Cayirci E Wireless sensor networks: a survey. Comput Netw 38:393–422. 1389-1286/02/$, PII: S1389-1286 (01)00302-42002
3. Kurs A, Karalis A, Moffatt R, Soljacic M (2007) Wireless power transfer via strongly coupled magnetic resonances. Science 317(5834):83–86
4. Zou T, Xu W, Liang W, Peng J, Cai Y, Wang T (2017) Improving charging capacity for wireless sensor networks by deploying one mobile vehicle with multiple removable chargers. Ad Hoc Netw. Under the CC BY-NC-ND license, 79–90
5. Xie L, Shi Y, Hou YT, Lou W, Sherali HD, Midkiff SF (2015) Multi-node wireless energy charging in sensor networks. IEEE Trans Netw 23
6. Shankar A, Jaisank N (VIT University) (2016) A novel energy efficient clustering mechanism in wireless sensor network. In: IMCIP, vol 89, pp 134–141
7. Singh V (Punjabi University) (2016) A survey of energy efficient clustering algorithms in wireless sensor networks. In: IJEC, vol 5, pp 17961–17966. ISSN:2319-7242
8. Mahajan S, Malhotra J, Sharma S (2014) An energy balanced QoS based cluster head selection strategy for WSN. EIJ 15:189–199

9. Nuray AT, Daraghma SM (Anadolu University, Turkey) (2015) A new energy efficient clustering-based protocol for heterogeneous WSN. JEES 1–7. ISSN: 2332-0796
10. Islam Md F, Kumar Y, Maheshwari S (2014) Recent trends in energy-efficient clustering in WSNs. Int J Comput Appl (0975–8887) 95(20):44–48
11. Baroudi U (2017) Robot-assisted maintenance of wireless sensor networks using wireless energy transfer. IEEE 17:4661–4671
12. Monti G, Dionigi M, Mongiardo M, Perfetti R (2017) Optimal design of wireless energy transfer to multiple receivers: power maximization. IEEE 65:260–269
13. Sarikaya Y, Ercetin O (2017) Self-sufficient receiver with wireless energy transfer in a multi-access network. IEEE 6:442–445
14. Shah RC, Rabaey JM Energy aware routing for low energy ad hoc sensor networks. DARPA on grant no. F29601-99-1-0169
15. Gour S, Sharma S A survey of security challenges and issues in Manet. IISTE 6: 21–28. ISSN 2222-1719
16. Hasan MZ, Al-Rizzo H, Günay M (2017) Lifetime maximization by partitioning approach in wireless sensor networks. EURASIP Wirel Commun Netw 1–18
17. Kaur I, Kaur N, Tanisha, Gurmeen, Deepi (2016) Challenges and issues in adhoc network. IJCST, vol 7. ISSN: 2229-433
18. Goldsmith AJ (2002) Design challenges for energy-constrained ad hoc wireless network. Office of Naval Research (ONR) under grants N00014-99-1-0698 and N00014-02-1-0003. 1070-9916/02, pp 1–21

Leeway of Lean Concept to Optimize Big Data in Manufacturing Industry: An Exploratory Review

Hardik Majiwala, Dilay Parmar and Pankaj Gandhi

Abstract Implementation of lean concept is most recent trend in manufacturing industries. Enterprise resource planning (ERP) solutions such as SAP, Oracle, and BAAN IV are used to carry out day-to-day life activities for operational convenience. All these activities recorded and generated very big data, and the managers or strategic decision-makers heavily rely on them for decision-making. The challenging task here is to manage such type of big data of manufacturing company using lean concept. Present study comprises two: lean principle and data optimization concept of manufacturing activities. Here, we focus to integrate lean principles for the optimization of big data for efficient and effective decision-making, and also attempt to summarize the experience gain from the study. Big data generally stands for datasets that may be recorded and analyzed computationally to generate trends and pattern. These data quantities stored are indeed required large space and have a valuable cost in present era of development. For manufacturing unit, the big data can help to improve the product quality and impart lucidity in the work practices, which are having an ability to untangle uncertainty such as inconsistence availability and performance of machines and assembly shop as a system. Desirable transparency and predictive manufacturing as application approach required large amount of data and advanced tool for prediction to use this big data as useful information. Lean term is applicable to minimize the waste generated from the big data collection. Application of lean principles for managing big data of manufacturing process is a kind of minimizing

H. Majiwala (✉) · D. Parmar · P. Gandhi
School of Engineering, P. P. Savani University, Gujarat, India
e-mail: hardik.majiwala@ppsu.ac.in; hardikmajiwala@gmail.com

D. Parmar
e-mail: dilay.parmar@ppsu.ac.in

P. Gandhi
e-mail: pankaj.gandhi@ppsu.ac.in

D. K. Mishra et al. (eds.), *Data Science and Big Data Analytics*,
Lecture Notes on Data Engineering and Communications Technologies 16,
https://doi.org/10.1007/978-981-10-7641-1_16

the GIGO (garbage in garbage out) to reduce the data cost and also reduce the time of data processing for the decision-making of the managers. Present study gives new approach to manage big data very accurately using lean principles.

Keywords ERP · Big data · Lean principles · Manufacturing industry

1 Introduction

Day by day, lots of data are being generated. This data is not necessarily of structured nature, which we can handle with our existing data handling methods. But the data is either of semi-structured or non-structured nature. The definition of big data is governed based on the 3Vs model [1]: "*Big data is high-volume, high-velocity, and/or high-variety information assets that demand cost-effective, innovative forms of information processing that enable enhanced insight, decision-making, and process automation.*"

ERP is the backbone of any manufacturing plant and most likely choices used to achieve competitive advantages [2]. ERP systems are intended to offer all-in-one integration of developments across all functional areas with amended workflow, standardization of many business practices, and permit to real-time matched data [3]. The fundamental assistances of ERP systems cannot come from inherent planning abilities but somewhat from their capabilities to process transactions powerfully and to keep organized data [4]. Lean-ERP matrix has been developed to integrate and take advantage of both ERP and lean [5]. The 15 keys to ERP support for lean production have been given to lean practitioners and researcher to integrate ERP and lean paradigm to implement the concept effectively [6]. It has been found that lean manufacturing process in industry optimizes using ERP-based system by integrating manufacturing activity with ERP module. Dell canceled an ERP system after spending 2 years and $200 million on its implementation. Hershey Food Corp. filed highly publicized lawsuits against its ERP vendors for a failed implementation [7].

India across the nation, there is a movement for digital economy. It is expected that due to this, there will be a surge in data growth to support the digital movement [8]. Now in the context of data mining, it is known fact whenever there is data growth, the concept of GIGO crystalized in the data system. We believe that to meet these needs of GIGO and large volume of data in digital economy, lean principles will be the tool to reduce the cost of data transfer, storage, exchange, and process.

2 Literature Review

Review of literature has been done on basis of collecting books, research publication, and document regarding the big data, ERP, and lean. Following subsequent headings describe different views of researcher and provide worth inputs.

2.1 Identification of Waste in Big Data

There are eight types of waste in big data generated in ERP-based manufacturing industries [9]. The same can be described as below:

(i) **Transportation (Unnecessary Transfer)**: Scuffle to share data via email, word press, spreadsheet, and presentation and resend synchronize document while losing track of data is under the category of transportation waste. Collecting, processing, and storing data of data in manufacturing shop is another excessive transmission of data.

(ii) **Inventory (Excess Storage)**: Scrutinize data through spreadsheet, email, and database even though filing cabinet searching for data. Data is stored electronically forever, while people generate data where no one needs. Organization needs to buy new big data tools to keep every record to make sense of business. Such excess data is one type of data entropy that the usability of data become diminishes over a time.

(iii) **Motion (Unnecessary movement)**: The situation of repeated problems within big data application and infrastructure cause improper layout, defects, excess data, reprocessing of data, and excessive motion. Obviously, such situation leads to production losses.

(iv) **Rework**: Fix error that can be avoided; summarize and relabel data as different time bounds, inaccurate data of testing, cut, and copy/paste; and remove unnecessary data in manufacturing shop that is kind of waiting.

(v) **Non-value-added processing (Over-Processing)**: Data generation during reporting managerial data in manufacturing shop is not added any value in production. Over-processing of data is another work, reworked for customer expectations and problem and capitals spent on mining and cleansing data.

(vi) **Over-production (excess Flow)**: Generate more data than required, ERP-based system force employee to make data that does not need to be used. It is non-value-added activity concomitant with creating and collecting data.

(vii) **Waiting**: Data congestion due to inadequate processing capability is reason for waiting in shop, waiting for data needed for actions in manufacturing shop. People are working on data and show up they are working long hours.

(viii) **Defect**: Defects in manufacturing data are incorrect decision-making for different levels of business model. This results in poor decision-making.

It is suggested to identify such big data and apply big data tool to minimize such big data in manufacturing industries.

2.2 Lean Principle from Big Data Perspectives

Womack and Jones have proposed five steps toward lean thinking [10]. Lean approach in manufacturing industry is a kind of large transformation or reformation to the company. Some of the researches have been carried out to cognize and apply lean practice in manufacturing industries [11].

Five principles from the perspective of big data in manufacturing industries are as follows:

(i) **Value Stream**: It is process of identifying the value from the view of customer. Value for big data is information provided by particular dataset to the different players of manufacturing shop: Supplier, security, supervisor, manager, deputy general manager, vice president, chief executive officer, and other stakeholders. The purpose of lean big data is to collect right data at right time and transfer such data to right person. The data which is not created value to service or product is waste.

(ii) **Value Stream Mapping (VSM)**: In the second point of lean principle, all the requirement of customer to the perspective of big data is identified and scrutinized for every step of manufacturing process. Apart from every action, step and practice which is not add value in the process for improvement are eliminated. Value stream mapping is exceedingly effective tool to seizure such non-value-added activity in process and try to eliminate.

(iii) **Flow**: Before attempting to implement the third principle flow, lean practitioner suggested going more in detail only after stipulating value and connecting the value stream. All activity examined properly by concern person in manufacturing plant for continuous flow of work using data analysis in ERP system. Generation of all data in manufacturing system and segregation of such data can be eliminated by identified the data waste in process. Data at different levels of manufacturing shop can be recognized and transfer to concern person for better decision-making process. So identification of such data plays a vital role in this principle. Flow of data can be classified by requirement of different players of manufacturing shop. Mura and Muda terms in Japanese mean unpredictability and same needs to be identified in flow of process. In order to achieve the continuous flow of process, such unpredictability has to reduce. So it is necessary to develop methodology to catch such waste and inconsistency in system by making flow of process smooth.

(iv) **Pull**: Womack and Jones state, instead of push the product to customers, customer pulls the product or service from you [10]. No concern department and any upstream functions can produce product or goods until customer ask for same. In big data, pull system eliminates unnecessary waste of generation of data and transportation of data associated with manufacturing shop. Pull system can only happen when service or product provide better value to customer need. Example of pull system in big data use cases is cited here. From the past data of manufacturing shop, one can identify future market, specific customer need, prediction of manufacturing, supply chain analysis, predictive maintenance task, and so on.

(v) **Perfection**: The process developed for system is said to be perfect when it produced only value-added activities and minimum waste. The last lean principle perfection ensures that there should not be flaw in system; one lean process is developed. But Womack and Jones state, "When customer gets what they want, there is no need to get improvement for optimizing time, cost, effort, defect and space [10]".

2.3 Big Data Issues

There are many issues that have been identified based on data used in manufacturing company. Many issues have been given by different researchers [12–14]. Description of such issues generated in ERP-based system as big data is given below:

- **Storage, Processing, and Transportation Issues**: Data generated by sensors and social media sites are an example of data generators for large volume of data. For performing analysis on such data, data should be transferred on the processing machines. The amount of time taken to transfer the large amount of data from storage points to computing points will be more compared to time taken to analyze the data.
- **Privacy and Security**: For generating big data, data related to individuals is collected and used in predictive analysis. This can lead to serious privacy issues of an individual. Sometimes, it is difficult to handle this large amount of data by their own and may require employing third-party solution for handling it. It may lead to security violations of data.
- **Analytical challenges**: The amount of data generation is enormous, and there is no provision for storing such large volume of data with increasing dimensionality in future. Then, at later stage it will create problems related to scalability and expandability. The data collected may not be of structured nature, which we can handle with the help of traditional relational datasets. At the same time, there is a need to effectively identify the important data and get the great value extraction from it.
- **Data Access and sharing mechanisms**: For making accurate decisions based on big data, analytics requires high availability of complete data in timely manner. For doing this, the data must be kept open, and APIs built for accessing this data must also be standardized. Cooperation from different companies is required for building the robust datasets. Companies hesitate to make their data open, as it can help their competitors to get information of their clients and other details.

3 Challenges to be Encountered

3.1 Meager Lean Initiative

Insufficient case study to understand and analyze tools and techniques used in lean paradigm in big data issues makes this task thought-provoking. Informatics approach to identify the non-value-added activity in data transaction and optimizing big data efficiency involves utilizing process, people, and technology using lean management system [15]. Such system targets the waste to be eliminated across whole big data system across non-production and production environment. Lean startup in organization popularized by Eric Rises gives solution for lean management system in manufacturing industry [16]. Another data reveals so that 75% initiative fail when it starts.

3.2 Intransigence to Change

Lean big data may require many changes at different levels of management, behavior, and operation. It may happen that workers, managers, and senior management resist such changes in organization. Multi-skill person concept requires person to do one activity to another. Same may be resisted by concern department in-charge or person at own. It may be possible in case of multi-trained personal, one trained for welding operation, and he needs to go to ground where he cannot use his basic skill of welding. Such problem has been addressed by many HR professionals. For more problems regarding resistance to lean practice, more investigation is required.

3.3 Lack of Visualization

Lean application in big data is more intangible and digital compared to lean application in manufacturing unit. For the implementation of the lean in manufacturing, ERP is used but managing the data in ERP authors need to apply lean principle in big data generated in ERP systems. Value stream is difficult in lean big data; tough lean application in big data is discussed in the previous section (Fig. 1).

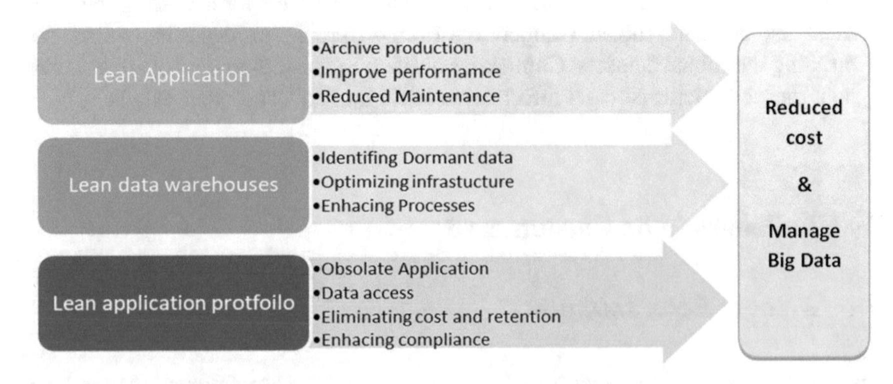

Fig. 1 Lean paradigm in big data

4 Lean Paradigm: A Case to Make Big Data Small

A big data case of one of the leading fabrication facilities of manufacturing unit has been studied for the present project. Company uses many modules in BAAN IV system for effective implementation of lean manufacturing. Later on, ERP-LN system has been implemented and journey toward lean is challenging. The problem and requirement of concern department have been collected through Kanban event. Kanban event is one of the tools to collect and identify the problems. All non-value-added activity in data collection has been identified and after rectifying all such activities, value stream mapping has been done. For the same statistical tool, Pareto analysis has been done to identify which waste is higher in generation of data. Pareto analysis uses 80–20 rules to give relative significance of explicit waste by ranking such waste in terms of priority.

Work Sampling: The authors have selected data sampling method for good judgment. IT department employee of manufacturing unit who was associated with big data of ERP system is the main source of all such information. Present case includes interviews, sampling of data, and document verification to measure activities in manufacturing unit. All activities carried out in the units are a kind of waste and have been identified and shown in Fig. 2 and Table 1.

Data Collection and analysis: Data is recorded with care and high accuracy in the form of primary data. Data analysis said that unnecessary transfer of data makes traffic in ERP system leads to delay in processing data. And storage of data leads to increase cost of units. Objective of this paper is to eliminate such waste and develop decision-making system from data segregation at different levels of management to minimize data traffic and additional cost to units.

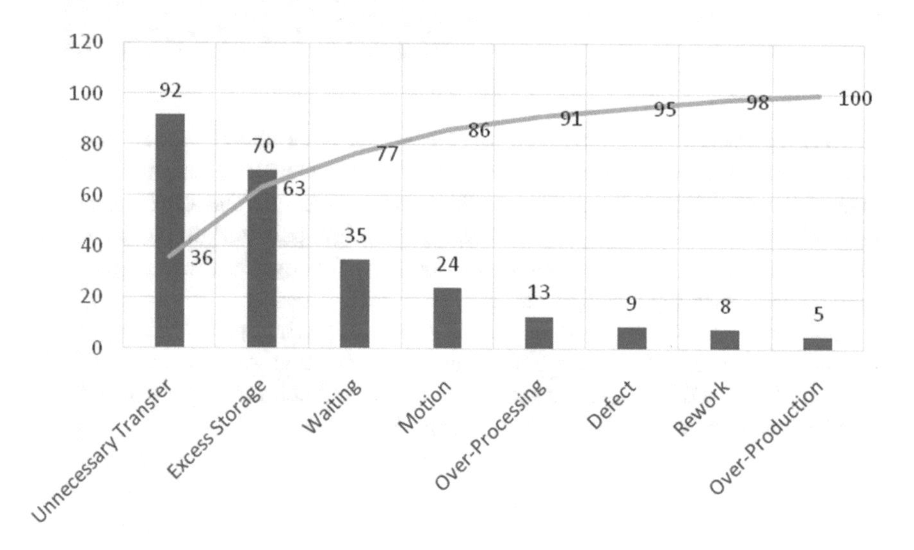

Fig. 2 Pareto diagram

Table 1 Analyzing big data waste

Sr. no.	Types of waste	Occurrence of waste during the month	Cumulative occurrence	% Cumulative
1	Unnecessary transfer	92	92	36
2	Excess storage	70	162	63
3	Waiting	35	197	77
4	Motion	24	221	86
5	Over-Processing	13	234	91
6	Defect	9	243	95
7	Rework	8	251	98
8	Over-Production	5	256	100
		256		

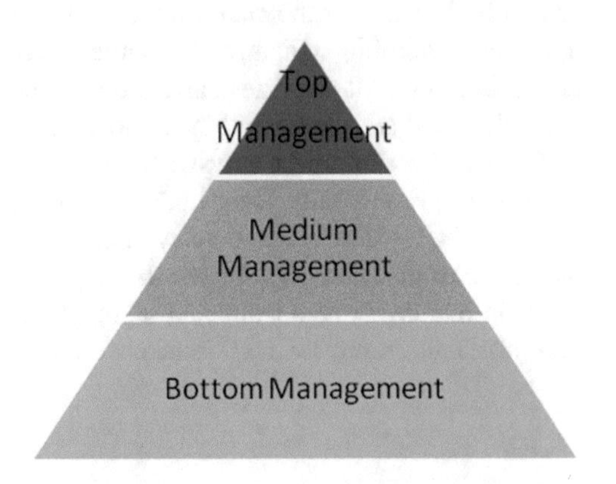

Fig. 3 Three levels of management

Decision-Making Support System: Once data is eliminated, other challenge is to develop decision-making support system for three levels of management: Top management, medium management, and bottom management. The lean concept such as Just in Time (JIT) has been used to transfer the right data to right person at right time. Other lean concept, Pokayoke, can be used for mistake-proof operation to transfer the data at top management.

The concept is to transfer right and accurate data to right person and right time is called just in time. For the applicability of such concept, one has to identify the required data for three different management levels. Such three types of different data have need to be investigated further (Fig. 3).

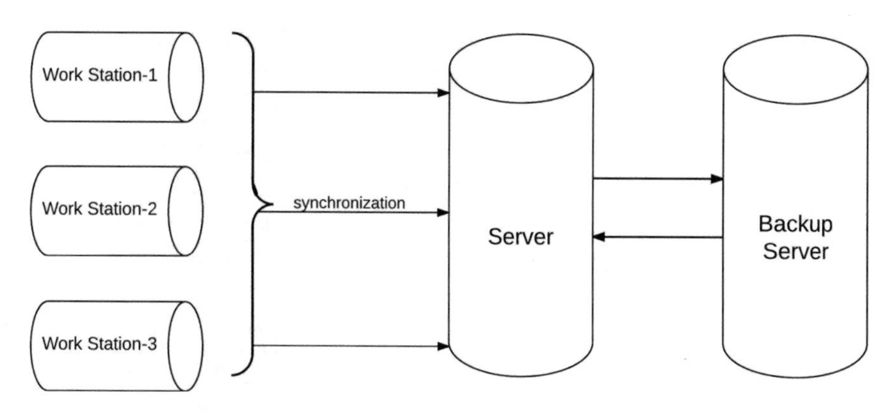

Fig. 4 Database model

Data has to be classified into three categories and transfers such data as suggested in requirement of different level managements. As shown in Fig. 4, algorithm has been developed to transfer and segregate the data as per management requirement. Decision can be trigger from the different levels of management. Top-level management needs to solve unstructured problem by executive management system. Similarly, medium management solves structured problem by decision support system and bottom management solve their routine problem by management information system (MIS). Data segregation plays vital role in this system. Data has been optimized using lean principles to eliminate waste generated in big data on manufacturing shop (Fig. 4).

Lean concept in big data is a kind of artificial intelligence where system itself takes decision. Data generated in workstations is stored in server. Once it is transferred to backup server, data stored in workstations and server will be automatically destroyed. So doing the same, unnecessary data travel, traffic and storage can be eliminated which is higher in the present study case. Management by objective has been set to get required data at different levels of management. Tacit knowledge of employee and workman on shop floor in manufacturing unit can be used to implement lean concept in big data management. Lean big data can be used to trigger the decision at different levels of management (Fig. 5).

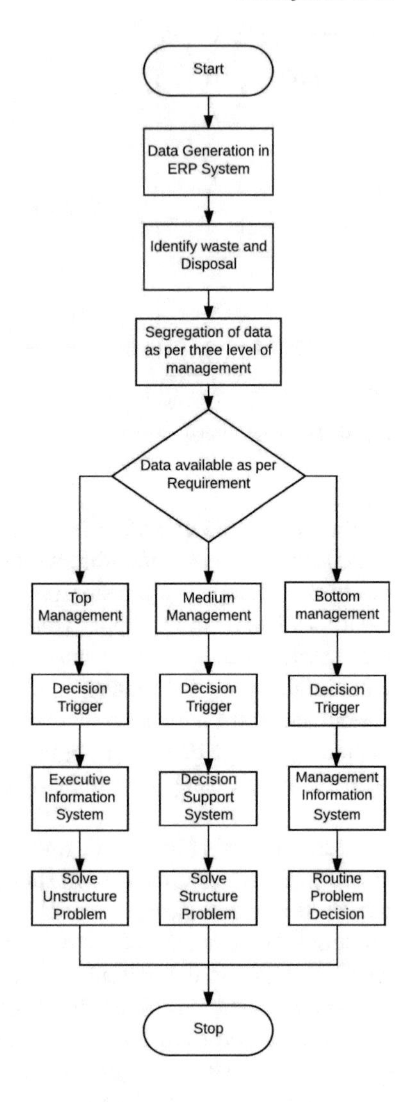

Fig. 5 Lean model for big data

5 Conclusion

Findings of the present study are based on preceding investigation of feasibility of study, current and future situation; researchers' work in the field shows that lean can be applicable in big data optimization in manufacturing industries and can be also used for strategic decision-making. Tough tools and methodology used in current practice are to be continued with lean applicability. Gradually, transformation in big data will take place in manufacturing industries. Lean implementation strategy and applicable principles have been studied in Sects. 1, 2, and 4 above. Lean process development and lean system design in big data are different from the production line. There are trials and experiments involved. In the present work, a case has been studied on big data issues in fabrication facility. The process of production in manufacturing unit demand high quality and as well for achieving excellence, research and innovation is needed. Toyota become world-class manufacturing facility just because of smooth synchronization of these two processes. By giving a training of lean concept to big data professional, one can synchronize and make bridge between these two processes for preeminent results.

References

1. Gartner IT Glossary, Big Data (definition), Gartner.com, Accessed September 24, 2017. http://www.gartner.com/it-glossary/big-data
2. Zhang Z, Lee MKO, Huang P, Zhang l, Huang X (2005) A framework of ERP systems implementation success in China: an empirical study. Int J Prod Econ 98:56–80
3. Mabert VA, Soni A, Yenkataramanan MA (2003) Enterprise resource planning: managing the implementation process. Eur J Oper Res 146:302–314
4. Jacobs FR, Bendoly E (2003) Enterprise resource planning: developments and directions for operations management research. Eur J Oper Res 146:233–240
5. Paper White (2007) Syspro. The When, Why and How of ERP support for lean
6. Steger-jensen K, Hvolby H-H (2008) Review of an ERP system supporting lean manufacturing. In: Koch T (ed) IFIP International federation for information processing: lean business systems and beyond. Boston, Springer
7. Wali OP (2016) Information technology for management. Wiley Publication, pp 279–280
8. Collective Envision (2017) Surge in data growth to fuel push towards digital economy. Economic Times, p 8, 28 Sept 2017
9. Cottyn J, Stockman K, Van Landegnhem H (2008) The complementarity of lean thinking and the ISA 95 Standard. In: WBF 2008 European conference. Barcelona, Spain, pp 1–8
10. Womack JP, Jones DT (2003) Lean thinking. Simon & Schuster
11. Rother M, Shook J (1999) Learning to see value stream mapping. Lean Enterprise Institute
12. Katal A, Mohammad W, Goudar RH (2013) Big data: issues, challenges, tools and good practices. In: 2013 sixth international conference on contemporary computing (IC3). IEEE, pp 404–409
13. Kaisler S, Armour F, Espinosa JA, Money W (2013) Big data: issues and challenges moving forward. In: 2013 46th Hawaii international conference on system sciences (HICSS). IEEE, pp 995–1004
14. Chen Min, Mao Shiwen, Liu Yunhao (2014) Big data: a survey. Mob Netw Appl 19(2):171–209
15. White Paper, Informatica (2011) The data integration company
16. Ries Eric (2011) The lean startup. Portfolio Penguin

Monitoring Public Participation in Multilateral Initiatives Using Social Media Intelligence

Ulanat Mini, Vijay Nair and K. Poulose Jacob

Abstract Governments, multilateral agencies like the World Bank, United Nations, and Development Banks as well as other nonprofits are involved in a variety of developmental activities across the world. A lot of resources are spent to ensure proper consultations and post-implementation verification of results. But this does not completely ensure whether the objectives are achieved. The new web technologies provided methodologies and developed tools that allow the users to pool resources on projects over the Internet. Social media allowed real-time feedback for citizens, monitoring developmental initiatives of Governments and multilateral agencies. The role of technology ensures that the consultations and ongoing feedback can be captured, analyzed, and used in understating the stakeholder reactions to the project and its implementation. This helps in making necessary course corrections avoiding costly mistakes and overruns. In this paper, we model a tool to monitor, study, and analyze popular feedback, using forums, social media, surveys, and other crowdsourcing techniques. The feedback is gathered and analyzed using both quantitative and qualitative methods to understand what crowd is saying. The summation and visualization of patterns are automated using text mining and sentiment analysis tools including text analysis and tagging/annotation. These patterns provide insight into the popular feedback and sentiment effectively and accurately than the conventional method. The model is created by integrating such feedback channels. Data is collected and analyzed, and the results are presented using tools developed in open-source platform.

Keywords Natural language processing · Content analysis and indexing
Crowdsourcing · Social media intelligence · Text mining · Sentiment analysis

U. Mini (✉) · K. Poulose Jacob
Cochin University of Science and Technology, Cochin 682022, India
e-mail: mini.ulanat@gmail.com

K. Poulose Jacob
e-mail: kpj@cusat.ac.in

V. Nair
Assyst International, Cochin, India
e-mail: vnair67@gmail.com

D. K. Mishra et al. (eds.), *Data Science and Big Data Analytics*,
Lecture Notes on Data Engineering and Communications Technologies 16,
https://doi.org/10.1007/978-981-10-7641-1_17

1 Introduction

1.1 Collaboration and Consultation Portal

Collaboration is the key aspect in every organization where opinions of their policies and programs (or products and services) are collected from a wide variety of custodians like the society (or customers), policy stakeholders (or employees), experts (business partners), external stakeholders (General public), etc. Organizations face challenges in terms of analyzing and reporting contents coming from various mediums including but not limited to feedbacks, reviews, comments, social media interactions, blogs, and surveys that are hosted internally or on the cloud. Manual evaluation of these is impossible, and there is a need to automate the review process and bring out the message in a format which stakeholders can understand and act upon appropriately. We are living in information age and the content is getting generated through various sources dynamically. Sourcing all the unstructured information into a common platform and processing are the challenges faced by the data engineers. Named entity recognition (NER), relationship extraction, and sentiment analysis (Polarity) are the key aspects of a text analysis process. Many research institutions are contributing heavily into this platform to enable the machine learning from the human-generated content.

Governments and multilateral agencies work on a variety of projects which impact the society. A large majority of mankind is living in a conflict-affected environment. The upkeep of peace, improving the living conditions, includes investment in destroyed and weak infrastructure. This is apart from various challenges including lack of food and essential services. This involves expenditure of large amount of money with multiple stakeholders. The beneficiaries of these projects are mostly communities of economically backward and less developed countries having limited access to communication and consultation. The project aims at benefiting the community in terms of social and economic upliftment, better infrastructure, education, health, environment, and sustainable practices. There is often a disconnect due to the long time frames, number of stakeholders, and the distance and communication lag between the affected community, the project implementation team, and the funding agencies. In the current context, greater transparency and open communications with all are very much essential for the involvement of stakeholders during all the stages of a project. Computer and communication technologies promise a greater role in peace-building efforts. The ubiquity and seamless communication capability of social media channels help to tide over these problems of traditional media.

1.2 Social Media and Public Participation

Traditional print media and communication like radio were the standards in peace-building efforts. But the more transformational nature of the promising technologies enhances the current process. There are a variety of tools currently available with public to express their experiences and opinions. These expressions further influence on the building up of opinions of others. Dialog is a critical component in this scenario. Social media can support and set stage for broader participation. Tapping into these sentiments will make it easier to implement projects. Policy decisions can be shaped more easily. Monitoring of media has been in process for long time whether for public relations, understanding public opinion or general market intelligence. But with fragmenting media and changing public sentiments over long-term projects, traditional monitoring methods like press clippings, field surveys, or ad hoc research are highly insufficient. Thus, involving the participation of the public and analyzing of public sentiment, much of which is expressed online and becomes a necessary part of public engagement.

The collaboration engine supports six channels for data collection: Review/Feedbacks, comments, social media interactions (LinkedIn, Facebook, Twitter, and Google+), blogs, and surveys. These channels act as data consumers and store those in data repositories for analysis. This also allows submissions of additional materials pertaining to any channels, which allows content extraction from documents (PDF, Excel, and Documents) as well. Collaboration hub is the platform where all the individual consultation-specific analysis gets added [1].

The solution enables the stakeholders to pick and choose the source of information (channels) depending on the interest or the need of the data collection practice. For example, someone wants to compare meeting comments versus social media comments. Sometimes, the user wants to filter analysis based on the geographical or sector or topic or theme based. The reason is that, when we compare the sentiments of two datasets if both are from different sectors or topics or geography, those results may be logically different. Technically, the data may be correct, but contextually that may be incorrect. The solution should enable the custodians to identify and define the logical relationships depending on the context of the business function. Organizations using this tool can build their own vocabulary (Keywords, Entities, and Concepts) for processing the data collected through various sources. The institutions follow specific vocabulary to avoid confusions or different interpretations on the results. This tool also facilitates integration with industry proven data analysis platforms like Alchemy and Open Calais along with the analysis engine. The custom solution benefits advanced optimization to narrow down the results and fine-tune the process according to the business requirements.

1.3 Challenges

1. Effective feedback at appropriate time;
2. Collecting and collating the feedback;
3. Intended beneficiaries and residents affected by these projects are distributed and vast;

4. Beneficiaries having no direct channels of conveying their opinions, and the extended logistics of such an exercise make it difficult to study and incorporate the feedback;
5. With multiple channels of social media of expression, there is an information overload and skewing tendency; and
6. With more data points, with all and sundry giving opinions and feedback on the multitude of media collating, curating all this data, and making sense of it, to achieve the desired goals is challenging.

2 Related Work

Sentiment analysis has become a hot research topic in recent years because of the variety of applications. The main focus areas are lexicon construction, feature extraction, and then determine the polarity which can be taken as a feedback for improvements in many cases. The emergence of crowdsourcing created new opportunities in data collection and annotation methods. Dave et al. [2] introduced the term opinion mining and further explains opinion mining tool capable of processing search results highlighting the attributes, combining all the opinions with qualitative featuring of product attribute [3]. User reviews are analyzed using machine learning methods by Pang et al. [4]. They also express the challenges in the analysis especially due to the presence of noise in the text and complexities of natural language processing. The synonyms and antonyms in WordNet are used by the set of opinion words tagged. Hu and Liu [5] used overall sentiment to classify documents. The well-known and much-cited papers of Liu [6, 7] represent component (which are product, person, event, etc.) and associated set of attributes as aspects. The majority of the algorithms for aspect-level sentiment analysis use machine learning classifier. Hoogervorst et al. [8] employ a discourse parser implementing Rhetorical Structure Theory (RST) [9]. In this case, context of each aspect is determined from the parser and expressed sentiment is computed with respect to the weightage of the discourse relations between word. Determine the polarity of comments whether it is positive, negative, or neutral by extracting features and components of the object that is opinion mining [10]. Hu and Liu [5] investigate the effectiveness of sentiment classification of documents by machine learning techniques. It is demonstrated that human-produced baseline for sentiment analysis on movie review data is found to be inferior than by machine learning techniques, but accuracy is better. The experiment was review of movie corpus classified using SVM, maximum entropy classification, and Naive Bayes, and features based on unigrams and bigrams.

3 Related Technologies

3.1 *Social Media and Public Participation*

Social media monitoring makes use of text mining and natural language processing (NLP), where the user-generated contents are analyzed to understand the awareness, mood, and emotions with relation to a particular topic. Text data mining (or opinion mining) is the process of obtaining relevant and high-quality information from text, typically by studying and coming up with patterns and trends using statistical pattern learning. By using known statistical patterns and keywords or taxonomies, it is possible to devise parsing techniques which can highlight certain words and ignore the others. By deriving patterns from the text data so structured, it is then examined for relevance and categorized, keywords extracted, and relationships between words or concepts are analyzed. Social media monitoring and analysis is a universal term used to accommodate brand monitoring, buzz monitoring, and online anthropology, to market influence analytics, conversation mining, and online consumer intelligence. Sentiment analysis attempts to extract sentiments or opinions, associated with positive or negative feelings or polarities toward a specific subject within one or more documents, rather than just branding and entire document as positive or negative. So it attempts to identify how sentiments are expressed in texts, and whether those expressions are favorable or unfavorable to the subject. Sentiment analysis involves the identification of the following:

1. Sentiment expressions or the statement containing sentiments,
2. Polarity and strength of the expressions (negative or positive), and
3. The relationship to the subject (the sentiment expressed about a car may not be relevant while considering opinions about a road construction).

3.2 *Lexical and Quantitative Analysis*

Lexical analysis has traditionally been used to design compilers. A modified lexical analysis tool is used to study the word distributions. Quantitative text analysis is used to extract semantic or grammatical relationships between words in order to find meaning or patterns, as in profiling. A variety of techniques support these models including pattern recognition, tagging and annotation, link and association analysis, and predictive analysis. The recognition of specific related keywords (e.g., country names or abbreviations, names of political or social leaders, or organizations) or patterns (email addresses, phone numbers, addresses, numeric or price data, etc.), within the context of textual data is also possible. While decoding such data, context becomes very important. For example, Ford can refer to a former US president, a car model, a movie, or some other entity. Disambiguation will require understanding of the context.

3.3 Text Analysis, Semantic Tagging, and Analysis

Statistical models and tools, lexical tools, and models for quantitative processing have been around for some time. But due to the complexity of use of these models, the processing power required, and the lower reliability made these more of tools for research or large organizations. Sentiment analysis and qualitative tools are computing intensive tools and hence was not very popular earlier, though algorithms were available. There was difficulty in creating the taxonomies that improve the quality of analysis. However, now with the advent of various commercial and open-source tools, text analysis is available to a much larger user base. Governments and multilateral agencies, as well as nonprofits and for profits, are all adopting these tools. The usage of tools can be illustrated with an example of analysis of comments. While the number of tools which work on semantic and text analysis is increasing, with a lot of open-source and proprietary tools, it should be understood that each tool does have limitations. For our needs, we have combined various tools and built part of our own unique taxonomies for the domain we have selected.

3.4 Other Analytic Techniques

Several other analytic techniques may be combined to extract further information. Relationship, fact, and event extraction involve identifying relationships between entities and other information in the text. Concept and entity extraction, and production of granular taxonomies are also important to the text mining tasks. The final objective is understanding the relationships between named entities, using natural language processing and analytic methods to populate a database or search index with the information extracted.

3.5 Open-Source and Commercial Tools Used

Some of the commonly available and useful tools include the following:

Open Calais: Open Calais [11], by Reuters, provides a web service allowing you to submit text for analysis and receive results of the analysis. With a free API, a lot can be done; however, the service is chargeable for usage beyond certain limits. Also, you need to consider that the data will be processed on their servers, which may raise a flag about privacy concerns. On the bright side, use can be made of their extensive taxonomy and semantics library.

TextRazor: Another service with free and paid services. TextRazor [12] uses natural language processing and artificial intelligence to provide content analysis. Other similar tools available provide free or paid services, and include Alchemy [13], Saplo [14], etc.

atlas.ti: Standalone tools like ATLAS.ti [15] and nVivo [16] provide many of these features and more. These tools are collectively called CAQDAS (Computer Assisted Qualitative Data Analysis Tools).

Stanford NER: The Stanford Named Entity Recognizer (NER) [17] provides a set of libraries, which can be used, among other things for text analysis. This tool has well-engineered feature extractors for named entity recognition. The advantage is that it has many options for defining feature extractors also.

Apache UIMA and TIKA: Apache UIMA or Unstructured Information Management Applications and TIKA [18] provide tools for text analysis and parsing. It extracts metadata from structured text content and unstructured content like audio and video. It is easy to integrate with Lucene/SOLR for indexing and searching.

4 Research Methodology

Most public policy, planning, and projects implementation address the inter-relationships between different stakeholders including the public in the region, the regional governments, civil society, the implementing agency, and the funding agencies. Stakeholders can have differing views on the project or policy, based on their understanding and the effect of the project on their daily life. Project seeks inputs from all stakeholders during all phases of the project. The levels of public participation help in ensuring transparency, by bringing about public accountability on issues and decisions that affect the community.

The aim of this paper is to discuss the methodology to model as follows:

(1) Promote public participation into an engagement,
(2) Bringing forth the public opinion and tracking it,
(3) Building tools to quantitatively and qualitatively assess interactions between people, and
(4) Building a feedback system to respond to the initiatives.

4.1 Case Study

Discussion forums, surveys, rating/polls, to more simpler forms like thumbs up/down, or five-star ratings are all popular methods used for getting feedback from sites. Increasingly Facebook, Twitter, and other social media sites are also used to track user feedback. Since all these platforms are now commonly used and prevalent, the model should include the discussions from all the channels. Our framework is specifically built to collect, aggregate, and analyze multiple social media streams. This is a media monitoring tools which does real-time keyword extraction, text analytics, and topic visualization. The tool helps in directly connecting the expression with real-time response platforms. This helps in reducing the elements of violence and respond to the peacekeeping and conflict response platform.

Tool is built for multilateral development agencies for management and monitoring of multiple consultations. All consultations of the agency and partner agencies across the world are managed by consultation hub. When a consultation starts, an entry is made, demarcating geography involved, stakeholders whether public or private, and key themes or areas of consultation. These key areas can be environment, health, power, etc. Once these parameters are decided, the tool starts collecting the inputs.

5 Architecture of the Automated Monitoring and Evaluation Tool for Multilateral Development Agency

The portal of consultation hub (Fig. 1) helps anyone to find, share, and participate in consultations that interest him. Users can be involved at any stage. Anyone including public can be a part of the consultation and generate content. This creates large volume of data. The tool needs to exploit opinion mining and promote public participation for better decision-making. This helps in the participating organizations to improve the delivery mode of services and resource optimization. The expressions are analyzed with opinion mining tools understanding the natural language using knowledge bases. A machine-readable ontology is defined to provide a unified schema for interpretation.

1. Tool listens to multiple data sources hiding the complexities involved. Inputs are collected from all channels including blogs, forums, Facebook posts, etc.
2. Semantic tagging done using the services includes Apache TIKA and UIMA and Text Razor. Tags are populated to SOLR indexing.
3. Natural language processing algorithms are applied to every post, interpreting emotions as positive and negative. The machine learning algorithms run across the data identifies hidden characteristics and bring out previously unnoticed patterns.

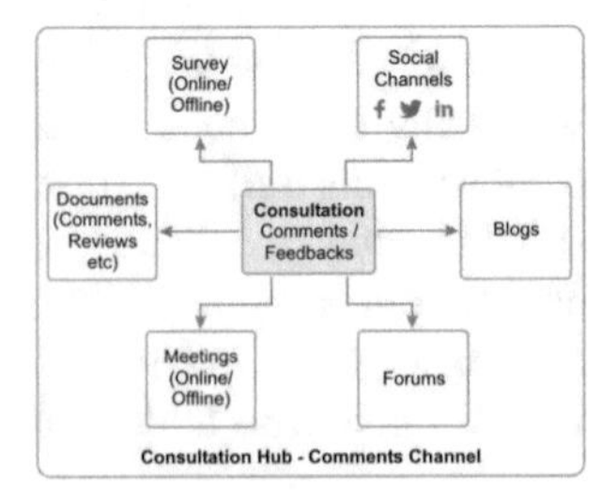

Fig. 1 Consultation hub

4. The tool automatically identifies the topic of every post and give insights into their authorship. Semantic context understood from ontology and taxonomy. Knowledge base is continuously updated.
5. Visualization tools help in understanding and acting on complex information, thus providing intelligence from unstructured data.

6 Solution Architecture

A screenshot (Fig. 2), from a sample implementation, is collected and is used to illustrate the usage of one of these mechanism comments or forum. This is further taken up for analysis. This is a text-based comment in the feedback system. Apart from the simple textual data, other methods like star ratings, polls, and online survey queries are collected regularly at data collection points. Compared with text data, these are easier to analyze and collate.

Apache UIMA (Unstructured Information Management Applications) an ad-don to Solr for indexing unstructured content has been identified. It is an API which connects easily with Solr and allows connectivity with various open text analysis engines (Alchemy and OpenCalais). It was a successful story and we were able to gather keywords, concepts, and entities out of the text passed to the analysis engine.

The critical aspect of the whole solution is that how best we portray the processed information in a way it connects to all sort of people easily and effectively, through strong self-depicting visualizations. Data visualization tools to consume those results and portray those effectively. It enabled the result visualization effectively irrespective of any devices or software solutions.

In this case study, we are taking text data, not concentrating on categorical data or numerical where results are easily obtainable for the purposes of analyzing feedback and measuring popular sentiments.

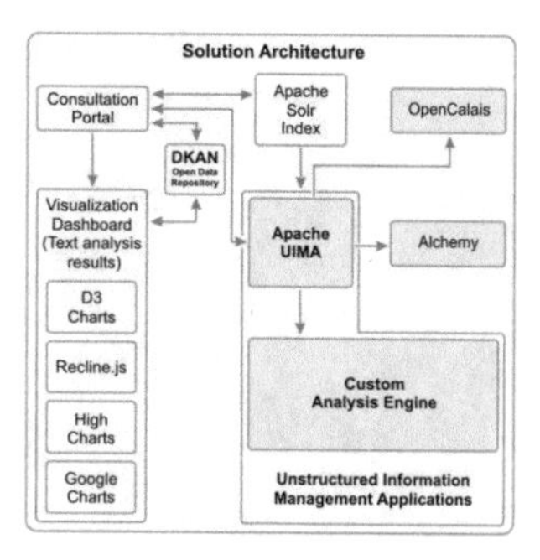

Fig. 2 Solution Architecture

7 Technical Architecture

The tool developed has exciting information extraction features. These features are extracted through the integration of open-source libraries like Apache TIKA tool kit, open NLP, and Apache UIMA project. Metadata is extracted from source document.

8 Working of Consultative Hub

In the below example, the first step is to properly identify the semantic relationships between the sentiment expressions and the subject (Fig. 3). Applying semantic analysis with syntactic parser and sentiment taxonomies, sentiments are found. It has been possible to get 80% or more precision in finding sentiments within multiple documents. The concept can be further explained with the help of an example of what is involved. Banks' effort to review and update safeguard policies is to be welcomed, but linking up clearly with other reforms within the bank and elsewhere will be critical.

We see that there are two statements being made here. The first statement indicates a favorable sentiment, while the second one shows a negative opinion. Thus, to do a proper analysis, we would need to identify the individual statements and present results accordingly.

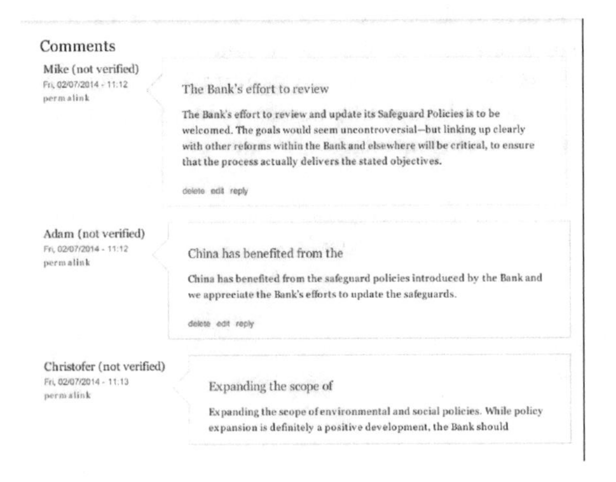

Fig. 3 Comments on a consultation

9 Text Analysis

We run an analysis of the comments, using relevant taxonomy. The screen in Fig. 4 shows the analysis being started.

Public conversations are tracked from the different channels. Understand what public is talking about. The sentiments can be tracked based on the written message. NLP offers the intelligence. The text is tagged as positive, negative, or neutral. Pipe the result into the tool, which assigns a sentiment score. NLP analyzes the language patterns. This analysis helps in much more than sentiment analysis. It can also monitor the malicious acts like attacks (Fig. 5).

Our experience in collecting, processing, and managing the open data accelerated our thoughts and actions to arrive at an apt solution to this business requirement. The core ideology was to reuse the solution stack as much possible and remain in the open space to avoid additional financial burden. We were successful in that to a great extent; UIMA (Unstructured Information Management Applications) solution is used. The API connects easily with Solr and allows connectivity with various text analysis engines (Fig. 6). Keywords, concepts, and entities derived out of the text were successfully gathered.

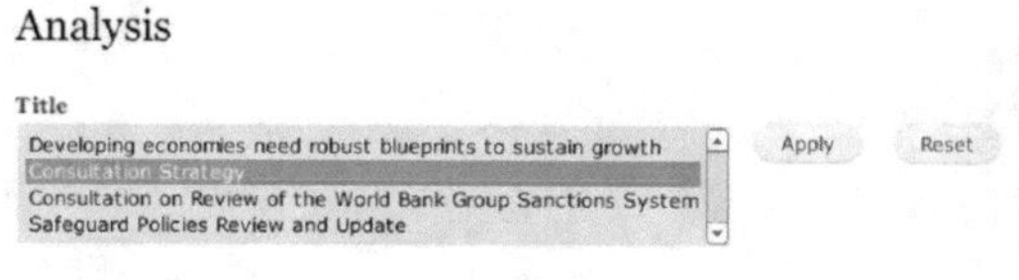

Fig. 4 Analysis of comments

Keywords

World Bank, world bank group, World Bank officials, social safeguard principles, Parliamentarians.\nThe best indicator, so-called \Development projects.\ , \Financial Intermediary Lending\ , best test, current exercise, genuine effort,

Concept

Concept	Relevance
World Bank Group	96.75 %
World Bank	71.89 %

Entities

Entity	Count
World Bank	1
World Bank Group	2
IFC	2
India	1
World Bank	7
Principle of Intergenerational Equity	1
FPIC	1

Fig. 5 Result analysis

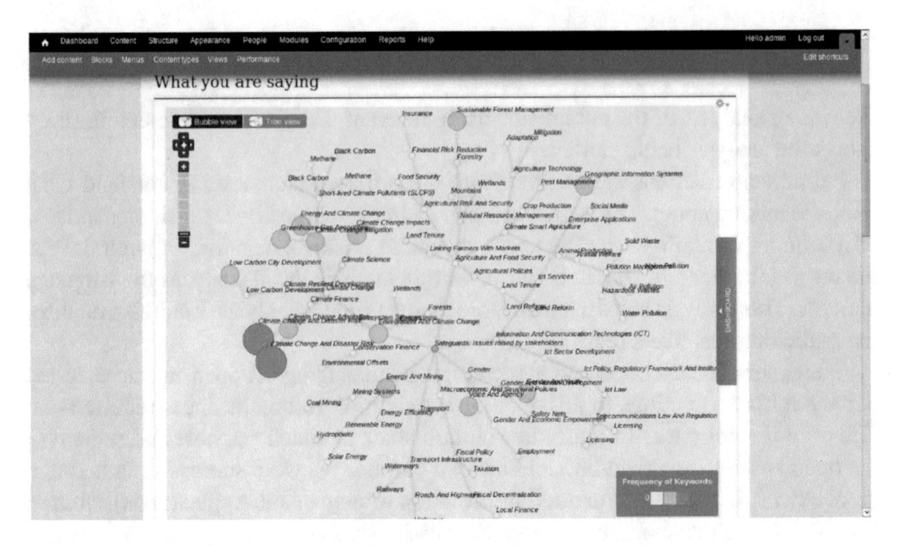

Fig. 6 Screenshot of crowdsourcing

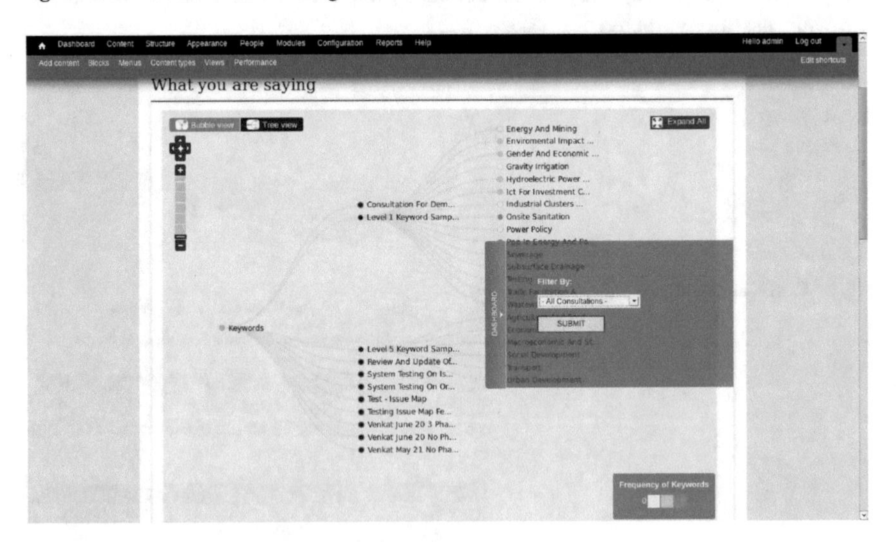

Fig. 7 Crowdsourcing in tree structure

10 Conclusion

Content analysis as well as text analysis is just coming into its own. With huge amounts of data exciting range of tools and usages, we will see more of analysis. Digital dashboards showing project performance, collection, and analysis of stakeholder feedback and comments are crucial along with tools like social media, open data, and open governance in the case of consultation hubs (Figs. 7 and 8).

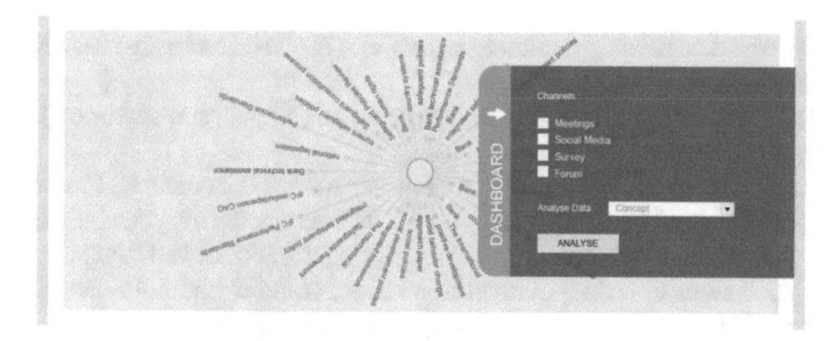

Fig. 8 UI for channel monitoring

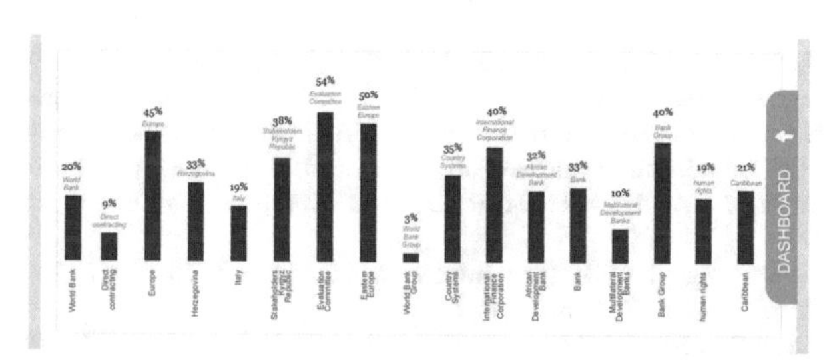

Fig. 9 Graphical view of dashboard

These tools will help usher a more engaging and democratic policy and help governments and other agencies to improve delivery capability, meet expectations, and increase transparency and accountability. The world is moving toward more and more transparency in actions, multilateral development agencies in the front to lead the world toward that goal. They are looking forward to publish the citizen-centric data online and encourage the civil society to participate in the policy-building or decision-making (Fig. 9). It moots the goals of each organization to activate the movement of transparent governance and collective decision-making.

References

1. Choo E, Yu T, Chi M (2015) Detecting opinion spammer groups through community discovery and sentiment analysis. In: IFIP annual conference on data and applications security and privacy. Springer International Publishing, pp 170–187
2. Dave K, Lawrence S, Pennock DM (2003) Mining the peanut gallery: opinion ex-traction and semantic classification of product reviews. In: Proceedings of the 12th international conference on World Wide Web. ACM, pp 519–528
3. Fang X, Zhan J (2015) Sentiment analysis using product review data. J Big Data 2(1):5
4. Pang B, Lee L, Vaithyanathan S (2002) Thumbs up?: sentiment classification using machine learning techniques. In: Proceedings of the ACL-02 conference on Empirical methods in natural language processing-Volume 10. Association for Computational Linguistics, pp 79–86
5. Hu M, Liu B (2004) Mining and summarizing customer reviews. In: Proceedings of the tenth ACM SIGKDD international conference on Knowledge discovery and data mining, pp 168–177

6. Liu B (2007) Web data mining: exploring hyperlinks, contents, and usage data. Springer Science & Business Media
7. Liu B (2015) Sentiment analysis: mining opinions, sentiments, and emotions. Cambridge University Press
8. Hoogervorst R, Essink E, Jansen W, van den Helder M, Schouten K, Frasincar F, Taboada M (2016) Aspect-based sentiment analysis on the web using rhetorical structure theory. In: International conference on web engineering. Springer International Publishing, pp 317–334
9. Popescu AM, Etzioni O (2007) Extracting product features and opinions from reviews. In: Natural language processing and text mining. Springer, London, pp 9–28
10. Turney PD (2002) Thumbs up or thumbs down?: semantic orientation applied to unsupervised classification of reviews. In: Proceedings of the 40th annual meeting on association for computational linguistics. Association for Computational Linguistics, pp 417–424
11. Open Calais. Retrieved Feb 16, 2017, from http://www.opencalais.com/
12. Extract Meaning from your Text. Retrieved Feb 16, 2017, from http://www.textrazor.com/
13. IBM Watson—AlchemyAPI. Retrieved Feb 16, 2017, from http://www.alchemyapi.com/
14. Text Analytics from Saplo. Retrieved Feb 12, 2017, from http://www.saplo.com/
15. The Qualitative Data Analysis & Research Software. Retrieved Feb 12, 2017, from http://www.atlasti.com/
16. NVivo product range | QSR International. Retrieved Feb 16, 2017, from http://www.qsrinternational.com/products-nvivo.aspx
17. Software ? Stanford Named Entity Recognizer (NER). Retrieved Feb 16, 2016, from http://nlp.stanford.edu/software/CRF-NER.shtml
18. Apache Tika, Retrieved March 16, 2016, from http://tika.apache.org/

An Efficient Context-Aware Music Recommendation Based on Emotion and Time Context

C. Selvi and E. Sivasankar

Abstract The enormous growth of Internet facilities, the user finds difficulties in choosing the music based on their current mindset. The context-aware recommendation has turned out to be well-established technique, which recommends a music based on the mindset of the user in various contexts. To enhance the potential of music recommendation, the emotion and time interval are considered as the most important context. Emotion context is not explored due to the difficulty in acquisition of emotions from user's microblogs on the particular music. This paper proposes an algorithm to extract the emotions of a user from microblog during a different time interval and represented at different granularity levels. Each music piece crawled from online YouTube repository is represented in a triplet format: (User_id, Emotion_vector, Music_id). These triplet associations are considered for developing several emotion-aware techniques to provide music recommendations. Several trial of experimentation demonstrates that the proposed method with user emotional context enhances the recommendation performance in terms of hit rate, precision, recall, and F1-measure.

Keywords Emotion-aware · Triplet association · Music-aware recommendation
Emotion analysis · Collaborative filtering

1 Introduction

With the easy use of smartphones and personal computers, posting microblogs frequently is a necessary part of many people's daily life. User post their opinion based on, what they have seen, heard, and thought about the music. Due to the overwhelmed

C. Selvi (✉) · E. Sivasankar
Department of Computer Science and Engineering, National Institute of Technology, Tiruchirappalli, Tiruchirappalli, Tamil Nadu, India
e-mail: selvichandran.it@gmail.com

E. Sivasankar
e-mail: sivasankarelango@gmail.com

D. K. Mishra et al. (eds.), *Data Science and Big Data Analytics*,
Lecture Notes on Data Engineering and Communications Technologies 16,
https://doi.org/10.1007/978-981-10-7641-1_18

music contents available on the Internet, it is very difficult for the user to choose the music based on their own interest. The largest online music stores such as Apple iTunes and Amazons MP3 have more than 20 million heterogeneous songs on their online repository. It makes the online user in confusion state to make the decision for their own preferences and also times consuming. There are many music Recommender System(RS)s proposed to provide the favorite music content to the target user with the reduced search time [1, 2]. Generally, user's own taste and various preferences are based on various contexts like location [3–5], time [6–8], activity [9–11], emotion [12–17], weather [18], and hybrid [19–21] context. Thus, to provide an efficient music recommendation, it is better to consider the various contexts of the user.

In context-aware recommendation, the target user's taste is analyzed based on user's past music history. Usually, music contents are described by emotions and there is a direct link between music emotion and user preferences [22]. But the emotional context of user's feeling on particular music is difficult to derive explicitly. Emotion-aware music RS is used to extract the emotion context from user's microblogs which contain user's personal opinion and preferences about the music on a particular time period. In [23], an emotion-aware music RS is proposed to extract the user's emotion from microblogs at different granularity levels in different time windows. Since the microblog gives the user's emotion about the music at the certain time either by implicitly or explicitly. In [17], an Affective RS (ARS) is developed by considering user behavior, facial expressions, user factors, mood, body gesture, senses, emotions, and physiological with human–computer interaction to provide the recommendation. None of the works focuses on how to extract the emotions in an effective manner from microblogs and used for providing recommendations.

In this work, we propose an algorithm that extracts emotions from microblogs at various granularity levels in the different time windows and how these emotions are used for prediction to provide an efficient recommendation to the target user. To validate the performance of the proposed emotion-aware RS, emotions at different granularity levels are extracted in different time windows. The experiment is conducted on online YouTube music dataset crawled from YouTube data API v3.

The main contributions of this work are as follows: (1) An emotions are extracted from microblog at different granularity levels in the different time intervals. (2) The extracted user's emotional contexts from their microblog are used to develop emotion-aware music recommendation methods based on traditional collaborative filtering (CF) memory-based approach. (3) Based on the prediction value, an appropriate music is recommended to the target user based on his/her current emotional context. (4) The extensive experimentation is conducted from user's coarse-grained emotions (2d) to fine-grained emotions (10d) in different time windows to improve the performance of music RS.

The remaining of the work is arranged as given below. The work related to context-aware RS is discussed in Sect. 2. Section 3 explains the proposed RS which includes the preprocessing algorithm that extracts emotions and prediction. The dataset used for experimental analysis, a design of experimentation, performance measures used for evaluating the proposed RS, and the analysis of results are clearly described in Sect. 4. Finally, Sect. 5 concludes the work with the summary and future work.

2 Related Work

Generally, people prefer to hear music on the sad situation than in happy situation. The music-aware RS mainly depends on emotion context of the user in a particular situation to provide the recommendation. The main RS algorithms are CF and content-based filtering. The CF algorithm works on the explicit rating given to the music, whereas the content-based filtering algorithm works by matching the characteristics of music with other music. The main goal of music-aware RS is to provide music that satisfies the user's current mindset. So emotion-aware music RS plays a significant role to attract attention to provide efficient music recommendation. This section deals with work related to context-aware music RS. Music-aware RS has been proposed in the various contexts like the environment and user-based approach.

Sometimes, the user's emotional behavior may be influenced by context related to environment and therefore the music suggestions were provided to the target user based on the environment [22]. Generally, people prefer to hear songs based on the weather conditions like summer and winter [24]. Thus, the music-aware RS based on environment-related context performs better. Some of the environment-related context includes time [6–8], location [3–5], climatic conditions [18], and hybrid context [19–21]. The performance of music-aware RS is enhanced in [7] by including temporal information in session-based CF method. Also, the work was proposed in [4] based on the Places Of Interests (POIs) that the target user is visiting, and the music is recommended to the user with maximum possibilities. The mobile music RS was proposed in [25] which suggest music playlist based on the present context of the target user such as weather condition, location, and noise or traffic level. Another context-aware music RS was proposed to use contexts such as time, light level, weather, and noise to provide recommendations [18]. The main drawback of the environment-based music-aware RS is, the researcher who has to label the music with contextual information to process it further and it is applicable only for smaller scale dataset and time-consuming process.

In user-based context method, the user-related context is having the closer relationship with user's opinion/feeling on a music piece. So it provides the satisfiable music suggestions to the user based on their preferences. The context related to user includes demographical information, emotions [12–16], and activity [9, 10]. The music-aware RS was developed in [26] based on user's activity context and the suggestions are provided by matching the user's walking steps-per-minute with the music tracks Beats-Per-Minute (BPM). In [15], the music similarity between the vectors is calculated by considering vector comprising emotional context. Those emotional context information of music include harmonics, rhythm, and scale. The recommendations are provided in [14] based on the user's present emotional state and emotional changes. The emotion match between the web document read by the user and the emotions of music has been done in [12], and the suggestions were provided to the target user. All the existing related works are based on the input contextual information and restricted based on its application. In [23], the authors have extracted the user's emotion at various granularity levels in the different time intervals that influence music recommendations. None of the work has been done to extract emotions of a user in an efficient way to provide recommendations.

3 Proposed Methodology

The proposed RS consists of two phases: preprocessing and prediction. In preprocessing stage, an algorithm says how the emotions are extracted and the triplets (User_id, Emotion_vector, Music_id) are framed from the user's microblog at different granularity levels in a different time interval. At the prediction phase, based on the user's present emotional state, the ratings of music are predicted and the recommendations are made accordingly.

3.1 Preprocessing

The main aim of the preprocessing stage is to form triplet tuple (User_id, Emotion_vector, Music_id) from the user's microblog. Table 1 shows the microblog of the single user at various time periods for the music "entertainment". This microblog contains the lines of the sentence with emotions, normal text like URL and music title, user-generated tags, etc. Each sentence in the microblog represents the opinion or feelings of the user about the music at a particular time period.

3.1.1 Triplet Emotion Tuple Identification from Microblog

Initially, the user's microblog is collected from online YouTube music repository. The collected microblogs are processed further to extract the emotions of a user by the following steps given in proposed Algorithm 1.

Table 1 Example microblog

User_ ID	Text content	Time period	Music name/ID
531325526	[smile][smile]so eager	2015-04-02 03: 15: 19	Entertainment
531325526	Very good day	2015-04-02 03: 18: 26	
531325526	Entertainment music is shared form Last.fm, http://x.yz/RWNSdQs	2015-04-02 03: 30: 16	

Algorithm 1 Proposed algorithm to extract emotions and finding emotion vectors for a microblog

1: Collect user's microblog from online YouTube repository.
2: Construct a parse tree for each sentence in a microblog.
3: Extract the emotion 1-gram words adverb and adjective from the parse-tree and create a emotion_list .
4: Let nd-emotion={emotion_word_1, emotion_word_2, emotion_word_3, ..., emotion_word_n}
5: **for** i=1 to n **do**
6: $Count_i$=0
7: **for** emotion ∈ emotion_list **do**
8: **for** i=1 to n **do**
9: **if** emotion = $nd - emotion[i]$ **then**
10: $Count_i$+=1
11: **if** emotion ∈ WordNet synset library **then**
12: Repeat step 8 to 10

Initially, the microblogs are extracted from the online music repository. Since the collected microblog contains user's feelings at different time periods, the sentences in the microblog are processed and finally combined. The emotions from each microblog are collected by constructing a parse tree. The emotions in the parse tree are considered as adverb and adjective. For each microblog, the emotion_list is framed. These steps are represented from steps 1–4 in Algorithm 1. For nd-emotion in Algorithm 1, the 2d-emotion vector is considered as (Positive, Negative) and the 10d-emotion vector is considered as (Joy, Surprise, Trust, Constructive, Anger, Anticipation, Disgust, Fear, Sadness, Pessimistic) and it is given in Table 2. For each emotion in the emotion_list, the count of number of emotion word in each nd-emotion type is identified, and the vectors are framed as the steps 8–10 in Algorithm 1. If the emotion word is not in the nd_emotion, then the synonym of emotion word is identified using WordNet synset library [27] and the steps 8–10 in Algorithm 1 are carried out again.

Table 2 Emotion granularity levels

2d-emotion	10d-emotion
1. Positive	1. Joy
	2. Surprise
	3. Trust
	4. Constructive
2. Negative	5. Anger
	6. Anticipation
	7. Disgust
	8. Fear
	9. Sadness
	10. Pessimistic

Based on the example microblog given in Table 1 and the different emotion granularity levels as in Table 2, the emotion vectors are extracted. The emotions extracted from example microblog are smile, eager, and good. The 2d-emotion vector is (4, 0) and the 10d-emotion vector is (2, 1, 1, 0, 0, 0, 0, 0, 0, 0). The triplet tuple for the example microblog is (531325526, ((4, 0), (2, 1, 1, 0, 0, 0, 0, 0, 0, 0)), Entertainment).

3.2 Prediction

In the prediction phase, first, the similarity between the target user with all other users are identified from the extracted emotional vectors. Second, the similarity values are arranged in descending order and the top k similar users with the target users are taken for prediction. Finally, the value of a music for target user is identified based on the extracted current emotional state of the target user and the retrieved top k similar users. Generally, the music piece with largest prediction value is recommended first and so on. The subsection shows the CF methods for calculating similarity and the prediction method for calculating the rating of music using traditional user and music-based CF methods.

3.2.1 User-Based CF with Emotion(UCFE)

Traditional user-based CF [23] find the similarity between the target user emotion vector with all other user's emotion vector and selects the subset of users who have top similarity values. Then, the interest of target user's music values is predicted based on the current emotions of target user by using the top similar users. The formula for calculating the similarity among the users is given in Eq. 1.

$$sim(l, m) = \frac{\sum\limits_{p \in P_l \cap P_m} cos(e_{lp}, e_{mp})}{\sqrt{|P_l| \times |P_m|}} \tag{1}$$

where

- l is the target user and m is the other user in music dataset.
- P_l is the set of music heard by user l, while P_m is the set of music heard by user m.
- e_{lp} is an emotional context of user l when hearing the music p, while e_{mp} is an emotional context of user m when hearing the music p.
- $cos(e_{lp}, e_{mp})$ represents the cosine similarity value between two emotional vector e_{lp} and e_{mp}.

Based on the calculated similarity value using Eq. 1, the prediction formula is represented in Eq. 2.

$$pre(l, p) = \sum_{m \in L_{l,k} \cap L_p} sim(l, m) \times cos(e_l, e_{mp}) \tag{2}$$

where

- l is the target user.
- $L_{l,k}$ is the set of top k users close to l.
- L_p is set of users heard the music p.
- e_l is the present emotion vector of target user l and e_{mp} represents an emotional context of user m when hearing the music p.

3.2.2 Music-Based CF with Emotion (MCFE)

Traditional music-based CF method [23] computes the similarity between the target music with all other music emotion vectors. Then, the top k similar music pieces are extracted, and the predictions are made to the target user's music piece and the recommendations are made. The formula for calculating the context-aware music-based CF is given in Eq. 3.

$$sim(p, q) = \frac{\sum_{l \in L_p \cap L_q} cos(e_{lp}, e_{lq})}{\sqrt{|L_p| \times |L_q|}} \tag{3}$$

where

- p and q are the music pieces.
- L_p is set of users heard the music p, while L_q is set of users heard the music q.
- e_{lp} represents an emotional context of user l when hearing the music p and e_{lq} represents an emotional context of user l when hearing the music q.
  Based on the calculated music-based similarity value using the Eq. 3, the prediction formula is defined in Eq. 4.

$$pre(l, p) = \sum_{q \in P_{p,k} \cap P_l} sim(p, q) \times cos(e_l, e_{lq}) \tag{4}$$

where

- l is the target user.
- $P_{p,k}$ is set of top k music piece similar to p.
- P_l is set of music piece heard by l.
- e_l is the present emotion vector of target user l and e_{lq} represents an emotional context of user l when hearing the music q.

4 Experimental Evaluation

This section is described with the dataset details and the performance analysis of proposed context-aware RS with traditional CF algorithms (User-based CF (UCF) and Music-based CF (MCF)) by considering with and without emotions. The experimental analysis is done in Python language.

4.1 Dataset Description

To evaluate the performance of proposed emotion-aware RS, microblogs from YouTube data API v3 are crawled. The experiment dataset contains the average of 5000 online users and 2700 music pieces in the triplet form (User_id, Emotion_vector, Music_id).

4.2 Experimental Design and Performance Measure

The performance of proposed emotion-aware music RS is evaluated based on the top *N* recommendation provided to the target user. At first, the music dataset is divided into, training and testing datasets, by using 10 fold cross-validation method. The test dataset is used to validate the performance of RS method based on the model developed using training dataset. Second, the traditional CF methods are used and few combinations of experimentation are conducted to compare them. At last, the evaluation measures are used to measure the recommendation results. The evaluation measures used for the result analysis include the following.

4.2.1 Hit Rate

Hit rate represents the fraction of hits. That is, a number of music pieces in the recommended list satisfy the target user's interest under the present emotional context. For example, triplet for test user u_1 is (u_1, e_{u_1}, p). If the recommended music list (N) of user u_1 contains the music p under the emotional context e_{u_1}, then it is a hit. If not, it is not a hit. The Eq. 5 represents the definition of hit rate.

$$\text{Hit rate} = \frac{Number\ of\ hits}{N} \tag{5}$$

where

- N times of recommendation.

4.2.2 Precision

Precision is a percentage of recommended music piece that is relevant. The formula for precision is given in Eq. 6.

$$Precision = \frac{\sum_{l \in L} |R(l) \cap T(l)|}{\sum_{l \in L} |R(l)|} \tag{6}$$

where

- $R(l)$ is the music list that is recommended to user l.
- $T(l)$ is the recommended music list to user l, which is listened by l in the test data.

4.2.3 Recall

The recall is a percentage of the relevant music piece that is recommended. The definition of a recall is given in Eq. 7.

$$Recall = \frac{\sum_{l \in L} |R(l) \cap T(l)|}{\sum_{l \in L} |T(l)|} \tag{7}$$

4.2.4 F1-Measure

F1-measure is the harmonic mean of precision and recall. It is defined in Eq. 8.

$$\text{F1-measure} = 2 \times \frac{Precision \times Recall}{Precision + Recall} \tag{8}$$

4.3 *Experimental Analysis of Proposed Method Using Traditional CF Methods*

Two ways of the traditional CF methods are compared using the proposed CF method: based on time window and based on a number of nearest neighbors. The traditional CF methods UCF and MCF methods are considered, and experiment is done without and with considering emotions under 2d and 10d granularity levels. UCF with 2d and 10d emotional context is represented as UCFE-2d and UCFE-10d respectively. Similarly MCF with 2d and 10d emotional context is represented as MCFE- 2d and

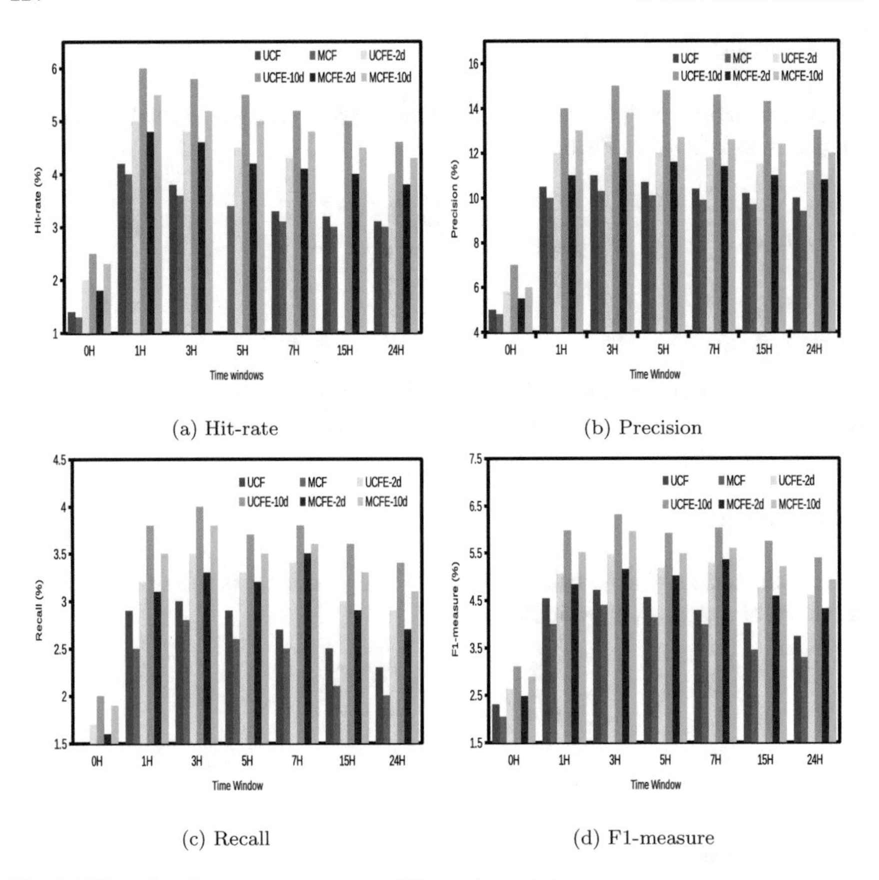

(a) Hit-rate (b) Precision

(c) Recall (d) F1-measure

Fig. 1 Effect of performance measures on different time windows

MCFE-10d respectively. Figures 1, 2, and 3 show the results obtained on various comparison levels.

Figure 1 shows the performance effect of CF methods under different time windows. From Fig. 1(a) to Fig. 1(d) represents the result of evaluation measures hit rate, precision, recall, and F1-measure, respectively. The experimentation is conducted from 0–24h duration with the certain interval. Generally, user's taste on the music piece is expressed at first when they start hearing the music. We found that the evaluation measures yields improved results in the early time duration between 1–7h and results are reduced in the later time period after hour 7. This is due to 95% of user's emotional context that is represented at the earlier stage and remaining 5% of emotions are in the later stage. This remaining 5% of data will not affect the result much. From Fig. 1, it is clearly understood that the proposed UCF on fine-grained 10d-emotions gives improved result than the MCFE-10d method. Since the emotional context of a user has direct linkage with emotions of a music piece, same

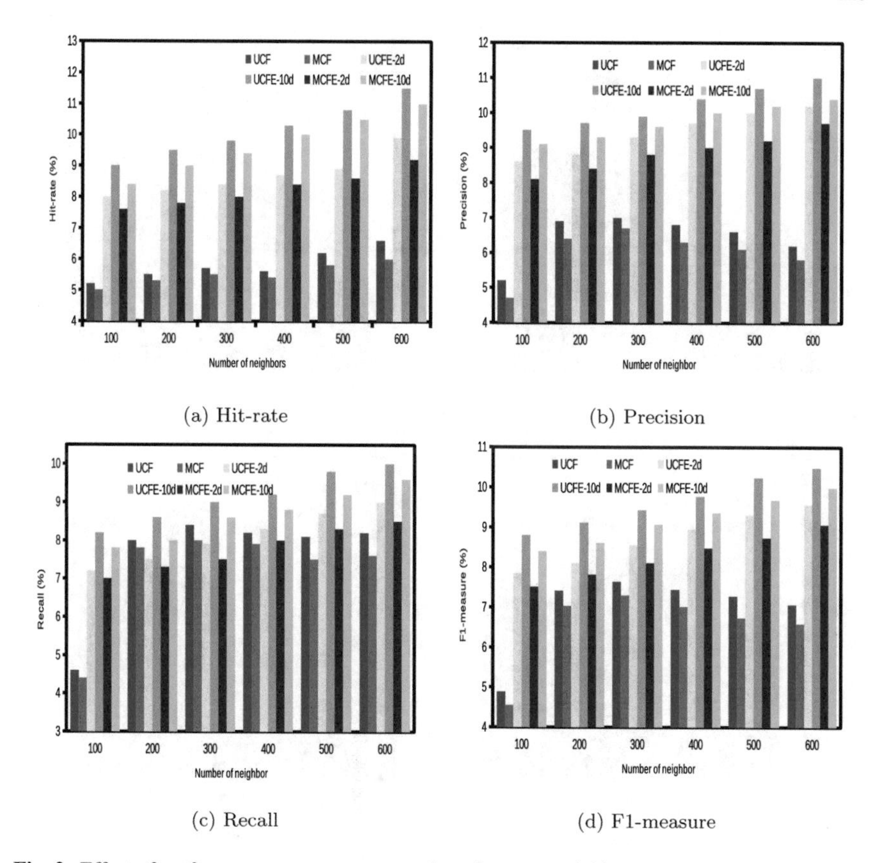

(a) Hit-rate

(b) Precision

(c) Recall

(d) F1-measure

Fig. 2 Effect of performance measures on number of nearest neighbors

way, the proposed UCF on coarse-grained 2d emotions obtains improved result than MCFE-2d method.

These proposed UCF and MCF methods with 2d- and 10d-emotions perform better than traditional UCF and MCF methods. Figure 1 clearly concludes that the performance of CF methods with considering user's emotions on a music piece gives enhanced result than the CF methods without considering emotions. Overall, the results from Fig. 1 clearly shows the time window 3H, and the proposed UCFE-10d gives improved performance result than all other methods taken for comparison.

Figure 2 represents the experimental result on various evaluation measures under different nearest neighbors considered. The results of evaluation measures hit rate, precision, recall, and F1-measure are shown in Fig. 2(a), Fig. 2(b), Fig. 2(c), and Fig. 2(d), respectively. The number of nearest neighbors is ranging from 100 to 600 with the interval of 100. Since the analysis of dataset gives the average number of music piece listened by each user as 638, the interval of 100 is taken to show the high-performance difference among considered neighbors.

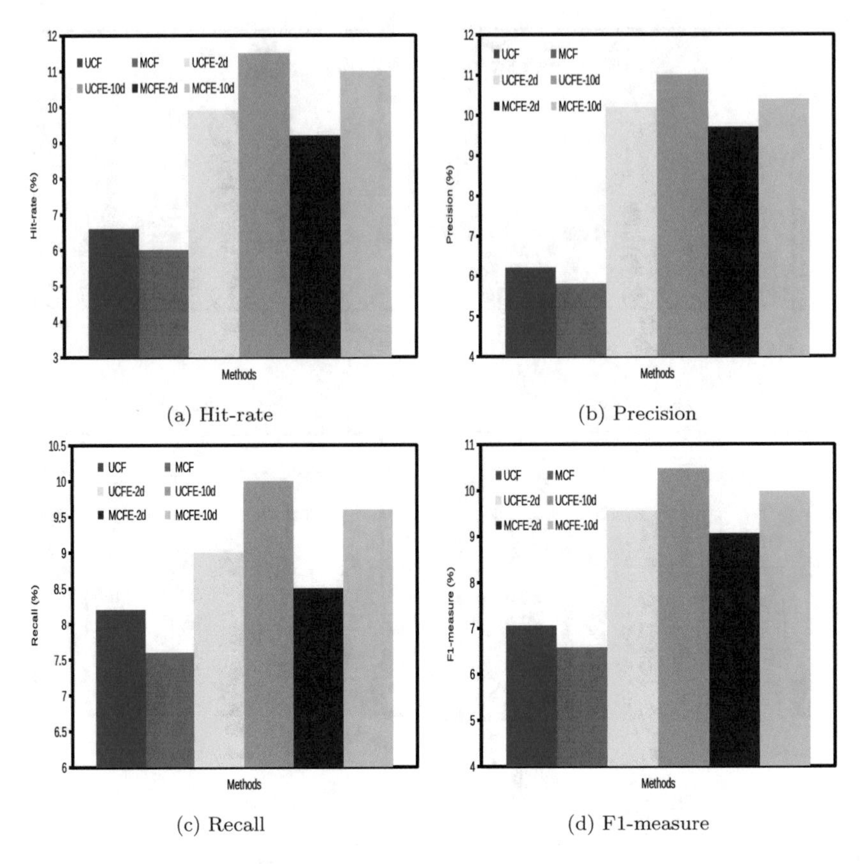

Fig. 3 Comparison of CF methods

From Fig. 2, it is clearly observed that all the considered evaluation measures show improved result for the maximum number of neighbors 600, because the similarity among the more number of nearest neighbors is considered for prediction, which will automatically satisfy the target user's taste. After 600 users, the results are started reducing since the similarity among the users is not strong. From Fig. 2(a) to Fig. 2(d), it is clearly understood that the UCF under 10d fine-grained emotional vector performs better than MCFE-10d, coarse-grained UCFE-2d, MCFE-2d, traditional measures UCF and MCF. Overall from Fig. 2, it is concluded that the UCF performs better than other methods taken for experimental analysis.

Figure 3 shows the overall comparison of CF methods taken for experimental analysis. Figure 3(a), 3(b), 3(c) and 3(d) represents the results of performance measures hit rate, precision, recall, and F1-measure, respectively. From Fig. 3(a) to Fig. 3(d), it is clearly understood that the fine-grained emotions with 10d-emotional vector with UCF performs better than other emotional vectors considered. And, also it is observed that the UCFE-10d and UCFE-2d emotions yield improved result than MCFE-10d and MCFE-2d methods, respectively. Overall, fine-grained emotional vector UCF is considered as a better emotion-aware music RS under the dataset which is considered for analysis.

5 Conclusion and Future Work

Context-aware Recommender System (RS) is considered as one of the best RSs in the various contexts like location, activity, time, emotion, etc. The recommendation based on music is having the direct link with the emotion of the target user. Though the lot of work has been done already, they are still lagging to extract meaningful emotions from the user's microblog. The effective preprocessing algorithm is proposed that extracts meaningful emotions under different granularity levels (2d and 10d) in different time windows. These extracted user's emotional vectors are used to provide recommendations to the target user based on its present emotional context. The proposed algorithm is tested with traditional Collaborative Filtering (CF) methods with and without considering the user's emotional vector. The performance measures hit ratio, precision, recall, and F1-measure of proposed methods are tested on the real-world YouTube music dataset that is crawled online. From the results, it is observed that the user-based fine-grained CF method with 10d-emotional vector outperforms than coarse-grained 2d-emotional vector. Overall, 10d-emotions with user-based CF method is considered as the better method for context-aware RS than other CF methods considered for experimentation. The work will be extended further, which deals about converting the emotion vectors into rating to enhance the performance of RS.

References

1. Ziwon H, Kibeom L, Kyogu L (2014) Music recommendation using text analysis on song requests to radio stations. Expert Syst Appl 41(5):2608–2618
2. Jose A, Mocholi VM, Jaen J, Catala A (2012) A multicriteria ant colony algorithm for generating music playlists. Expert Syst Appl 39(3):2270–2278
3. Cheng Z, Shen J (2014) Just-for-me: an adaptive personalization system for location-aware social music recommendation. In: Proceedings of international conference on multimedia retrieval, pp 185. ACM
4. Kaminskas M, Ricci F, Schedl M (2013) Location-aware music recommendation using auto-tagging and hybrid matching. In: Proceedings of the 7th ACM conference on recommender systems, pp 17–24. ACM
5. Schedl M, Vall A, Farrahi K (2014) User geospatial context for music recommendation in microblogs. In: Proceedings of the 37th international ACM SIGIR conference on research & development in information retrieval, pp 987–990. ACM
6. Cebrián T, Planagumà M, Villegas P, Amatriain X (2010) Music recommendations with temporal context awareness. In: Proceedings of the fourth ACM conference on recommender systems, pp 349–352. ACM

7. Dias R, Fonseca MJ (2013) Improving music recommendation in session-based collaborative filtering by using temporal context. In: 2013 IEEE 25th international conference on tools with artificial intelligence (ICTAI), pp 783–788. IEEE
8. Su J-H, Yeh H-H, Yu Philip S, Tseng VS (2010) Music recommendation using content and context information mining. IEEE Intel Syst 25(1)
9. Ning-Han L, Szu-Wei L, Chien-Yi C, Shu-Ju H (2009) Adaptive music recommendation based on user behavior in time slot. Int J Comput Sci Netw Security 9(2):219–227
10. Wang X, Rosenblum D, Wang Y (2012) Context-aware mobile music recommendation for daily activities. In: Proceedings of the 20th ACM international conference on multimedia, pp 99–108. ACM
11. Lee W-P, Chen C-T, Huang J-Y, Liang J-Y (2017) A smartphone-based activity-aware system for music streaming recommendation. Knowl Based Syst
12. Cai R, Zhang C, Wang C, Zhang L, Ma W-Y (2007) Musicsense: contextual music recommendation using emotional allocation modeling. In: Proceedings of the 15th ACM international conference on multimedia, pp 553–556. ACM
13. Chang C-Y, Lo C-Y, Wang C-J, Chung P-C (2010) A music recommendation system with consideration of personal emotion. In: 2010 international computer symposium (ICS), pp 18–23. IEEE
14. Rho S, Han B, Sanghoon J, Hwang E (2010) Music emotion classification and context-based music recommendation. Multimed Tools Appl 47(3):433–460
15. Rho S, Han B, Hwang E (2009) SVR-based music mood classification and context-based music recommendation. In: Proceedings of the 17th ACM international conference on multimedia, pp 713–716. ACM
16. Man-Kwan S, Fang-Fei K, Meng-Fen C, Suh-Yin L (2009) Emotion-based music recommendation by affinity discovery from film music. Expert Syst Appl 36(4):7666–7674
17. Katarya R, Verma OP (2016) Recent developments in affective recommender systems. Physica A: Stat Mech Appl 461:182–190
18. Park H-S, Yoo J-O, Cho S-B (2006) A context-aware music recommendation system using fuzzy Bayesian networks with utility theory. In: International conference on fuzzy systems and knowledge discovery, pp 970–979. Springer
19. Hong J, Hwang W-S, Kim J-H, Kim S-W (2014) Context-aware music recommendation in mobile smart devices. In: Proceedings of the 29th annual ACM symposium on applied computing, pp 1463–1468. ACM
20. Knees P, Schedl M (2013) A survey of music similarity and recommendation from music context data. ACM Trans Multimed Comput Commun Appl (TOMM), 10(1):2
21. Lee JS, Lee JC (2007) Context awareness by case-based reasoning in a music recommendation system. In: International symposium on ubiquitious computing systems, pp 45–58. Springer
22. North A, Hargreaves D (2008) The social and applied psychology of music. Oxford University Press
23. Shuiguang D, Dongjing W, Xitong L, Xu G (2015) Exploring user emotion in microblogs for music recommendation. Expert Syst Appl 42(23):9284–9293
24. Pettijohn TF, Williams GM, Carter TC (2010) Music for the seasons: seasonal music preferences in college students. Current Psychol 29(4):328–345
25. Reddy S, Mascia J (2006) Lifetrak: music in tune with your life. In: Proceedings of the 1st ACM international workshop on human-centered multimedia, pp 25–34. ACM
26. Elliott GT, Tomlinson B (2006) Personalsoundtrack: context-aware playlists that adapt to user pace. In: CHI '06 extended abstracts on human factors in computing systems, pp 736–741. ACM
27. Miller GA (1995) Wordnet: a lexical database for english. Commun ACM 38(11):39–41

Implementation of Improved Energy-Efficient FIR Filter Using Reversible Logic

Lavisha Sahu, Umesh Kumar and Lajwanti Singh

Abstract The demand for high-speed processing has been increasing as a result of expanding computer and signal processing applications. Nowadays reducing the time delay and power consumption main factor of the circuit. One of the main advantage of reversible logic gates is to reduce the heat dissipation and improve the performance of circuit. Reversible logic gate is used for building complex circuits like multiplier, adder, FIR, and much more and reduce heat dissipation. FIR (finite impulse response) filter is used in various range of digital signal processing applications. This paper describes reversible Vedic FIR filter and compared with irreversible Vedic FIR filter.

Keywords Reversible logic · FIR · CNOT · HNG · PERES gates · Low power Multiplier

1 Introduction

In signal processing, a filter response is used for finite duration and pass low power signal is known as FIR (finite impulse response) filter. As the technology increase

L. Sahu (✉)
Department of Computer Science and Engineering, Government Women Engineering College, Ajmer, India
e-mail: sahulavisha@gmail.com

U. Kumar
Department of Information Technology and Engineering, Government Women Engineering College, Ajmer, India
e-mail: ume2222@gmail.com

L. Singh
Department of Electronics and Communication Engineering, Banasthali Vidyapith, Newai, Tonk, India
e-mail: urslaj@gmail.com

D. K. Mishra et al. (eds.), *Data Science and Big Data Analytics*,
Lecture Notes on Data Engineering and Communications Technologies 16,
https://doi.org/10.1007/978-981-10-7641-1_19

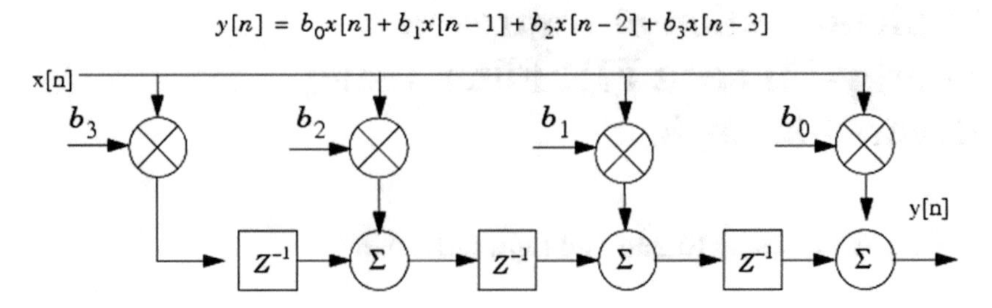

Fig. 1 *N*th order discrete time FIR filter

according to Moore's law [1, 2] the number of transistors per square inch on integrated circuits had doubled every year since their invention. It leads to increase the dissipation of heat resulting degradation in performance so as to maintain this heat dissipation reversible logic can be used and according to Launder's [3] research "the amount of energy dissipated by every irreversible bit operation is at least KTln2 joules, where $k = 1.3806505 * 10^{23}$ 23 m^2 kg^{-2} K^{-1} (J/K^{-1}) is the Boltzmann's constant and T is the temperature [4] at which operation is performed".

In this paper Verilog implementation of 4-bit FIR filter implemented using reversible logic [5] and Vedic multiplier [6]. In 4-tab FIR filter having four coefficient b[0], b [1], b [2], b [3]. FIR filter equation can be expressed as

$$\mathrm{Y[N]} = \sum_{i=0}^{N} b_i \mathrm{X[i]} \tag{1}$$

- "Y" is the output
- "X" is the input signal
- "b" is the filter coefficient
- N is the filter order (Fig. 1).

2 Vedic FIR Filter Using Irreversible Logic

FIR filter [7] is implemented using Vedic mathematics for achieving higher speed performance. Urdhva Tiryakbhayam sutra is used to implement the Vedic multiplier.

2.1 Urdhva Tiryakbhayam

"Urdhva" and "Tiryakbhayam" sutra comes from Sanskrit literature which means "cross and vertical" [8]. This method generates all partial products and sums and it generalized up to N * N multipliers.

2.2 *Reversible Urdhva Tiryakbhayam Multiplier*

A basic block diagram of 2 ∗ 2 multiplier is as shown in Fig. 2. In this diagram a0, a1 are the bits of first digit and b0, b1 are the bits of second digit. In this multiplier [9] we are taking half adder of respective bits as shown in block diagram. The multiplication is done on the sutra of Urdhva Tiryakbhayam. The result obtained is of 4 bits (Fig. 3).

Digital logic of 2 * 2 Vedic multiplier is implemented using Urdhva Tiryakbhayam sutra. Vedic 16 * 16-bit multiplier is built using 2 * 2 Vedic multipliers. In 4-tab FIR filter implementation we use the 16 * 16-bit Vedic Multipliers and 8-bit, 16 bit, 24-bit Adders.

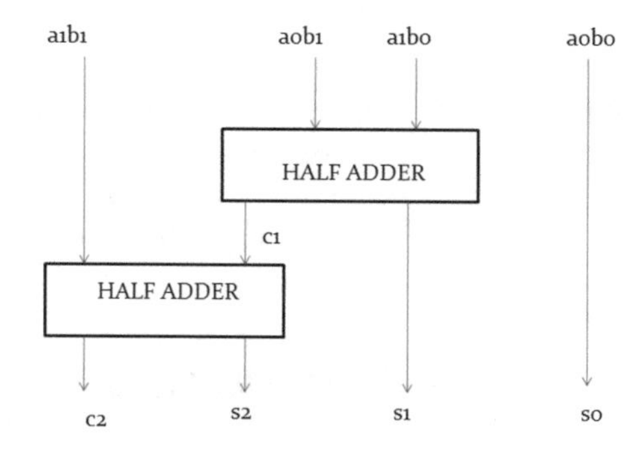

Fig. 2 Block diagram of 2 ∗ 2 vedic multiplier

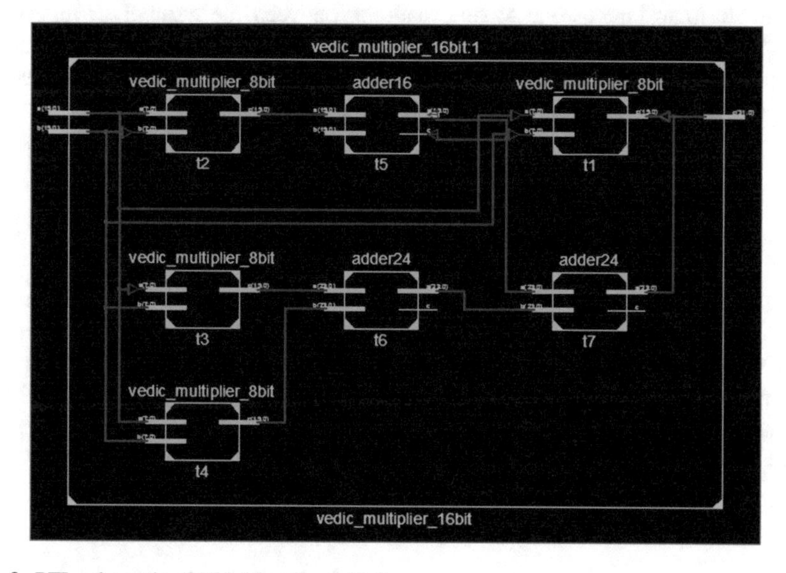

Fig. 3 RTL schematic of 16 * 16 vedic multiplier

2.3 Ripple Carry Adder

In multipliers basic building is adder so in that procedure we specify the ripple carry adder using a series of full adder. In ripple carry adder 2 outputs are generated one is sum and another is carry as follows:

$$s_i = \text{a xor b xor c;}$$
$$c_{i+1} = \text{ab + bc + ca;}$$

3 Reversible Vedic FIR Filter

Reversible Vedic FIR filter is implemented through Vedic multiplier, adder and delay using reversible logic gate. In this paper we describe the Peres gate, CNOT gate for 2 * 2 reversible Vedic multiplier because it is best suitable based on quantum cost and ancillary inputs and HNG gate is used for ripple carry adder.

3.1 Reversible Logic Gates Used for Multipliers

Reversible computing is required so as to reduce the power dissipation of the circuit. Reversible logic having one to one mapping between the number of inputs and outputs. Reversible circuit achieving the reversibility having the functionality outputs and garbage outputs. Some of the required reversible logic [10] gates are described peres Gate, CNOT Gate, HNG Gate [11], etc. (Tables 1 and 2).

3.2 Reversible Logic Gates Used for Adders

In reversible ripple carry adder all the full adders are replicated through reversible HNG gate. Ripple carry [12] is sequence of standard full adder in irreversible logic we can show in Fig. 4. In Table 3 it can be clearly seen that HNG gates have minimum garbage and gate counts. 32-bit adder is implemented using Toffoli [13] gate and this adder is used when we implement the 16-bit 4-tab FIR filter (Figs. 5, 6 and 7; Table 4).

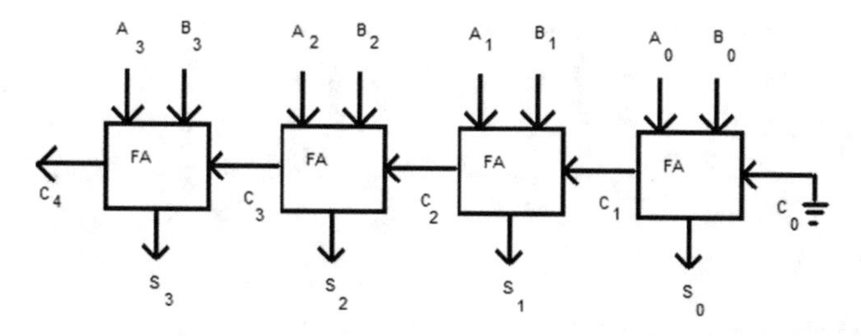

Fig. 4 Ripple carry adder using full adder

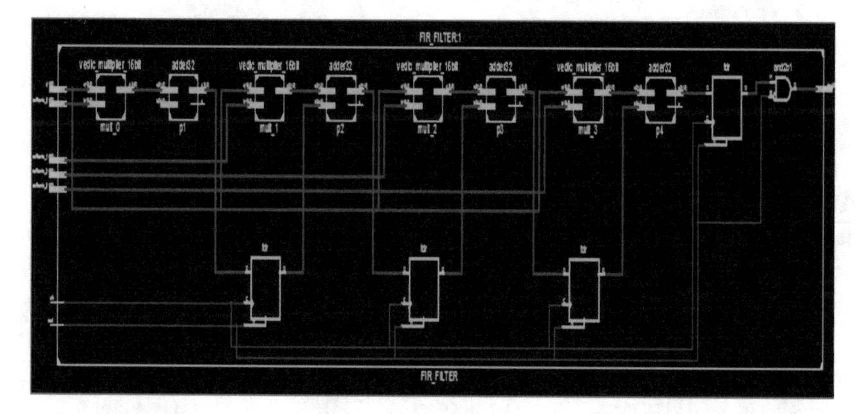

Fig. 5 RTL schematic of irreversible vedic FIR filter

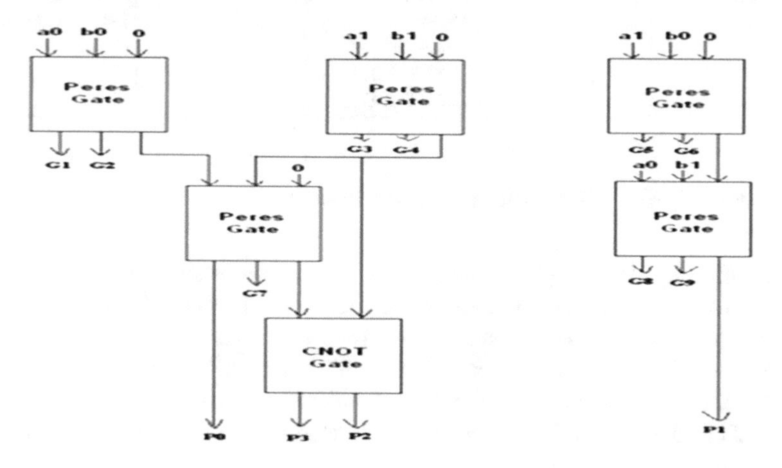

Fig. 6 2 * 2 reversible vedic multiplier

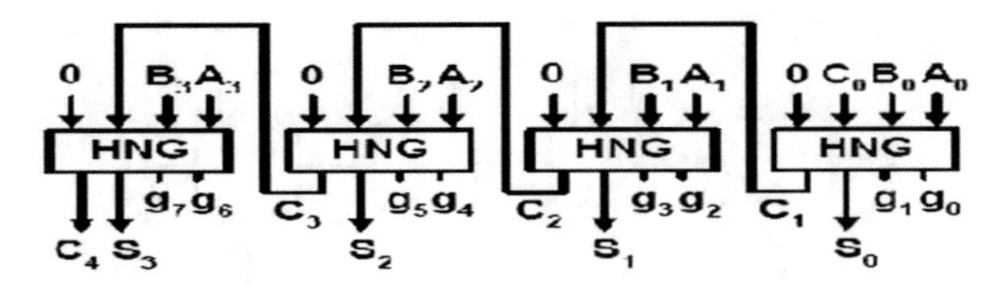

Fig. 7 Reversible adder using HNG gate

Table 1 Peres gate

Reversible logic	Gate specification	Expression	Q.C	Feature
Peres gate/NTG	3 * 3	$L = A$ $M = A \oplus B$ $N = AB \oplus C$	4	Lowest quantum cost

Table 2 CNOT Gate

Reversible logic	Gate specification	Expression	Q.C	Feature
CNOT/Feynman gate	2 * 2	$L = A$ $M = A \oplus B$	1	Copying gate, fan-out gate

Table 3 HNG gate

Reversible logic	Gate spec.	Expression	Q.C	Feature
HNG gate	4 * 4	$L = A$ $M = B$ $N = A \oplus B \oplus C$ $O = (A \oplus B)$ $C \oplus AB \oplus D$	6	Minimize the garbage and gate counts

4 Results, Analysis, and Comparison

Power consumption [14] is important factor in battery-powered devices, considering this fact in mind this paper describes about use of reversible logic to reduce power consumption. Power consumption of proposed system (FIR filter using Vedic reversible logics) is better than Irreversible FIR filter (FIR filter using Vedic irreversible logics) shown in Figs. 10 and 11. Synthesis report showing that number of slice LUT and number of occupied slice reduce compare to existing system design

Table 4 Toffoli gate

Reversible logic	Gate spec.	Expression	Q.C	Feature
Toffoli gate/CCNOT gate	3 * 3	$L = A$ $M = B$ $N = AB \oplus C$	5	Universal reversible logic gate

Device Utilization Summary			
Slice Logic Utilization	Used	Available	Utilization
Number of Slice Registers	128	12,480	1%
Number used as Flip Flops	128		
Number of Slice LUTs	2,270	12,480	18%
Number used as logic	2,270	12,480	18%
Number using O6 output only	1,681		
Number using O5 and O6	589		
Number of occupied Slices	943	3,120	30%
Number of LUT Flip Flop pairs used	2,271		
Number with an unused Flip Flop	2,143	2,271	94%
Number with an unused LUT	1	2,271	1%
Number of fully used LUT-FF pairs	127	2,271	5%
Number of unique control sets	1		
Number of slice register sites lost to control set restrictions	0	12,480	0%
Number of bonded IOBs	114	172	66%
Number of BUFG/BUFGCTRLs	1	32	3%
Number used as BUFGs	1		
Average Fanout of Non-Clock Nets	3.95		

Fig. 8 Synthesis report of vedic FIR using reversible

as well Average Fan-out of Non-Clock nets is also found better than the Irreversible FIR filter shown in Figs. 8 and 9. Simulation results of reversible 16-bit FIR Filter are shown in Fig. 12.

5 Conclusion

Reversible logic based FIR filter required less power consumption and reduce number of occupied slice, number of slice LUT compare to existing system design (Vedic FIR Filter using irreversible logic). We can see from Figs. 10 and 11 that power consumption is reduced from 0.327 to 0.324 W and dynamic power supply also reduce from 0.006 to 0.003 (Fig. 12).

Average fan-out of non-clock nets is reduced from existing design (5.28) to proposed design (3.95). In future reversible FIR filter is used to improve the performance

Device Utilization Summary			
Slice Logic Utilization	Used	Available	Utilization
Number of Slice Registers	128	12,480	1%
Number used as Flip Flops	128		
Number of Slice LUTs	2,441	12,480	19%
Number used as logic	2,441	12,480	19%
Number using O6 output only	2,441		
Number of occupied Slices	1,132	3,120	36%
Number of LUT Flip Flop pairs used	2,441		
Number with an unused Flip Flop	2,313	2,441	94%
Number with an unused LUT	0	2,441	0%
Number of fully used LUT-FF pairs	128	2,441	5%
Number of unique control sets	1		
Number of slice register sites lost to control set restrictions	0	12,480	0%
Number of bonded IOBs	114	172	66%
Number of BUFG/BUFGCTRLs	1	32	3%
Number used as BUFGs	1		
Average Fanout of Non-Clock Nets	5.28		

Fig. 9 Synthesis report of vedic FIR using irreversible logic

of the system and reducing the power consumption of battery powered system. Synthesis and simulation report is generated through Xilinx 13.4 and power consumption also generated through power analyzer report from power supply to system.

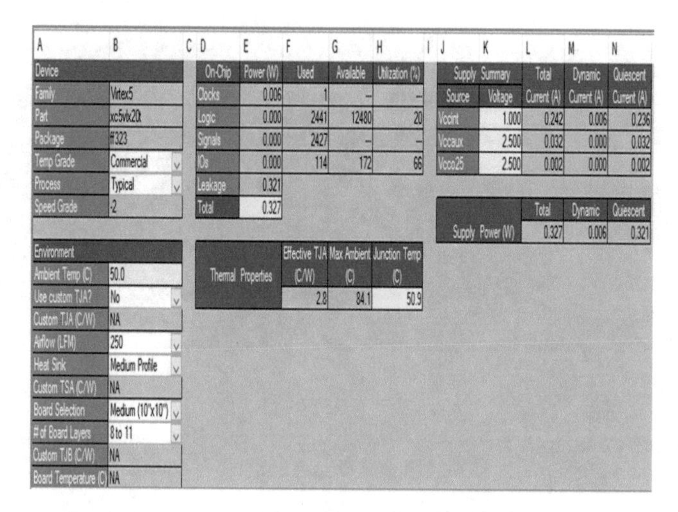

Device	
Family	Virtex5
Part	xc5vlx20t
Package	ff323
Temp Grade	Commercial
Process	Typical
Speed Grade	-2

Environment	
Ambient Temp (C)	50.0
Use custom TJA?	No
Custom TJA (C/W)	NA
Airflow (LFM)	250
Heat Sink	Medium Profile
Custom TSA (C/W)	NA
Board Selection	Medium (10"x10")
# of Board Layers	8 to 11
Custom TJB (C/W)	NA
Board Temperature (C)	NA

On-Chip	Power (W)	Used	Available	Utilization (%)
Clocks	0.006	1	–	–
Logic	0.000	2441	12480	20
Signals	0.000	2427	–	–
IOs	0.000	114	172	66
Leakage	0.321			
Total	0.327			

Thermal Properties	Effective TJA (C/W)	Max Ambient (C)	Junction Temp (C)
	2.8	84.1	50.9

Supply Summary Source	Voltage	Total Current (A)	Dynamic Current (A)	Quiescent Current (A)
Vccint	1.000	0.242	0.006	0.236
Vccaux	2.500	0.032	0.000	0.032
Vcco25	2.500	0.002	0.000	0.002

	Total	Dynamic	Quiescent
Supply Power (W)	0.327	0.006	0.321

Fig. 10 Power analyzer of 16-bit vedic FIR using irreversible logic

Device	
Family	Virtex5
Part	xc5vlx20t
Package	ff323
Temp Grade	Commercial
Process	Typical
Speed Grade	-2

Environment	
Ambient Temp (C)	50.0
Use custom TJA?	No
Custom TJA (C/W)	NA
Airflow (LFM)	250
Heat Sink	Medium Profile
Custom TSA (C/W)	NA
Board Selection	Medium (10"x10")
# of Board Layers	8 to 11
Custom TJB (C/W)	NA
Board Temperature (C)	NA

On-Chip	Power (W)	Used	Available	Utilization (%)
Clocks	0.003	1	–	–
Logic	0.000	2270	12480	18
Signals	0.000	2942	–	–
IOs	0.000	114	172	66
Leakage	0.321			
Total	0.324			

Thermal Properties	Effective TJA (C/W)	Max Ambient (C)	Junction Temp (C)
	2.8	84.1	50.9

Supply Summary Source	Voltage	Total Current (A)	Dynamic Current (A)	Quiescent Current (A)
Vccint	1.000	0.239	0.003	0.236
Vccaux	2.500	0.032	0.000	0.032
Vcco25	2.500	0.002	0.000	0.002

	Total	Dynamic	Quiescent
Supply Power (W)	0.324	0.003	0.321

Fig. 11 Power analyzer of 16- bit vedic FIR using reversible logic

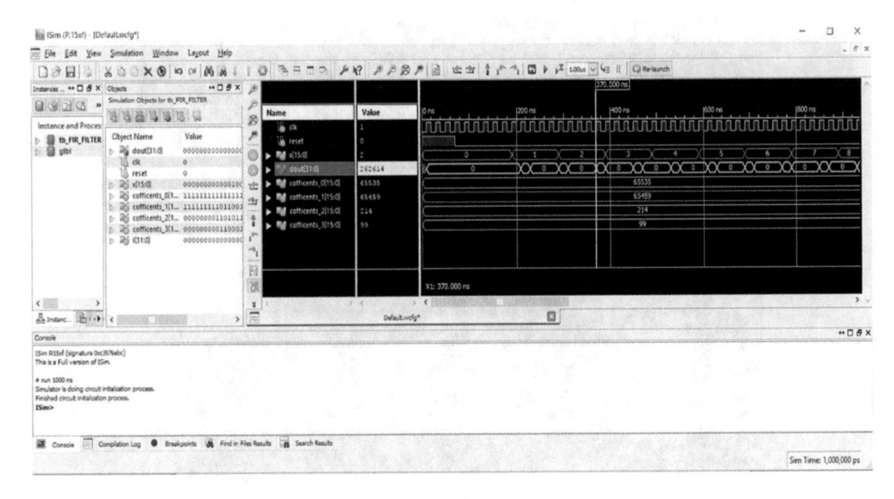

Fig. 12 Simulation results of reversible 16-bit FIR filter

References

1. Moore GE (1965) Cramming more components onto integrated circuits. Electronics 38(8):1–14
2. Moore GE (1975) Progress in digital integrated electronics. In: Proceedings technical digest international electron devices meeting, vol 21, pp 11–13
3. Launder R (1961) Irreversibility and heat generation in the computing process. IBM J Res Devel 5(3):183–191
4. De Garis H, Batty T (2004) Robust, reversible, nano-scale, femto-second-switching circuits and their evolution. In: Proceedings of the 2004 congress on evolutionary computation. IEEE. ISBN 0-7803-8515-2
5. Joy K, Mathew BK Implementation of a FIR filter model using reversible Fredkin gate. In: Control, instrumentation, communication and computational technologies (ICCICCT), international conference on IEEE explore, 22 Dec 2014
6. Thakral S, Bansal D (2016) Comparative study and implementation of vedic multiplier for reversible logic based ALU. MR Int J Eng Tech 8(1)
7. Chaudhary A, Fatima N (2016) Survey paper on FIR filter using programming reversible logic gate. Int J Comput Appl 151(11):0975–8887
8. Nagaveer T, Neelima M (2014) Performance analysis of vedic multiplier using reversible logic in Spartan 6. Int J Res Comput Commun Technol 3(10)
9. Shams M, Navi K, Haghparast M Novel reversible multiplier circuit in Nanotechnology. https://doi.org/10.1016/j.mejo.2011.05.007
10. Kumar U, Sahu L, Sharma U (2016) Performance evaluation of reversible logic gate. Print-ISBN:978-1-5090-5516-6, ICTBIG
11. Garipelly R, Madhu Kiran P (2013) A review on reversible logic gates and their implementation. ISSN 2250-2459, ISO 9001:2008 Certif J 3(3)
12. Gowthami P, Satyanarayana RVS (2016) Design of digital adder using reversible logic, vol 6, Issue 2, (Part—1), pp 53–57. ISSN: 2248-9622
13. Toffoli T (1980) Reversible Computing. Tech memo MIT/LCS/TM-151, MIT Lab for Computer Science
14. Vasim Akram R, Rahmatullah Khan M, Raj Kumar B (2014) Design of high speed low power 32-bit multiplier using reversible logic: a vedic mathematical approach, 02(08):0624–0629. ISSN 2322-0929

A Study on Benefits of Big Data for Healthcare Sector of India

Komal Sindhi, Dilay Parmar and Pankaj Gandhi

Abstract Big data has taken the world by storm. Due to the tremendous amount of data being generated in each and every field, the use of big data has dramatically increased. Health is the heart of a nation, and thus healthcare is one of the unavoidable and best examples to be given when discussed application of big data in today's era. Similar to western countries like US leveraging the benefits of big data starting from every simpler thing to handling the most complicated tasks, India can also utilize the potential of big data. In the present paper, we have started with the overview of healthcare sector of India in urban as well as rural areas, followed by general merits of big data in healthcare as well as domain-specific uses and ended with a broad framework depicting big data in context to healthcare sector of India.

Keywords Big data · Healthcare · India

1 Overview of Healthcare Sector of India

Healthcare sector is a part of economic sector of a country which is an accumulation of all the organizations that provide services and treatments to patients. It is one of quickest growing sectors of the world. The major contenders involved are doctors, hospitals, clinics, insurance companies, pharmaceutical companies, industries manufacturing medical instruments, laboratories, medical data analysts, and many more.

K. Sindhi (✉)
S V National Institute of Technology, Surat, Gujarat, India
e-mail: sindhikomal93@gmail.com

D. Parmar (✉) · P. Gandhi
School of Engineering, P. P. Savani University, Gujarat, India
e-mail: dilay.parmar@ppsu.ac.in

P. Gandhi
e-mail: pankaj.gandhi@ppsu.ac.in

D. K. Mishra et al. (eds.), *Data Science and Big Data Analytics*,
Lecture Notes on Data Engineering and Communications Technologies 16,
https://doi.org/10.1007/978-981-10-7641-1_20

The revenue generated by this raging sector was recorded to be USD 45 billion in the year 2008, increased to USD 110 billion in 2016, and is expected to reach USD 280 billion by 2020 as per the statistics given by Indian Brand Equity Foundation on July 27, 2017 [1]. Thus, the CAGR (Compound Annual Growth Rate) is measured to be 16.5%. The healthcare industry of India accounts for approximately 5.25% of the total GDP (Gross Domestic Product) of India according to the World Health Organization. On the other hand, number of patients is also expanding in an unexpected manner. Speaking of diversity in the type of data generated in this sector, the medical information appears to be arriving in many different types like handwritten prescriptions, reports (in the form of multiple image formats), information obtained as a result of monitoring devices and wearables, medical claims of patients and insurances, Pharmaceutical Research and Development, discussion about health on social media platforms, and some other electronic records. Hence, the source data in the present case has obviously increased with the increasing number of diseases and patients; in which our population increase of course plays a prominent role and it has reached to zettabytes of information which is equal to 10^{21} GB. This massive amount of data comes up with the concept of big data in healthcare industry.

Now focusing on the issues in the present healthcare sector, it is said that the healthcare system in rural areas suffers far more than any other sector in our country [2]. Nearly, 86% of the total patients visiting hospitals or getting admitted come from rural regions. WHO has defined certain norms/ratios for the accessibility of healthcare services which are very higher than the ones measured in India, especially those in the rural regions. Most of the Public Health Centers and Community Health Centers are situated far away from rural areas, eventually demanding higher wages. This in turn gives rise to villagers to consult private doctors who are unregistered [3]. Moreover, statistics say that 2% of total number of doctors are available in rural regions of India, whose population accounts for 68% of the entire people [4]. As recently reported by India today about the statement from Medical Council of India, the number of doctors to the number of patient (i.e., doctor to patient ratio) is 1:2000, whereas the ideal ratio decided by WHO is 1:1000 [5].

Out of total population of India, 75 % people live in rural areas while 25% people live in urban areas. Majority of people living in rural areas are uneducated and having poor basic infrastructure facilities. India as an economy is in developing stage, where funding is limited and weather is also in favor of epidemics diseases like malaria, typhoid, and hepatitis. Despite having these weaknesses, the strength of India lies in IT. India is a software hub, where youth is inclined for Computer/IT education. Maximum techno-savvy people of USA and EU are from India. This indicates we need to focus on our strength to overcome the weaknesses. We believe that the talent of India especially can be utilized in the domain of big data to resolve the issues of healthcare sector.

2 Advantages of Big Data in Healthcare

Starting with the discussion about the applicability of big data in any healthcare unit on a general basis, the major benefits are as follows:

- **Betterment in the Type of Treatment Given**: With the advent of big data, nowadays the data being collected comes from a variety of sources like previous hospital visits, concerned laboratories, and heredity information records as well as the social media profiles of the patient. It helps the doctor to determine more precisely the cause of the disease the patient is going through. This is because it might happen sometimes that the treatment may be based on some daily personal activity (which can be known through social media profile), rather than just the fundamental health information at that point of time.
- **Forgery Detection**: There has been a huge increase in the number of people producing false medical claims to get more money. Big data researchers and analysts being able to access enormous amount of data have been quite successful in detecting such fraudulent people. After medical practitioners started using big data for checking the patient claims, there has been a saving of $1 billion according to [6].
- **Electronic Health Records (EHRs)**: EHRs are nothing but the paper medical records stored in electronic devices. Big data has made it possible for the patients to access their medical prescriptions given by different doctors, laboratory records, etc., which can also be modified by doctors on online basis. These records also store the patients, scheduled for next regular check-ups or laboratory tests so as to remind them about that.
- **Early Involvement**: Taking the advantages of predictive analysis techniques in big data involves learning on the basis of past experiences as well as related databases; it can be efficiently used to diagnose a disease before it reaches a worst stage.
- **Decreased Costs**: Costs in healthcare have been decreased in a number of ways due to big data. One of them is reduced paper costs, second one is the cost saving due to reduced number of visits to the hospital because of remote monitoring facilities, third is the cost saved by unnecessary tests undertaken, and so on.
- **Flexible and Improved Analysis**: The vast data handling and controlling characteristic of big data can be utilized by analysts in medical foundations who work on studying the progress of various healthcare sectors or hospitals in terms of quality of healthcare provided and revenue generated by that hospital, during a long span. For example, for a particular hospital under consideration, a record can be maintained of the bill paid by every patient till date, salary given to various doctors, nurses, and other workers, equipment and maintenance expenses, and so on. All these can be combined to analyze the overall profit generated by that hospital.

In addition to the benefits mentioned above, there are many other ones that have revolutionized the healthcare industry in recent times. In terms of applications of big data in particular domains of healthcare, some of them are discussed below:

- **Physiotherapy**: Dugani et al. [7] have taken a real-time dataset of a large number of X-ray images into consideration. It is a general practice that a doctor recommends a particular physiotherapy treatment to a patient on the basis of his/her X-ray reports. Here, the author has explained the mapping of various injuries to the physiotherapy treatment given to the patient suffering from it. A number of test cases have been generated, among which the first one was used to find out the number of male and female patients admitted; second one was used to find out the injury/(s) that has been found to be present in maximum number of patients; third one was used to find out which injury occurred maximum separately in males and females; and the last one showed the cost for the individual physiotherapy treatment given. Hence, combining all the abovementioned results, whenever a patient is admitted it can be smoothly predicted which injury might have happened and which treatment should be given, and finally the authorities or patients can also check that only the appropriate amount is paid as bill and not more or lesser.

- **fMRI (functional Magnetic Resonance Imaging)**: Functional MRI basically determines the activity of brain by measuring the changes in blood flow. These are specifically 3D and 4D images produced in very large volume and at the same time differ from each other in terms of format, resolutions, etc. Here, in [8], the authors have laid their emphasis on flexible processing, storage, and analytics of fMRI images. The big data tool utilized was the big data Spark platform because the former used tool Hadoop supports lesser number of programming languages and has a messy installation process. On the other hand, Spark tool has advantages like improved speed, greater memory processing abilities, and wider support to programming languages. In both [8] and [9], a pipeline is proposed and tested in which template matching and sum of square differences was used to obtain brain networks from fMRI data given as input. Lastly, the Spark tool used in [9] showed 4 times better performance as compared to Python used in [8].
- **Chronic Diseases Monitoring**: A chronic disease is one lasting 3 months or more, by the definition of the U.S. National Centre for Health Statistics [10]. These diseases tend to be common according to the age and they are mostly found in elderly people according to [11]. Here, an architecture has been proposed in compliance to a European project named Virtual Cloud Carer Project, which is implemented using wireless sensors and big data. The target was to perform remote monitoring of the chronic patients at in-house (controlled places like hospital) as well as outside environments. The patients were supposed to carry a device which was a combination of various sensors capable of measuring vital information, and these huge data were managed using big data services like NoSQL. The interface to the other end where this data was forwarded could be shared by patients close relatives, doctors, and other concerned people. Thus, the emergency situations could be communicated in an easy manner and also the decision-making power for the future conditions was improved.
- **Emotion-Aware Healthcare**: According to [12], physical or vital characteristics of a patient are not the only important factor for diagnosing a patient but his/her emotions also play a prominent role. Keeping this as a motive, a system known as BDAEH (Big Data Application in Emotion-Aware Healthcare) was proposed in which input data was collected using wearable devices; this data was transmitted to

data centers using 5G technologies, processed on a cloud platform, and the finally results were conveyed to doctor after extensive analysis.

- **Diabetic Data Analysis**: Saravana Kumar et al. [13] laid emphasis on a commonly prevailing non-contagious disease, i.e., diabetes. Three types of diabetes have been categorized till date. In [13], a predictive analysis algorithm was used and it was implemented in Hadoop/ MapReduce environment, which was used to (i) estimate the type of diabetes observed; (ii) intricacies arising due to that, and (iii) treatment that should be given to the respective circumstances. Thus, the proposed system could be used to provide improved treatment and better cure.
- **Progressive Lung Diseases**: Lung diseases that progressive gradually, also known as chronic obstructive pulmonary diseases, have been the center part of research in [14]. These diseases are often found to be difficult to be present on the basis of patients' present conditions, but the past history of the patient can prove useful in this case. The repository of data taken into consideration stored the details of the patient uniquely according to the manner. In this manner, treatment undertaken by the patient can be known, even from different hospitals. The overall process includes some preprocessing steps at the beginning, preparing dataset (consisting of some particular attributes), and thereby training the system, finally implementing the classification by selecting J48 algorithm for decision tree using tenfold validation.
- **Heart Attack Prediction**: Effective prediction of heart attack using big data has been demonstrated in [15]. Heart attack, also known as acute myocardial infarction, is one of the most dangerous diseases anyone could ever have. This disease has multiple reasons to occur and it is always very necessary to identify the correct one out of them, which leads to the requirement of personalized solution to the problem instead of a general one. Hence, enormous datasets need to be analyzed which can be done with the help of big data. Big data analytics has been used in [15], thereby accomplishing better prevention of heart attack, its occurrence, and personalized treatment for heart attack. Data visualization and mining techniques have been used, and the tool used is Hadoop.
- **Parkinsons Disease**: Parkinson's disease (PD) is a neurodegenerative brain disorder that progresses slowly in most people. Most people's symptoms take years to develop, and they live for years with the disease [16]. Gradually, the brain loses its control over body movements and emotions as well. Dinov et al. [17] have utilized an expansive and heterogeneous dataset for Parkinsons disease maintained by Parkinson's Progression Markers Initiative. They have used classification techniques which are based on model-free big data learning as well as model-based big data approaches to classify the disease. It was being concluded that the model-free methods like SVM (Support Vector Machine, AdaBoost) surmounted the performance of the model-based machine learning big data methods.

3 Model of Big Data for Healthcare Sector of India

In this section, we discuss a model for healthcare sector of India. The healthcare big data is to be generated from different sources as shown in Fig. 1. On captured data, preprocessing should be applied for cleaning the data. This acquired data which is of structured, semi-structured, and unstructured nature should be stored in the storage area. This data generation and information consumption is in the context of India, where data is to be collected from government healthcare centers, private hospitals, census data, weather data, and social media data of patients. After that, analysis of data should be done, and useful information must be derived for different information consumers. The details regarding how the gained information can be used by different information consumers are given below:

- **Doctors**: The doctor can see the past health history of patient with the help of electronic health records. The patient who has shifted to new place can continue his treatment in new hospital in hassle-free manner, because the doctor is able to access all the diagnosis details and reports from the electronic health records. Healthcare big data can help doctors in their decision support systems also, where they can use the potential use of big data in disease prediction for patients.
- **Government**: The gained information can help government in giving subsidies to different private hospitals based on the patients' records of that area where government health centers are not available. It can help planning commission in estimating and planning funds for different health projects in a better way.
- **Insurance Companies**: Insurance companies can use big data to identify and prevent false claims. The information gained can also help insurance companies in improving their health products.
- **Pharmaceutical Companies**: Pharmaceutical companies can use big data in their research and development work. They can make customized drugs for certain disease based on the gained information. They can use the information of clinical trials done for large segment of patients and can use this information in their research work for drug invention.

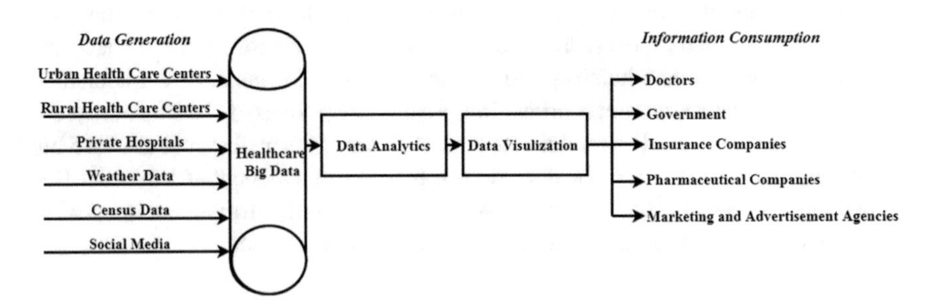

Fig. 1 Model of Big data for healthcare sector of India

- **Marketing and Advertisement Agencies**: They can use big data information about different diseases prevalent in society. They can then make appropriate training programs for educating the rural people of India.

4 Conclusion

There is large skill pool of IT in India. If these youth can be utilized for the development of social engineering arena, then we will be able to resolve social issues and the issues in healthcare sector are one of them. Health services to rural and poor people is a big challenge to government of India, which can be addressed by using the big data concept. Here, an attempt has been made to develop basic architecture of big data in healthcare domain. In addition to that, the preliminaries of big data and their generalized benefits in case of any healthcare sector have been summarized. Lastly, the advancements or applications of big data in particular domains of medicine have been discussed.

References

1. Healthcare Industry in India. https://www.ibef.org/industry/healthcare-india.aspx. Accessed 20 Sept 2017
2. Singh S, Badaya S (2014) Healthcare in rural India: a lack between need and feed. South Asian J Cancer 3(2):143. https://doi.org/10.4103/2278-330X.130483
3. Iyengar S, Dholakia R (2012) Access of the rural poor to primary healthcare in India. Rev Mark Integr 4(1):71–109. https://doi.org/10.1177/097492921200400103
4. Britnell M (2010) In search of the perfect health system. Palgrave Macmillan
5. Grim picture of doctor-patient ratio, 1 doctor for 2,000 people. http://indiatoday.intoday.in/story/grim-picture-%20of-doctor-%20patient-ratio/1/654589.html. Accessed 20 Sept 2017
6. Big data could mean big problems for people's healthcare privacy. http://www.latimes.com/business/lazarus/la-fi-lazarus-big-data-healthcare-20161011-snap-story.html. Accessed 26 September 2017
7. Dugani S, Dixit S (2017) Physiotherapy data analysis of big data in healthcare applications. In: International conference on innovative mechanisms for industry applications (ICIMIA), pp 506–511. https://doi.org/10.1109/ICIMIA.2017.7975666
8. Sarraf S, Saverino C, Ghaderi H, Anderson J (2014) Brain network extraction from probabilistic ICA using functional magnetic resonance images and advanced template matching techniques. In: 27th Canadian conference on electrical and computer engineering (CCECE), pp 1–6. https://doi.org/10.1109/CCECE.2014.6901003
9. Sarraf S, Ostadhashem M (2016) Big data application in functional magnetic resonance imaging using apache spark. In: Future technologies conference (FTC), pp 281–284. https://doi.org/10.1109/FTC.2016.7821623
10. Chronic disease. http://www.medicinenet.com/script/main/art.asp?articlekey=33490. Accessed 24 Sept 2017
11. Pez D, Aparicio F, De Buenaga M, Ascanio J (2014) Chronic patients monitoring using wireless sensors and big data processing. In: 8th international conference on innovative mobile and internet services in ubiquitous computing (IMIS), pp 404–408. https://doi.org/10.1109/IMIS.2014.54

12. Lin K, Xia F, Wang W, Tian D, Song J (2016) System design for big data application in emotion-aware healthcare. IEEE Access 4:6901–6909. https://doi.org/10.1109/ACCESS.2016.2616643
13. Saravana kumar NM, Eswari T, Sampath P, Lavanya S (2015) Predictive methodology for diabetic data analysis in big data. Proced Comput Sci 50: 203–208. https://doi.org/10.1016/j.procs.2015.04.069
14. Koppad S, Kumar A (2016) Application of big data analytics in healthcare system to predict COPD. In: International conference on circuit, power and computing technologies (ICCPCT), pp 1–5. https://doi.org/10.1109/ICCPCT.2016.7530248
15. Alexander C, Wang L (2017) Big data analytics in heart attack prediction. J Nurs Care 6(393):1168–2167. https://doi.org/10.4172/2167-1168.1000393
16. What is Parkinson's? http://www.parkinson.org/understanding-parkinsons/what-is-parkinsons. Accessed 24 Sept 2017
17. Dinov I, Heavner B, Tang M, Glusman G, Chard K, Darcy M, Foster I (2016) Predictive big data analytics: a study of Parkinsons disease using large, complex, heterogeneous, incongruent, multi-source and incomplete observations. PloS one 11(8):e0157077. https://doi.org/10.1371/journal.pone.0157077

Handling Uncertainty in Linguistics Using Probability Theory

Annapurna P. Patil, Aditya Barsainya, Sairam Anusha, Devadas Keerthana and Manoj J. Shet

Abstract Uncertainty is the lack of knowledge, or insufficient information. In this paper, we will be majorly discussing uncertainty occurring in natural language. Numerous natural language processing techniques can be applied to minimise linguistic ambiguities. We discuss one of the most widely used techniques—probability theory. An attempt is then made to solve the linguistic uncertainty using the theory.

Keywords Uncertainty · Probability theory · Bayes' theorem · Natural language processing · Knowledge base · Learning agent

1 Introduction

Uncertainty is formally defined as the lack of knowledge, or insufficient information. The vagueness, ambiguity, lack of specificity or conflict in information makes it hard for the user to take decisions or fix solutions under uncertainty. Probability theory

A. P. Patil · A. Barsainya · S. Anusha · D. Keerthana (✉) · M. J. Shet
Department of Computer Science and Engineering, Ramaiah Institute of Technology, Bengaluru, India
e-mail: keerthana.d1@gmail.com

A. P. Patil
e-mail: annapurnap2@msrit.edu

A. Barsainya
e-mail: aditya.barsainya16@gmail.com

S. Anusha
e-mail: anushaisnow@gmail.com

M. J. Shet
e-mail: manojjs168@gmail.com

D. K. Mishra et al. (eds.), *Data Science and Big Data Analytics*,
Lecture Notes on Data Engineering and Communications Technologies 16,
https://doi.org/10.1007/978-981-10-7641-1_21

and Fuzzy Set Logic help in dealing with uncertainty under various circumstances in spite of not having a common accepted theory in academia.

Uncertainty can be seen in all fields and aspects. Here, we talk about the uncertainty existing in Linguistics that are produced by statements in natural language. For example, in English, one word can have different meanings and represent different parts of speech (POS) contextually in various places. The word "water" assume different meanings in "Give me some water." and "Did you water the plants?". It is a noun and a verb respectively in the above example. These ambiguities are usually dealt with the knowledge of the context or common sense.

Handling uncertainty means trying to get the best possible solution for a given situation by the agent without any confusion. Though it is easily dealt with by humans, it is not the same for machines. Sarcasms, spoken language, excessive use of negative words, and the likely are sometimes not understood and ends up in misclassification of the POS for the same. To make the task simpler and the verification easier, we have fixed the domain to a food ordering system where the ambiguity in the customers' request is to be handled. Our attempt is to find a solution for this and make the machine learn.

Among many ways of handling uncertainty, like fuzzy logic, hidden Markov models, certainty factors, neural networks and likewise, Bayes' theorem can be used to resolve the uncertainty in Parts of Speech (POS) tagging of words in English. We propose the use of a prior knowledge base (KB) and a continuous learning agent that enables the machine to classify using regression and Bayes' logic and overcome the confusion. This can be achieved by storing the frequencies of occurrence of a word in a particular form in KB and assigning the most probable one in case of uncertainty. The learning agent learns with each incoming example and updates the knowledge it already has, i.e., occurrence frequency.

2 Literature Survey

Dutt and Kurian [1] studied the need of handling uncertainty using different techniques. Authors also give a classification of the different types of uncertainty prevailing in the real world and various techniques such as, probability analysis, fuzzy analysis, Bayesian Network analysis, etc., to handle uncertainty. The paper also suggests that probability analysis and fuzzy analysis as the best to reduce linguistic uncertainty.

Meghdadi et al. [2] presents the fuzzy logic for modelling uncertainty. They also give a comparison of the probability theory and fuzzy logic which are the best for handling uncertainty. The authors write that fuzzy logic is used for non-statistical uncertainty and probability theory for statistical uncertainty. They also formulate that the combination of these theories can be used for handling uncertainties that exist at the same time in the real-world systems. The authors present the mathematical framework named Discrete Statistical Uncertainty (DSU) model for representing probabilistic fuzzy systems.

Dwivedi et al. [3] states the different forms of uncertainty in the real-world problems. Authors mention the use of probability theory, certainty factor, and possibility theory to solve the problems of uncertainty. The comparison of the three methods to solve uncertainty is also mentioned by the authors.

Auger and Roy [4] account the expression of uncertainty in textual data. The paper depicts the ambiguity in the natural language which can be resolved by linguistic context. The authors solve the linguistic ambiguity by lexicons, grammars, dictionary, and algorithms. The authors mention that the method of automatic tagging of expression used explains how humans access certainty.

Zadeh [5] states the basic approach to quantifiers in natural language such that it can be interpreted as fuzzy numbers. The author also explains the application of semantics to the fuzzy possibilities that provide inference from knowledge bases which can contain proposition similar to the real-world knowledge bases. The paper states a method to reduce the uncertainty in natural language knowledge bases.

Groen and Mosleh [6] explain Bayesian inferences as a process in which representation of a state is assigned a possible true or false based on the observations. The author has also recognised that the Bayes theorem is not capable of handling uncertainty since the uncertain observation prevent from computing the likelihood of the observation. The authors, in their work, have overcome this restriction by redefining the likelihood function of probability to probability of whatever is not contradicting in the observation. The theory proposed in this paper is applicable to situations where the uncertainty does not depend on the prior knowledge of the environment.

Earlier fuzzy theorem was used to solve the problem of linguistic uncertainty as mentioned by Zadeh [5], but we will use Bayesian (knowledge base) view of probability [3] theory to solve linguistic uncertainty.

3 Linguistic Uncertainty and Ways of Handling Uncertainty

This section provides review of research work and theories in the field of linguistics, uncertainty, probability along with the intersections of the fields.

3.1 Linguistic Uncertainty

Linguistic uncertainty refers to indeterminacy in language. Inexact information is often an inherent characteristic of human reasoning. It arises from the statements that are spoken or through direct communication. Linguistic uncertainty is of five major types [3]:

- Ambiguity
- Vagueness

- Context Dependency
- No specificity, and
- Indeterminacy of theoretical terms.

Ambiguity occurs due to the lack of relevant information that could be known and aid in eliminating the uncertainty in probability. Ambiguity uncertainty majorly arises from the concept that a word can have more than one meaning. For example, the word "place" has different meanings in "I want to visit that place." and "I want to place an order."

Anything not precise is said to be vague [7]. For example, in the context of food ordering, "I want to have something nice." does not clearly state what he/she wants. "Nice" can mean different things to different people depending on his/her liking. The uncertainty in vagueness can be eliminated by giving it an operational meaning or defining the vague term.

Context dependency refers to the confusion among the various meanings a word can take depending on the context. Sarcasms and puns come under this category. No specificity refers to insufficient specification as to what the word means and indeterminacy of theoretical terms is not knowing what the theoretical terms mean in real world.

In this paper, we are majorly trying to resolve uncertainty in ambiguity and try to assign correct POS tags to each word.

3.2 Ways of Handling Uncertainty

There are numerous ways of handling uncertainties—Probabilistic reasoning, Fuzzy logic, Certainty Factor, Hidden Markov model, Neural networks, etc. We here discuss the two most widely used techniques, Probability theory, and fuzzy logic.

3.2.1 Probability Theory

The theory of probability has been in existence since the sixteenth century, when attempts to analyse games of chance was made by Gerolamo Cardano. Probability in simple terms is the likeliness/degree of belief that an event will occur. Probability theory is a means of analysis of random events. It is a value between [0, 1], which gives the chances of an event occurring and all of it adds up to 1 for a given experiment.

Conditional Probability:

One of the major extensions of probability theory is the Bayes' logic. It is derived from Conditional probability which defines the possibility of occurrence of one event given that another evidence/event has occurred. The relation is given as

$$P(A|B) = \frac{P(A^{\wedge}B)}{P(B)},$$

where,

P(A|B)	probability of A happening given that B has already happened
P(A^B)	probability of both A and B happening together
P(B)	probability of B happening.

Bayes' Theorem
In probability theory, Bayes' logic defines the probability of an event based on the prior knowledge of conditions related to it. The knowledge base is continuously built through observation, monitoring and validation of previous experiment or existing facts about the experiment. The relation between two events by Bayesian theorem is given as

$$\boldsymbol{P(A|B) = \frac{P(B|A) \cdot P(A)}{P(B)}},$$

where

P(A) and P(B)	probabilities of A and B independent of each other
P(A|B)	the conditional probability of A being true given that B is true
P(B|A)	probability of B given that A is true.

The main significance of Bayes' theorem is that it is based on theoretical proof and uses conditional probability for classifying events. It is one of the few methods capable of forward uncertainty propagation when little or data is available, or a statistical inference when it is [1]. It is widely used in the field of medical diagnosis.

3.2.2 Fuzzy Logic

Unlike Probability theory, where the outcome is atomic, i.e. True/False or 0/1, Fuzzy logic deals with "degree of truth" or "degree of belongingness". It assigns any real number between [0, 1] which depicts "how true" the condition is.

The inventor of fuzzy logic, Lotfi Zadeh, observed that human reasoning involves a range of possibilities between a complete Yes and a No like, possibly no, cannot say, certainly yes, etc. [8]. Fuzzy logic resembles human reasoning in a way that it assigns a degree of truth involving intermediate possibilities.

While building a fuzzy system, the designer has to clearly define all the membership functions considering how average people use concepts. Here, expert knowledge is coded into fuzzy rules [5]. A Fuzzy Logic System can deal with vagueness and uncertainty residing in the human knowledge base, and allows us to represent linguistic terms.

3.2.3 Which Is Better? A Comparison

In general, probability theory and fuzzy logic have different domains of applicability and distinct agendas. Though closely related, the key difference lies in what they mean.

Probability handles with occurrence or non-occurrence of events, and not facts. Probability theory has nothing to reason. It is either entirely true or false (atomic). On the other hand, fuzzy logic is all about the degree of truth and captures the vagueness concept.

Though Bayesian network is one of the most efficient approaches of probabilistic reasoning, it has a few drawbacks. Unknown values, inconsistent knowledge and computational expense being the major ones.

The main drawback of fuzzy logic is that it completely depends on context and general-purpose rules are quite hard to get. It also requires all membership functions to be clearly defined pre-experiment.

Classical probability theory has been successful in mechanistic systems where the interactions and dependencies are clearly defined, like—quantum mechanics, communication systems, statistical mechanics, and related fields. But, when it comes to fields where human emotions, perceptions, and reasoning are to be involved, it fails. The best possible way to deal with this is explained in [5], which says that both probability theory and fuzzy logic can be infused. This approach helps in generalising the former and widen its scope of application.

In simple words, probability is the chances of an event being true, while fuzzy logic defines how true it is. As Zadeh [5] says, the two are complementary rather than competitive.

3.2.4 When Will the System Fail?

Though Bayesian theory is very efficient, it does have pitfalls

- It requires all dependencies of variables to be clearly defined. Thus, ceases to be practical when the domain is too big.
  This can be resolved by restricting the domain and solving smaller uncertainties.
- Bayesian approach totally depends on the prior probabilities defined in the KB. For example, the word "duck" is usually a noun. If a sentence, "I ducked to avoid getting hit by the bullet." is given, the machine will end up classifying "duck" as a noun rather than a verb because the prior probability says so. This is called the Danger of strong priors [9]. As Mark Twain succinctly puts it, "It ain't the unknown that puts you in danger. It's what you know for sure that just ain't so."
  This problem can be overcome by assigning a small probability value to every outcome or use normal distribution as priors.
- Sometimes the reasoning requires the user to make a few assumptions about the results. This contradicts the whole idea of learning by measuring [10].
- Bayesian logic cannot handle boundary values and fails to rightly classify.

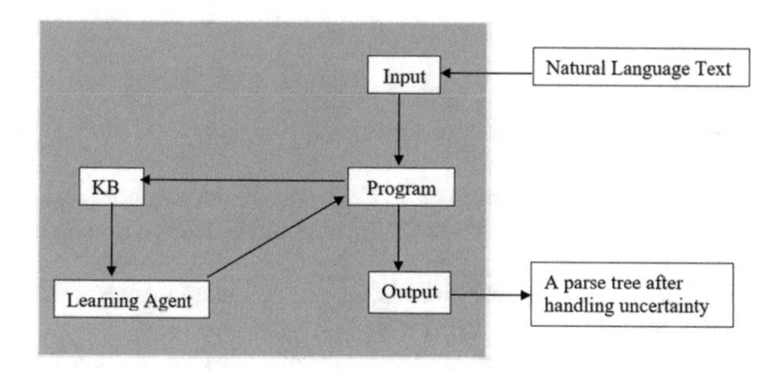

Fig. 1 Agent block diagram

4 System Architecture

The structure of the agent that we use to linguistic uncertainty is depicted as a block diagram (Fig. 1).

In this agent structure, the agent receives input from an external source in the form of text in natural language. The domain of the input text is fixed to the context of food ordering. The input text is sent to programme which splits the text into a tree with a verb as root at each level of the tree.

The knowledge base (KB) stores the relevant word in connection with the domain. The words stored in KB are assigned a value, called the degree of belief, the prior probability.

The knowledge base updates the assigned degree of belief on accepting new inputs and using the Bayesian theorem, thus, forming a learning agent. The output of the agent is a parse tree with appropriate POS tagging which is expected to be a response to an order with the knowledge it has acquired from the KB and the updated probability.

5 Conclusion

Traditionally, linguistic uncertainty was resolved using fuzzy logic as it is best suited when human psychology is to be accounted for. In this paper, we attempt to solve the same using simple, straight-forward Bayesian approach of probability theory.

Here, the domain was restricted. This proposed idea can be expanded to various domains and finally integrated to resolve uncertainty in the whole of natural language.

References

1. Dutt LS, Kurian M (2013) Handling of uncertainty—a survey. IJSRP 3(1). Edition [ISSN 2250-3153]
2. Meghdadi AH, Mohammad R, Akbarzadeh T (2003) Uncertainty modelling through probabilistic fuzzy systems. In: Proceedings of the 4th international conference uncertainty modelling and analysis, pp 56–61
3. Dwivedi A, Mishra D, Kalra PK (2006) Handling uncertainties—using probability theory to possibility theory. Mag IIT Kanpur 7(3):1–12
4. Auger A, Roy J (2008) Expression of uncertainty in linguistic data. In: Published 2008 in 2008 11th international conference on information fusion
5. Zadeh LA (1983) A computational approach of fuzzy quantifiers in natural language. Comput Math Appl 9(1):149–184
6. Groen FJ, Mosleh A (2004). Foundations of probabilistic inference with uncertain evidence. Int J Approximate Reason 39(1):49–83, Apr 2005
7. How Bayesian inference works, data science and robots blog
8. An intuitive explanation of Bayes' Theorem—better explained
9. A summary on fuzzy probability theory: Michael Beer, Publication Year: 2010 GrC
10. Qin B, Xia Y, Prabhakar S, Tu Y (2009) A rule-based classification algorithm for uncertain data. In: Proceedings of the IEEE international conference on data engineering (ICDE), pp 1633–1640

Review of Quality of Service Based Techniques in Cloud Computing

Geeta and Shiva Prakash

Abstract Nowadays cloud computing is an emerging technology for enhancing our daily life. cloud computing provides the facility to pay for use. There are some important service providers in cloud computing such as IBM, Google, Amazon, Microsoft, etc., these service providers give the different services to the users. Cloud providers offer services to the users based on the expected quality requirements, So it is a big challenge of cloud to provide enhanced Quality of Service (QoS) to their users. In this paper, we are presenting main existing research work of different QoS management techniques in cloud computing and also compare them with their strength and weakness. This review will be helpful for new researchers in this field to get exposure of quality of service based management techniques.

Keywords Cloud computing · Service level agreement (SLA) · Trade-offs
Quality of service (QoS)

1 Introduction

In these days cloud computing plays a main role in IT sector. The aim of cloud computing is to provide several network services to the users so that they can access these services from anywhere, any time and from anywhere in the world over the Internet on the payment of users QoS requirement.

Geeta (✉)
Dr. APJ. Abdul Kalam Technical University, Lucknow, India
e-mail: geetasingh02@gmail.com

S. Prakash
Department of Computer Science & Engineering, M.M.M. University of Technology, Gorakhpur, India
e-mail: shiva.plko@gmail.com; shiva.pkec@yahoo.com

D. K. Mishra et al. (eds.), *Data Science and Big Data Analytics*,
Lecture Notes on Data Engineering and Communications Technologies 16,
https://doi.org/10.1007/978-981-10-7641-1_22

There are various service provider of cloud computing available in the market as Amazon, Google, IBM, Microsoft to give cloud computing services such as Software as a Service (SaaS), Governance as a Service (GaaS), Platform as a Service (PaaS), Infrastructure as a Service (IaaS), and Business as a Service (BaaS). This computing technique is the new and most challenging and important emerging technology defined by NIST. This technology has a high impact in the business because of several availability of cloud computing resources. It is also the latest very popular emerging technology in distributed and parallel computing that provides software applications as services [1] and hardware Infrastructure. cloud computing stacks have become popular in enterprise or business data centers, where hybrid and private Cloud frameworks are highly adopted [2]. In cloud system, resources [3] are hosted or provided over the Internet and delivered to users as services. Although users do not have control or manage over the underlying Cloud resources and they desire to ensure that the availability, quality, reliability, and performance of these resources are provided [4]. There are Challenges in QoS area. Qos represents the performance, reliability, and availability of the system. QoS represents the level of reliability, availability, and performance offered by the infrastructure by the platform or an application that hosts it. It represents the capability of a system or network for presenting higher services [5]. QoS in Cloud have represented qualitative features such as agility, accountability, cost, performance, usability, assurance, privacy, and security. These features are used for comparing several cloud services.

It is basic to cloud computing consumers, who anticipate providers to give the advertised features, and for the cloud providers, who require to get the appropriate trade-off between operational cost and QoS levels. Cloud Service providers require to acquiesce with SLA agreements which conclude the penalties and revenue based on the accomplish performance level. SLA's are marked between the user and cloud provider where Service Level Agreement violation acts as major constraints. SLA violation is minimized through technologies involving in monitoring [6]. Thus, many authors are investigating automated Quality of Service management techniques that can influence the great programmability of software and hardware resources in cloud. We have done detailed literature review and also provided the comparative study of well-known QoS-based approaches on the basis of author and years of publication, technique used and their strength and weaknesses.

In Sect. 2 we have given brief description of cloud service metrics which are used by researchers to enhance quality of the services. In this research paper we have described about different QoS management techniques in Sects. 3 and 4 have detailed comparison of QoS management techniques and we discussed the comparison in Sect. 5. Finally we provide conclusion in Sect. 6.

2 QoS Matrices in Cloud

Usually the cloud services generally distributed by the various cloud computing providers and some features are very important to compare several cloud computing services. Quality of Services represents the performance, reliability and availability of the system. In this section we are representing the QoS features with their respective metrics of cloud computing services [7–9].

This Table 1 shows the features of Quality of Service with respective matrices. These features are very helpful to do further research to improve the performance of Cloud system.

3 Related Work on Quality of Services Management Techniques

A monitoring system has proposed by Chitra et al. [10] to improve QoS at the time of SLA negotiation between user and cloud computing service provider. The negotiation sporadic voting is created and generated intelligences in a complete method. Afterward finding local changes, every network element has to released signals to guarantees that worldwide attributes may not be affected. Through monitoring system, the unsuccessful terminals should be observed and it enhanced the effectiveness of the cloud computing atmosphere and entice the users. Other quality of service attributes can be assumed.

A SoS method has proposed by Hershey et al. [11] to enable Quality of Services management, response and monitoring enterprise systems which deliver cloud as a service. Enterprise Monitoring, Management, and response Architecture in cloud computing system (EMMRA CC) enhanced previous researched work to support architecture from which to recognize points of the administrative domains where Quality of Service metrics may be managed and monitored. An actual example was given for applying the new SoS technology to a actual world scenario, i.e., distributed denial of service (DDoS). This technique is not applied to federated clouds in real time.

A generic Quality of Service framework has proposed by Liu et al. [12] for cloud computing workflow systems. This framework system consists of Quality of Service aware service selection, Quality of Service requirement specification, QoS violation handling and Quality of Service consistency monitoring. Although, knowledge sharing and data communication between the parameter for various Quality of Service dimensions is not relevant for solving difficult issues such as service selection based on multi-quality of Service, violation handling and monitoring.

A Quality of Service oriented federated framework in cloud computing has been proposed by Salam and Shawish [13] where several cloud service providers can coordinate seamlessly to provide scalable Quality of Service assured services. Federation Coordinators and cloud coordinators key elements are used to enable cloud federation.

A framework of ranking prediction (QoS) in [14] has proposed by Zibin Zheng for cloud computing services by considering previous service experiences of users. It is used to eliminate the expensive actual world service and time-consuming invocations.

Table 1 Features of quality of services with respective matrices

S. no.	Features	Matrices	Descriptions
1.	Efficiency	Resource utilization, ratio of waiting time, time behavior	The resources employed for services when providing the wanted operation and also therefore the amount of efficiency, underneath settled situation, that means it evaluate however nice the web services employs the sources
2.	Availability	Accuracy, flexibility and response time	In classic device users entry services from web browser via the net, however right here the accessibility to that services is not assured
3.	Reliability	Accuracy of service, service constancy, fault tolerance, recoverability and maturity	Is the power of a service to remain functional with time without malfunction. The capability of the service to keep on operating with a certain degree of efficiency timely
4.	Scalability	Average of assigned resources among the requested resources	Represent the potentiality of enhancing the calculating power of the service provider's program as well as the product ability to process a number of clients' demands at a certain time period
5.	Reusability	Coverage of variability, readability and publicity	Known as the level to which a application component or even other work system may be used in a number of program or application
6.	Usability	Attractiveness, operability and learn ability	The quantity to which a service could be used by particular consumers to gain certain aims with usefulness, effectiveness and also approval in a certain background of usage
7.	Composability	Service interoperability and service modularity	It is the interoperability characteristics
8.	Adaptability	Coverage of variability and completeness of different set	The level of efficiency in adjusting the solutions for the utilization of every service based software

Therefore, the proposed framework needs no extra invocations of cloud computing services when preparing Quality of Service ranking prediction. The Collaborative filtering approach is used to anticipate Quality of Services only for internet facilities may also be utilized for cloud computing amenities. Now the uniqueness among different users is computed by Pearson Correlation Coefficient approach.

A formal model has proposed by W. C. Chu to help not only the ECC services construction and design through IaaS, PaaS, SaaS but also concurrent dynamic analysis and monitoring on the Quality of Services elements for the guarantees from Quality of Service providers and the SLA for multiple ECC users. On the bases of this formal model, analysis and testing models were created to help runtime monitoring and automatic testing to assure the requirements/SLA constraints. The limitations of this approach are not adapting the properties and solutions of IOT into the framework in addition to field experiment.

A framework was proposed by Saurabh Kumar Garg to find the priority and quality cloud services in [15]. This framework creates an important impact and builds healthy competition in the cloud service providers to improve their Quality of Services (QoS) and to satisfy their SLA. They have likely an AHP (Analytical Hierarchical Process) position approach which may calculate the cloud computing amenities related on various tenders contingent on Quality of Services necessities. This approach is cast-off only for measurable Quality of Services parameters such as guarantee of amenity, agility, accountability, safekeeping, secrecy, performance, cost and utility. This is not acceptable for non-quantifiable Quality of Services parameters such as Sustainability, Suitability, Service Response Time, Interoperability, Transparency, Accuracy, Availability, Reliability, and Stability.

M. M. Hassan has represented and tested the workload of Big data in [16], by running a set of Big data on Amazon cloud EC2 and construct a big simulation schedule and contrasted the proposed mechanism with other methods. Represented method was cost effective, performance features such as throughput, delay and delay variable were not considered into application.

An Algorithm QoS-aware service selection has proposed by Ruozhou et al. [17] for comprising various services offered by cloud computing. Several kinds of resources required to ritualize as a group of cloud computing amenities by means of different virtualization methods. The cloud system user services are providing through modified cloud computing amenities that includes not only various types of cloud amenities, formerly it also the communicating networks those cloud services. Hence, a group of the networking services and cloud services has been designed as a combined customized computing service.

In [18] a mechanism was proposed by R. Karim to map the consumers Quality of Services requirements of cloud services to the appropriate Quality of Services specifications of Software as a Services and map them to the best infrastructure as a services that provides the optimal Quality of Services guarantees and end-to-end Quality of Service values was examined as a output of the mapping. The researchers proposed a group of rules to accomplish the mappings process. The Quality of Service specifications of cloud computing services were hierarchically designed using Analytic hierarchy process (AHP) technology. This technology helped to provide the mapping process throughout the cloud computing layers and to rank the users cloud computing services.

A model of cloud resource pricing balancing Quality of Service requirements and greater profits has proposed by Sharma et al. [19]. This prototypical does only active formation and does not contain utilizing in cloud system price. Therefore, it may supremacy to erroneous predictions. The usage is a higher aspects that has to be applicable for all computing price estimation.

In [20], architecture has proposed by S. Lee that occupied the negotiator method to control, the monitoring of requested QoS requirements and SLA, to help the validation and verification. Furthermore, the agent method dynamically analyzed the resources deployment and allocation. Therefore the weak point of this method was absence of self-learning to find the automatic allocation time of the system resources.

In opposite, composing service elements into a cloud computing service that assembles several Quality of Service is a challenge. Since such a quandary can be mainly addressed as Multi-Constraints Path problem (MCP), which is known as NP-hard [21].

A novel of Quality of Service aware dynamic data replicas delete mechanism proposed for maintenance cost saving and disk space purpose by Bin et al. [22]. Its investigational outcomes displayed that the DRDS technology the availability and outcome of quality of service necessities are established.

Linlin et al. [23] has considered admission control and several IaaS providers. Obtaining from several Infrastructure as a Service providers brings large-scale of capitals, Different cost architecture, and elastic cloud computing facilities concert. The researchers used scheduling algorithms and a creative cost-effective admission control to Increase the Software as a Service benefactor's return. Therefore, their planned resolutions are capable to increase recognized users number of complete the effective settlement of demand on Virtual Machines hired from different IaaS suppliers and they used different user's Quality of Service needs and dissimllar framework.

Xu et al. [24] has proposed a multiple QoS and multi-workflows (MQMW) constrained scheduling strategy to address the issue of various workflows with several Quality of Service requirements. This proposed method could schedule various workflows which were started at several times through QoS restraint such as reliability and availability were not included to workflows.

P. Zhang has presented a QoS framework et al. [25] for an adaptive Quality of Service management process and mobile computing to control QoS assurance in mobile computing system and also they proposed a QoS management approach which is based on Fuzzy Cognitive Map (FCM).

An efficient prominence based Quality of Service provisioning method has presented by Xiao et al. [26] which can reduce the cost of cloud resources, while fulfilling the required QoS metrics and they also considered the statistical contingency of the response time as a practical metric rather than the typical mean response time. This method is not used to privacy and security metrics.

In next Sect. 3, Table 1 demonstrates a comparison of the various reviewed algorithms on the basis of QoS management techniques with their strength and weakness.

4 Comparative Table of Different Existing QoS Management Techniques

This section presents comparative study of various quality of service based techniques in cloud on the basic of year and authors of publication, what techniques used, strength, weakness and reported by title.

5 Discussion

Different QoS techniques as shown in Table 1 have their strength and weakness which is very beneficial to us for further studies. For example a cloud monitoring system is very important technique, used to improve Quality of Service during compromise among users and facility benefactors intermittent voting is created and its intelligences produced in an complete progression. It gradually attracts the cloud users and increases the cloud efficiency. But it is failed to calculate communication cost. In "Enterprise Monitoring Management and Response Architecture" in cloud computing (EMMRA CC) and System of System approach it enhance QoS performance and prevents Distributed denial of service attack. The method is effective but it is not applied to federated clouds in real time. On the other hand, a framework (generic Quality of Service) for the cloud workflow systems has proposed which also have four parameters such as Quality of Service aware service selection, QoS requirement specification, Quality of Service violation handling and QoS consistency monitoring. Though, the knowledge sharing and data communication between the parameters for various QoS features is not applicable for solving complicated difficulties such as different Quality of Service selection, violation handling, and monitoring. A Quality of Services position forecast context for cloud computing amenities using with previous facility knowledges of users has proposed. This QoS context is used to circumvent the expensive real-world service invocations and time-consuming. This framework used Collaborative filtering approach to predict "Quality of Services" for the web amenities; this can also be helped for cloud computing amenities. In this framework one of the authors have used Pearson Correlation Coefficient approach very nicely to find the likeness among customers and correctness of status method has to be measured. Similarly, all the Quality of service based approaches have their strength and weakness which is very beneficial to researchers for further studies. This Table 2 concludes that all the existing approaches having their strength and weakness.

Table 2 Comparison of Qos techniques in cloud computing

S. no.	Author/Year	Title	Technique used	Strength	Weakness
1.	B. Chitra/2013	A "survey on optimizing QoS during service level agreement in cloud" [10]	Cloud monitoring system for quality of services	Periodic polling is operated and details are produced	It is failed to find communication cost
2.	P. C. Hershey/2015	System of system for QoS observation and response in cloud [11]	Enterprise management monitoring and response architecture in cloud computing (EMMRA CC) and system of system method (SoS)	It improved the performance of QoS and prevents distributed denial of service (DDoS) attacks	It is not applied to federated clouds because cloud providers servers are not integrated in the real time
3.	Xiao Liu/2011	A Generic quality of services framework for cloud workflow systems [12]	Generic QoS framework for cloud workflow systems	Demonstrated cloud system implementation and evaluated effectiveness of performance framework	Complex difficulties such as monitoring and violation handling occurs
4.	M. Salam/2015	A Quality of services oriented inter cloud computing federation framework [13]	Quality of services based cloud framework, federated coordinators (FC) and cloud coordinators (CC)	Enables service providers to dynamically work as backup for each other in peak times, Protect the Providers from any possible SLA violation	Complex services are not produced using a combination of cloud services from various cloud computing providers and No provision is made for DdoS attacks (distributed denial of service)
5.	Zibin Zheng/2013	"QoS ranking prediction for cloud services" [14]	Personalized QoS ranking prediction framework	Outperformed rating based methods and greedy techniques	It has to be considered accuracy of ranking technique
6.	W. C. Chu/2014	An approach of QoS assurance for enterprise cloud computing (QoSAECC) [27]	Used multi agent method	It Established an integrated cloud computing data services to help quality of services and service level agreement manipulation	The features & solutions of IOT is not approved into the framework and the field experiments
7.	Saurabh Kumar Garg/2011	SMICloud: a cloud framework for comparing and ranking cloud services [15]	AHP based ranking mechanism	QoS parameters are explained for both cloud service providers and consumers	Non-quantifiable QoS attributes are not used
8.	M. M. Hassan/2014	Quality of service aware resource provisioning for big data processing in cloud computing environment [16]	Used heuristic algorithms	Dynamic virtual machine allocation model and cost effective to handle big data tasks	Performance metrics (delay variation, throughput and delay) are not considered

(continued)

Table 2 (continued)

S. no.	Author/Year	Title	Technique used	Strength	Weakness
9.	Ruozhou Yu/2012	QoS-aware service selection in virtualization based cloud [17]	Algorithms based on quality of service aware selection	Dissimilar properties have been virtualized using virtualization technology	Amenity provisioning difficulties are not overcome
10.	R. Karim/2013	An end-to-end QoS mapping technique for cloud service selection [18]	Used ranking algorithm based on AHP (Analytic hierarchy process)	It Proposed a new technique of computing end-to-end values in cloud computing	There is no performance enhancement on actual QoS datasets of cloud services
11.	B. Sharma/2012	Pricing cloud computer commodities: a novel financial economic model [19]	Pricing model and profit balancing for QoS	Used realistic values as a major constraint	Consumption is not well-thought-out for computational price
12.	S. Lee/2012	A quality of service assurance middleware model for enterprise cloud computing [20]	Used agent technology	1. Automated resource allocation 2. Enhance transmission of cross layer control information with respect to some services performance such as SLA	Self-learning method is not used to find the timing of automatic allocation of cloud resources
13.	Z. Wang/1996	Routing techniques based on quality of service that supports multimedia applications [21]	Multi-constraints path problem	various quality of service requirements are optimized	Problem occurs based on NP-Hard
14.	L. Bin/2012	A Quality of service aware dynamic data replica deletion strategy for distributed storage systems under cloud computing [22]	Data replica deletion strategy (DRDS) algorithm	Save maintenance cost and disk space for distributed storage system	Increased overhead on inconsistency and update of data
15.	Linlin Wu/2012	SLA based admission control for a software as a service provider in cloud [23]	Admission Control and scheduling algorithm	Profit is maximized for service providers	Only fewer QoS constraints are considered
16.	M. Xu/2009	A Multiple quality of service constrained scheduling approach of multiple workflows for cloud computing [24]	Used scheduling algorithm	Gives good scheduling results	Availability and reliability was not added to workflows.
17.	P. Zhang/2011	A quality of services aware system for mobile cloud computing [25]	Fuzzy cognitive map and QoS prediction algorithm	Facilitates quality of service establishment, assessment, prediction and assurance	No good model with suitable configurations was generated

(continued)

Table 2 (continued)

S. no.	Author/Year	Title	Technique used	Strength	Weakness
18.	Y. Xiao/2010	Reputation-based quality of services provisioning in cloud environment via Dirichlet multinomial model [26]	Dirichlet multinomial techniques	The proposed management framework provides an efficient QoS provisioning scheme for cloud computing	This method is not used to privacy and security metrics

6 Conclusion

In this survey, we have studied various QoS techniques in cloud to ascertain the extent to which QoS challenges have been resolved and their advantages and disadvantages. We have compared various QoS management techniques with their advantages and disadvantages. Although many researchers have provided scheduling techniques, traffic control, admission control, and dynamic resource provisioning response time to handle the issue of QoS management in cloud. The important process in cloud computing is QoS which leads to faster response to requests of cloud consumers. Several researchers have been done the work in this area to enhance the QoS in cloud computing, but still lot of possibilities to enhance the performance as to improve the response time, decrease delay over the network. Various researchers have searched the ways to enhance it and their research work is very useful for further study.

References

1. Dubey S, Agrawal S (2013) Methods to ensure quality of service in cloud computing environment. Int J Advanc Res Comp Sci Softw Eng 3(6):405–411
2. Ardagna D, Casale G, Ciavotta M, Perez JF, Wang W (2014) Quality-of-Service in cloud computing: modeling techniques and their applications. J Internet Serv Appl 5(11):1–12
3. Yuchao Z, Bo D, Fuyang P (2012) An adaptive QoS-aware cloud. In: Proceedings of international conference on cloud computing, technologies, applications and management, pp 160–163
4. Stantchev V, Schropfer C (2009) Negotiating and enforcing QoS and SLAs in grid and cloud computing. In: Proceedings of international conference on advances in grid and pervasive computing, GPC 09, pp 25–33
5. Zhang Z, Zhang X (2010) A load balancing mechanism based on ant colony and complex network theory in open cloud computing federation. In: Proceedings of 2nd International conference on industrial mechatronics and automation (ICIMA), Wuhan, China, May 2010, pp 240–243
6. Rajesekaran V, Ashok AA, Manjula R (2014) Novel sensing approach for predicting SLA violations. Int J Comp Trends Technol (IJCTT) 10(1), 25–26
7. Schubert L (2010) The future of cloud computing, opportunities for European cloud computing beyond 2010. In: Jeffery Em K, Neidecker- Lutz B (eds) Mar 2013 http://cordis.europa.eu/fp7/ict/ssai/docs/Cloudreportfinal.pdf
8. Li ZH, O'Brien L, Zhang H, Cai R (2012) On a catalogue of metrics for evaluating commercial cloud services. In: 13th international conference on grid computing. ACM/IEEE, pp 164–173

9. Reixa M, Costa C, Aparicio M (2012) Cloud services evaluation framework. In: Proceedings of the workshop on open source and design of communication. ACM, pp 61–69
10. Chitra B, Sreekrishna M, Naveenkumar A (2013) A survey on optimizing QoS during service level agreement in cloud. Int J Emerg Technol Advanc Eng 3(3)
11. Hershey PC, Rao S, Silio CB, Narayan A (2015) System of systems for Quality-of-Service observation and response in cloud computing environment. IEEE Syst J 9(1):1–5
12. Liu X, Yang Y, Yuan D, Zhang G, Li W, Cao D (2011) A generic QoS framework for cloud workflow systems. In: Published in Ninth IEEE international conference on dependable, autonomic and secure computing
13. Salam M, Shawish A (2015) A QoS-oriented inter-cloud federation framework. IEEE Syst J 642–643
14. Zibin Z, Xinmiao W, Yilei Z, Michael RL, Jianmin W (2013) QoS ranking prediction for cloud services. IEEE Trans Parallel Distribut Syst 24(6)
15. Saurabh Kumar G, Steve V, Rajkumar B (2011) SMICloud: a framework for comparing and ranking cloud services. In: Fourth IEEE international conference on utility and cloud computing
16. Hassan MM, Song B, Shamin MS, Alamri A (2014) QoS aware resource provisioning for big data processing in cloud computing environment. In: Proceedings of international conference on computational science and computational intelligence, pp 107–112
17. Ruozhou Y, Xudong Y, Jun H, Qiang D, Yan M, Yoshiaki T (2012) QoS-aware service selection in virtualization-based cloud computing. In: China-EU international scientific and technological cooperation program (0902)
18. Karim R, Ding C, Miri A (2013) An end-to-end QoS mapping approach for cloud service selection. In: Proceedings of IEEE Ninth world congress on services, pp 341–348
19. Sharma B, Thulasiram RK, Thulasiraman P, Garg SK, Buyya R (2012) Pricing cloud compute commodities: a novel financial economic model. In: Proceedings of 12th IEEE/ACM international symposium cluster, cloud and grid computing, pp 451–457
20. Lee S, Tang D, Chen T, Chu WC (2012) A QoS assurance middleware model for enterprise cloud computing. In: Proceedings of IEEE 36th international conference on computer software and application workshops, pp 322–327
21. Wang Z, Crowcroft J (1996) Quality-of-service routing for supporting multimedia applications. J Sel Areas Commun 14(7):1228–1234
22. Bin L, Jiong Y, Hua S, Mei N (2012) A QoS-aware dynamic data replica deletion strategy for distributed storage systems under cloud computing environments. In: Proceedings of second international conference on cloud and green computing, pp 219–225
23. Linlin W, Saurabh Kumar G, Rajkumar B (2012) SLA-based admission control for a Software-as-a-Service provider in cloud computing environments. J Comp Syst Sci 78 1280–1299
24. Xu M, Cui L, Wang Y, Bi B (2009) A multiple QoS constrained scheduling strategy of multiple workflows from cloud computing. In: IEEE international symposium on parallel and distributed proceeding with applications, pp 629–633
25. Zhang P, Yan Z (2011) A QoS-aware system for mobile cloud computing. In: Proceedings of IEEE, pp 518–522
26. Xiao Y, Lin C, Yiang Y, Chu X, Shen X (2010) Reputation-based QoS provisioning in cloud computing via Dirichlet multinomial model. In: Proceedings of IEEE international conference on communications, pp 1–5
27. Chu WC, Yang C, Lu C, Chang C, Hsueh N, Hsu T, Hung S (2014) An approach of quality of service assurance for enterprise cloud computing (QoSAECC). In: Proceedings of international conference on trustworthy systems and their applications, pp 7–13

Skyline Computation for Big Data

R. D. Kulkarni and B. F. Momin

Abstract From a multidimensional dataset, a skyline query extracts the data which satisfy the multiple preferences given by the user. The real challenge in skyline computation is to retrieve such data, in the optimum time. When the datasets are huge, the challenge becomes critical. In this paper, we address exactly this issue focusing on the big data. For this, we aim at utilizing the correlations observed in the user queries. These correlations and the results of historical skyline queries, executed on the same dataset, are very much helpful in optimizing the response time of further skyline computation. For the same purpose, we have earlier proposed a novel structure namely Query Profiler (QP). In this paper, we present a technique namely SkyQP to assert the effectiveness of this concept against the big data. We have also presented the time and space analysis of the proposed technique. The experimental results obtained assert the efficacy of the SkyQP technique.

Keywords Skyline queries · Correlated queries · Query profiler

1 Introduction

Skyline queries are special types of data retrieval queries that accept multiple user preferences and produce the most ideal objects of the users' interest. Hence, these queries have a wide range of applications in decision support systems. Unlike many other decision support systems which use either of mathematical models, cumulative scoring functions, data mining, or statistical analysis, skyline queries are more understandable to the end users. As an example, consider that a person desiring to rent a house need to find out all such possible houses where rent is minimum,

R. D. Kulkarni (✉) · B. F. Momin
Department of Computer Science and Engineering, Walchand College of Engineering,
Sangli, Maharashtra, India
e-mail: rupaliwaje@rediffmail.com

B. F. Momin
e-mail: bfmomin@yahoo.com

D. K. Mishra et al. (eds.), *Data Science and Big Data Analytics*,
Lecture Notes on Data Engineering and Communications Technologies 16,
https://doi.org/10.1007/978-981-10-7641-1_23

distance to surrounding schools is minimum, and distance to hospitals in the vicinity is also minimum. This generates a skyline query for houses with minimum rent, minimum distance to schools, and minimum distance to hospitals. From the related large dataset, the desired data is returned which satisfies the preferences imposed by the user, which is *Skyline* of the houses. The task of decision-making gets simplified using the skyline.

In practice, the following observations are common in concern with the skyline computation. (1) To serve the best to the users, the datasets tend to store more and more data, and hence they are huge on the factors like the cardinality and the number of dimensions. (2) As number of user preferences grow, the response time of the skyline queries is impacted, and (3) Popular datasets get queried at large which affects response time of overall skyline computation. Exactly these issues have been addressed through this paper. We focus on optimization of the response time of the skyline computation against the practical observations mentioned above.

Through this paper, we make the following contributions:

1. We reutilize the concept of *Query Profiler* (*QP*) [1], a data structure earlier proposed by us and use it for maintaining the metadata of the skyline queries raised against especially the large dataset.
2. We present a technique namely *SkyQP* for skyline computation of the correlated queries which is based on *QP*.
3. We then present the exclusive time and space analysis, based on the experimental results obtained from the work which justify the efficacy of the proposed technique.

The rest of the paper has been organized as follows. Section 2 gives a brief review of the parallel skyline computing techniques. In Sect. 3, the technique *SkyQP* has been detailed. Section 4 discusses the experimental work along with the time and space analysis. The concluding Sect. 5 is about the inferences drawn from the experiments and the extensions possible for this work in near future.

2 Background and Related Work

S. Borzsonyi et. al. proposed the theme of skyline queries and a skyline operator [2]. Since the era was of centralized computing, all the algorithms evolved targeted the centralized, single dataset for query processing. Various techniques which evolved include the algorithms like *BNL*, *D&C* [2] which scan the entire dataset for producing the skylines. To reduce this burden of the complete scan of a dataset, the algorithms which tend to preprocess the datasets using sorting, data partitioning like *SFS* [3], *LESS* [4], *SaLSa* [5], *Bitmap* algorithm [6] were proposed. The efficient algorithm for the centralized environment has been *BBS* [7]. As the concept of data mining came up, the skyline computing techniques desired to seek the efficacy of the mining concepts. The techniques based on this include the techniques like *SkyCube* [8–10], skyline graphs [11], and *CSC* [12].

In the due course of time, technologies flourished and the era of parallel and distributed computing emerged. Then, the skyline computation algorithms were developed for the new infrastructural and processing requirements. Few of the techniques based on these technological improvements include *DSL* [13], *SSP* [14], *iSky* [15], *Skyframe* [16], and *SFP* [17]. Such approaches used the smarter data indexing structures for proper data partitioning, data pruning, and data selection from the proper data nodes. Exploiting the patterns observed in data and using such summarizations for the skyline computation has been another parallel approach which has been used in techniques like *DDS* [18, 19]. As the new programming paradigm of *MapReduce* evolved, few of the skyline computing algorithms were developed to utilize its attractive features like flexibility, scalability, fault tolerance, and user-friendly model of programming of *MapReduce*. These techniques include MR-SFS, MR-Bitmap and *MR-BNL* [20], *SKY-MR* [21], *MR-GPMRS* [22], and *MR-Angle* [23]. The emergence of modern, faster hardware like *GPU*s, multicore processors, *FPGA*s, and grids has offered faster memory operations and better degree of parallelism. To utilize these features, newer skyline computing techniques evolved. Few of them are *GGS* [24], *SkyAlign* [25], *GNL* [26], and *FPGA*-based skylines as in [27, 28].

In all the papers reviewed, none of the research effort except in [29] have used the correlations observed in the skyline queries raised by the users. The approach in [29] is limited by the features of cache and lack of cache control under heavy system loads. Hence, our research effort presents a *SkyQP* technique which differs in two ways. First, we analyze the correlations in the user queries for serving the subsequent, correlated skyline queries by making use of our earlier concept of *Query Profiler* (*QP*) [1] which is an efficiently maintained and properly indexed data structure. Second, the metadata of pre-executed skyline queries stored in *QP* efficiently cuts or minimizes the computational efforts of subsequent, correlated skyline queries. Unlike the approaches which tend to use complex memory or data indexing structures as in [8–12], our approach presents a much simpler, efficient solution which improves the response time of overall skyline computation. And unlike approaches as in [18, 19] which use correlations in the data to be processed, we focus on utilizing the correlations present in the queries raised by the users. The presented *SkyQP* approach is also free from any data preprocessing requirements.

In the next section, the proposed approach has been discussed in detail.

3 The SkyQP Technique

This section elaborates the proposed *SkyQP* technique. In the forthcoming discussions, we make the following assumptions: (1) A single large dataset gets queried by the users, (2) The minimum type of constraints (e.g., minimum rent, minimum distance, etc.) is raised by the users, and (3) The next elaboration assumes that a term query refers to a skyline query.

The next step is to understand the way in which correlated queries are processed. This has been discussed next.

3.1 Processing Correlated Skyline Queries

When large number of queries are raised against a dataset, chances are high that query dimensions overlap and hence generate a correlation each other and with the previously executed skyline queries against the same dataset. These correlations can be one of the following types: (1) an exact correlation: The dimensions of the current query are exactly same as some previous query, (2) a subset correlation: The dimensions of the current query happen to be subset of the dimensions of some of the pre-executed queries, (3) a partial correlation: The dimensions of the current query happen to be subset of one or few of the pre-executed queries and newer dimensions are also added to the query, and (4) a novel correlation: The dimensions of the current query do not match with dimensions of any of the historical query.

The *SkyQP* technique correlates every query raised by the user, with the pre-executed queries, strictly in above order and manner before it is considered for further computation. Now let us discuss how the correlated queries are processed for computing the skylines.

An Exact Correlation: The skyline for a subsequent query which has an exact type of correlation is returned immediately by returning the skyline of that query with which the current query has an exact type of correlation. Dataset access is avoided and the re-computational efforts are totally waved.

A Subset Correlation: If the subsequent query has a subset type of correlation, then its skyline is present in skylines of the previously executed queries with which the current query has a subset correlation. The intersection of these skylines of the parent queries serves as the skyline of the current query. Again, dataset access and the re-computational efforts are totally waved.

A Partial Correlation: If a subsequent query has a partial type of correlation, then the skyline is figured as described next. The skylines of the pre-executed queries with which the current query is partially correlated are called as the initial set. A query may possess a partial correlation with several pre-executed queries. Under this scenario, the union of the skylines of all such previous queries with which current query has a partial correlation formulates the initial set that assists the further computation. Also, a partially correlated query may carry a new dimension or a set of new dimensions, which is not confined in the dimensions of any of the previous queries with whom it is partially correlated. This mandates the dataset scan. Although, the first window that contains the filtering tuples is availed by the initial set and this assists in speeding up the further computations.

A Novel Correlation: If a subsequent query has a novel type of correlation, then its skyline is computed by using any of the skyline computation algorithms as this has not been computed earlier. The dataset access is obviously mandatory.

This elaboration highlights the fact that, if the skylines of the historical queries are preserved, such metadata about the queries serves best for the queries correlated by either an exact or a subset correlation as the scan of the dataset is totally avoided. And for the partially correlated queries, the metadata helps to gear up of the computation.

And in due course of time, with the continuous queries raised against the same dataset, the queries which were correlated by either a partial or novel type of correlation may correlate with an exact or a subset correlation and as a result their skylines can be served relatively faster. So, we conclude that a structure that keeps statistics of every query executed by the system is very much helpful to improvise the response time of the overall skyline computation related to a dataset. Such data structure is *Query Profiler* (*QP*) previously proposed by us. The structure of *QP* is: *QP* = {*QId*, *Att*, *S*, *Sb*, *Pr*, and *Qf*} where *QId*: a unique, numerical identifier for each query, *Att*: is the set of the dimensions present in the query which proves helpful in finding the correlations, *S*: is the skyline of the query, *Sb*: is the set of *QId*s that denote with which different queries, the current query has a subset type of correlation, *Pr*: is the set of *QId*s that denote with which different queries, the current query has a partial type of correlation, and *Qf*: indicates the frequency of each query occurrence. This *QP* is maintained by a query receiving machine and managed efficiently by hash indexing and sorting as explained in [1]. The proposed technique *SkyQP* is based on *QP*. The next section discusses the proposed technique *SkyQP*.

3.2 *The Overview of the* SkyQP *Technique*

The proposed technique *SkyQP* aims at optimization of the response time of skyline computation by utilizing the concept of *QP* and the correlations observed in the skyline queries. Figure 1 depicts the overview of the proposed technique.

The technique works on a machine or server containing the large dataset *D* which receives the skyline queries from users. This machine or server also maintains *QP* with itself and maintains the metadata of all skyline queries. Upon receipt of a

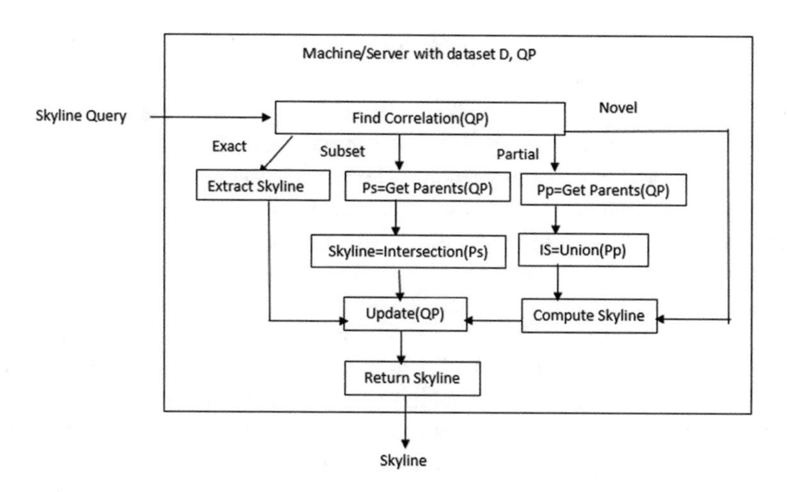

Fig. 1 The overview of SkyQP technique

query, the correlation of this current query is found with the existing queries in *QP*. Depending upon the type of correlation observed, the skyline computation steps vary. For the correlations of exact and subset type, the dataset access is totally avoided as their skylines are served from *QP* itself as explained in Sect. 3.1. This results in saving the re-computational efforts and the skylines are immediately returned to the users. For the partial type of correlations, their related parent queries are identified and an initial set *IS* is computed as explained earlier which helps to gear up the computations and reduces the computational efforts. With the help of *IS* and *D*, the skyline is returned to the user. Lastly, a novel type of correlation is served by accessing *D*.

This is how the *SkyQP* technique improves the response time of overall skyline computation for the correlated skyline queries raised against *D*. The efficacy of the technique is justified by the experimental results. These have been discussed in next section.

4 Experimental Work and Analysis

This section details the experimental work carried out and the analysis part. Totally, two experiments have been carried out to evaluate the performance of the *SkyQP* technique. The first one evaluates the technique on the parameter of speed and the other one intends to judge the memory requirements.

The work has been detailed next.

4.1 Experimental Work

The configuration of Intel Core i-3 2100 CPU, 3.10 GHz, 2 GB RAM having Windows 7 environment has been used for carrying out the experiments. For this work, a large dataset of a high-energy project called STAR has been used. The dataset is available at https://sdm.lbl.gov/fastbit/data/samples.html. This dataset has more than one lac records and thirteen dimensions. The terms assumed in the experiments are as follows: *n*: the dataset cardinality means the number of tuples in the dataset, *d*: the number of dimensions of the dataset, and *Q*: number of queries raised against the dataset. Totally, two experiments have been carried out. The first experiment observes effect on the response time upon variance of the number of queries, and the second one evaluates the memory efficacy. The experiments involve comparison between two methods: (1) *NQP*: This method does not implement *SkyQP* and computes the skylines of all the user queries without inspecting the correlations among the queries and without use of *QP*. (2) *QP*: This method implements the *SkyQP* technique. In both the methods, the skyline computation algorithm used is *BNL*. The results obtained have been shown in Fig. 2.

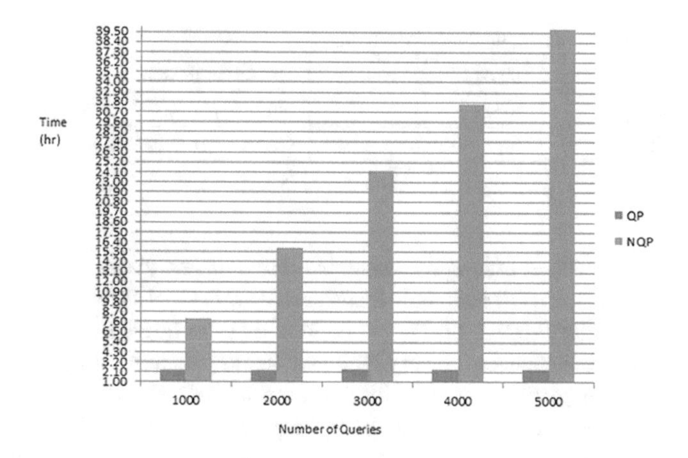

Fig. 2 Response time variance against number of queries

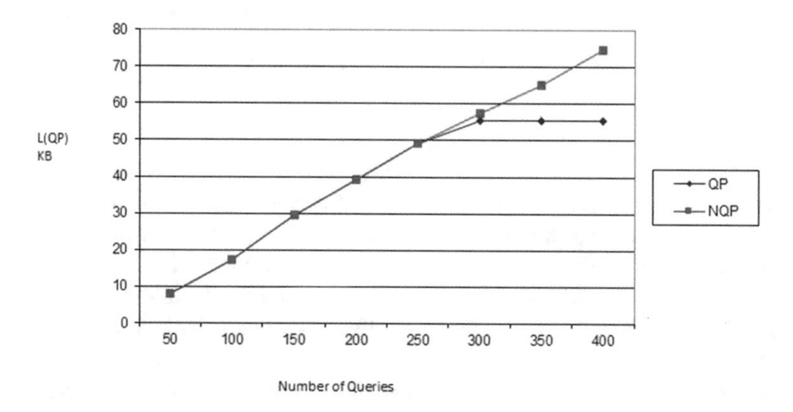

Fig. 3 Memory efficacy of the techniques

The parameter sets for this experiment are $n = 100{,}000$ and $d = 5$. The parameter Q is varied from 1000 to 5000. Totally, 28.6 % of novel queries have been used. The observations assert the optimal performance by the *QP* method showing 86% gain in the response time. This gain is high because, when very large number of queries are raised against the same dataset, a very close correlation is observed in the user queries. As the *QP* method has implemented *SkyQP*, the correlations in user queries get well exploited for achieving optimum response time.

The aim of the second experiment has been to study the memory requirements of the above two methods. The parameter sets for this experiment are $n = 100{,}000$ and $d = 5$. The parameter Q has been ranged from 50 to 400. The results obtained have been shown in Fig. 3.

It is observed that, for the *QP* method, the memory requirements are much lesser than the *NQP* method. This is because of the fact that the *QP* technique uses the

Table 1 Time and space complexities of the methods

	The *NQP* approach	The *QP* approach
Time complexity	$O(n \cdot q)$	$O(n \cdot (q_n + q_p)) + O(q_e + q_s)$
Space complexity	$O(q \cdot S_s)$	$O((q_n + q_p + q_s) \cdot S_{qp}) + O(S_{qp})$

SkyQP technique which efficiently maintains the metadata about the user queries. On the other hand, the *NQP* technique treats each query raised as the novel query and invests in the memory repeatedly. When large number of queries are raised against the same dataset, the chances are very high that the novel queries do get categorized as exact, subset, or partial queries and in turn, the subset and partial queries repeat. When such scenario occurs, no more memory requirement is put by the *QP* technique and hence after a linear shift, the *QP* method has demonstrated a steady behavior, as depicted in Fig. 3.

The inferences drawn from these experiments have been analyzed next.

4.2 Analysis

For the analysis, consider the following terms: n: total number of tuples, q: total number of skyline queries raised against the dataset, q_n: number of novel correlations, q_e: number of exact correlations, q_s: number of subset correlations, q_p: number of partial correlations, S_s: size of skylines generated, and S_{qp}: size of the *QP* for related single query. The observations derived from the experiments have been summarized in the table given (Table 1).

As per the observations mentioned above, it is found that $q = q_n + q_e + q_s + q_p$ and that $q_n \ll (q_e + q_s + q_p)$ and the due to the efficient management strategies of *QP*, $S_{qp} < S_s$. When the dataset gets queried at large, more and more queries raised by users get correlated and better performance on both the aspects of time and space is delivered by the proposed approach. With the help of these practical observations, efficiency of the proposed technique is justified.

The conclusions and future extensions possible to this work have been detailed in the last section.

5 Conclusions and Future Extensions

The proposed technique *SkyQP* has exploited the fact that correlations exist in user queries and they are very useful in either avoiding or minimizing the efforts involved in the skyline computation. When large number of queries are raised against a dataset, a tighter correlation is observed in the user queries. These correlations are maintained by *QP*. And both these facts benefit the *SkyQP* technique. The results obtained are encouraging and assert the effectiveness of the proposed technique.

In the near future, we have plans to extend this work for including the practical facts like parallel updates done on the dataset, various types of skylines demanded by the user (e.g., top-k or reverse skylines), implementation of the technique on multicore processors, etc.

References

1. Kulkarni RD, Momin BF (2016) Skyline computation for frequent queries in update intensive environment. J King Saud Univ Comput Inf Sci 28(4):447–456
2. Borzsonyi S, Kossmann D, Stocker K (2001) The skyline operator. In: Proceedings IEEE international conference on data engineering, pp 421–430
3. Chomicki J, Godfrey P, Gryz J, Liang D (2003) Skyline with presorting. In: Proceedings IEEE international conference on data engineering, pp 717–719
4. Godfrey P, Shipley R, Gryz J (2005) Maximal vector computation in large data sets. In: Proceedings IEEE international conference on very large databases, pp 229–240
5. Bartolini I, Ciaccia P, Patella M (2006) SaLSa: computing the skyline without scanning the whole sky. In: Proceedings IEEE international conference on information and knowledge management, pp 405–411
6. Kossmann D, Ramsak F, Rost S (2002) Shooting stars in the sky: an online algorithm for skyline queries. In: Proceedings IEEE international conference on very large databases, pp 275–286
7. Papadias D, Tao Y, Fu G, Seeger B (2005) Progressive skyline computation in database systems. ACM Trans Database Syst 30(1):41–82
8. Xia T, Zhang D (2005) Refreshing the sky: the compressed skycube with efficient support for frequent updates. In: Proceedings ACM SIGMOD International Conference on Management of Data, pp 493–501
9. Yuan Y, Lin X, Liu Q, Wang W, Yu JX, Zhang Q (2005) Efficient computation of the skyline cube. In: Proceedings IEEE international conference on very large databases, pp 241–252
10. Zhang N, Li C, Hassan N, Rajasekaran S, Das G (2014) On skyline groups. IEEE Trans Knowl Data Eng 26(4):942–956
11. Zheng W, Zou L, Lian X, Hong L, Zhao D (2014) Efficient subgraph skyline search over large graphs. In: Proceedings ACM international conference on conference on information and knowledge management, pp 1529–1538
12. Lin J, Wei J (2008) Constrained skyline computing over data streams. In: Proceedings IEEE international conference on e-business, engineering, pp 155–161
13. Wu P, Zhang C, Feng Y, Zhao B, Agrawal D, Abbadi A (2006) Parallelizing skyline queries for scalable distribution. In: Proceedings IEEE international conference on extending database technology, pp 112–130
14. Wang S, Ooi B, Tung A, Xu L (2007) Efficient skyline query processing on peer-to-peer net-works. In: Proceedings IEEE international conference on data engineering, pp 1126–1135
15. Chen L, Cui B, Lu H, Xu L, Xu Q (2008) iSky: efficient and progressive skyline computing in a structured P2P network. In: Proceedings IEEE international conference on distributed computing systems, pp 160–167
16. Wang S, Vu Q, Ooi B, Tung A, Xu L (2009) Skyframe: a framework for skyline query processing in peer-to-peer systems. VLDB J 18(1):345–362
17. Jensen HC, Lu H, Ooi HB (2006) Skyline queries against mobile lightweight devices in MANETs. In: Proceedings IEEE international conference on data engineering, pp 66–72
18. Hose K, Lemke C, Sattler K (2006) Processing relaxed skylines in PDMS using distributed data summaries. In: Proceedings IEEE international conference on information and knowledge management, pp 425–434

19. Hose K, Lemke C, Sattler K, Zinn D (2007) A relaxed but not necessarily constrained way from the top to the sky. In: Proceedings international conference on cooperative information systems, pp 339–407
20. Zhang B, Zhou S, Guan J (2011) Adapting skyline computation to the MapReduce framework: algorithms and experiments. In: Proceedings international conference on database systems for advanced applications, pp 403–414
21. Park Y, Min J-K, Shim K (2013) Parallel computation of skyline and reverse skyline queries using MapReduce. J VLDB Endow 6(14):2002–2013
22. Mullesgaard K, Pederseny JL, Lu H, Zhou Y (2014) Efficient skyline computation in MapReduce. In: Proceedings international conference on extending database technology, pp 37–48
23. Chen L, Hwang K, Wu J (2012) MapReduce skyline query processing with a new angular partitioning approach. In: Proceedings international conference on parallel and distributed processing symposium, pp 403–414
24. Bgh K, Aasent I, Maghni M (2013) Efficient GPU-based skyline computation. In: Proceedings international workshop on data management on new hardware, Article no. 5
25. Bgh K, Chester S, Assent I (2015) Work-efficient parallel skyline computation for the GPU. J Very Large Data Bases Endow 962–973
26. Choi W, Liu L, Yu B (2012) Multi-criteria decision making with skyline computation. In: Proceedings IEEE international conference on information reuse and integration, pp 316–323
27. Woods L, Alonso G, Teubner J (2013) Parallel computation of skyline queries. In: Proceedings IEEE international conference on field-programmable custom computing machines, pp 1–8
28. Woods L, Alonso G, Teubner J (2015) Parallelizing data processing on FPGAs with shifter lists. J ACM Trans Reconfig Technol Syst 8(2)
29. Bhattacharya A, Teja P, Dutta S (2011) Caching stars in the sky: a semantic caching approach to accelerate skyline queries. In: Proceedings international conference on database and expert systems applications, pp 493–501

Human Face Detection Enabled Smart Stick for Visually Impaired People

Shivam Anand, Amit Kumar, Meenakshi Tripathi and Manoj Singh Gaur

Abstract The present work enhances the capabilities of a newly developed smart stick (Sharma et al Multiple distance sensors based smart stick for visually impaired persons, Las Vegas, pp 1–5, 2017 [1]) by detecting human faces using the PI camera on Raspberry Pi board. Visually impaired people can use this stick developed by us (Sharma et al Multiple distance sensors based smart stick for visually impaired persons, Las Vegas, pp 1–5, 2017 [1]) to locate static and dynamic obstacles using multiple distance sensors and now can even detect the presence of a human if he/she is in front of the user. The problem of human face detection with simple and complex backgrounds is addressed in this paper using Haar-cascade classifier. Haar classifier has been chosen because it does not require high computational cost while maintaining accuracy in detecting single as well as multiple faces. Experimental results have been performed on the smart stick in indoor and outdoor unstructured environments. The stick is successfully detecting the human face(s) and generates alerts in form of vibration in the stick as well as audio in a headphone. OpenCV-python is used to implement Haar-cascade classifier and an accuracy $\approx 98\%$ is achieved with this setup.

S. Anand
Electronics Engineering, HBTU Kanpur, Kanpur, India
e-mail: shivam.anand936@gmail.com

A. Kumar (✉)
Computer Science and Engineering, Indian Institute of Information Technology Kota, Jaipur, India
e-mail: amit@iiitkota.ac.in

M. Tripathi · M. S. Gaur
Computer Science and Engineering, Malaviya National Institute of Technology Jaipur, Jaipur, India
e-mail: mtripathi.cse@mnit.ac.in

M. S. Gaur
e-mail: gaurms@gmail.com

D. K. Mishra et al. (eds.), *Data Science and Big Data Analytics*,
Lecture Notes on Data Engineering and Communications Technologies 16,
https://doi.org/10.1007/978-981-10-7641-1_24

Keywords Raspberry pi · OpenCV · Python · Haar-cascade · Face detection

1 Introduction

The goal of face detection is to locate the face regardless of the object's position, a number of objects, scale, rotation, orientation, and illumination. Human face detection problem considers more challenges compared to normal object detection due to its dynamic characteristics in terms of shape, texture, and geometric features. Additional challenges occur in the presence of glass, noise, lightning, and resolution. Particularly, face detection and tracking become infeasible when a human face is not clearly distinguished from background scene. Face detection, recognition, and tracking are found very helpful in numerous applications like biometric security, human–machine-interaction, surveillance systems, gender detection and classification and much more. Therefore, these problems have attracted researchers' attention and a lot of research has been reported in past few decades [2–6]. Recently, it has also become popular in commercial places for person identification.

Face detection techniques can be broadly categorized as (i) Feature-based approach and (ii) Appearance-based approach. First, feature-based approaches extract features in an image to detect faces. These features may be edges, color, geometry, Eigen value, and other vector-based invariant features [7–10]. In [9], face detection technique in color images with complex backgrounds has been proposed based on nonlinear color-transformation and fuzzy systems.

These methods detect the region of skin over the entire image and then produce face objects based on the spatial arrangement. Facial features have efficiently been extracted using Gabor filters in [7, 8] with some preprocessing to improve the performance of a facial recognition system. Learning techniques such as recurrent convolutional neural networks (RCNN), principal component analysis (PCA) and independent component analysis (ICA) methods have also been used to segment facial region in complex backgrounds even in the presence of noise and illumination [7–10]. Feature-based face detection techniques are quite robust and accurate but computationally heavy for real-time applications, especially for embedded systems. Further, these methods need more integration features to enhance the adaptability.

The appearance-based approaches have the capability to process the whole image simultaneously and detect the face very fast. Several human face detection results based on appearance method has been reported in the literature [11–14]. Recently this technique has gained momentum for practical applications [15–18]. Appearance-based approaches use the geometric structure and/or the intensity values of pixels in an image as the measurements. The geometric primitives such as points and curves can be used to locate distinctive features such as hair, eyes, nose, mouth, lips, and others [12, 18]. Appearance-based methods have also incorporated some preprocessing phase and constructed a rapid frontal face detection system [17, 18]. Further, the face detection technique based on Gabor transform and wavelet transform especially using Haar-cascade classifier have been adapted for speedy face detection without compromising the robustness [18]. Initially, Viola and Jones adapted the concept of Haar wavelet which is computationally light to process and developed Haar-like features. In [19], Viola and Jones have introduced a real-time face detection system using Haar-like features which are further analyzed and improved in [20–22]. The key advantage of Haar-like features over most of the other methods is its computational speed, while maintaining accuracy, in real-time applications. We have also adapted the concept of Haar-like features by considering its run-time benefits and developed a smart stick which is capable to detect single/multiple human faces. The face detection speed and accuracy are shown in the experimental results section.

The rest of the paper is organized as follows. First, the problem, which is addressed in this paper, is discussed in Sect. 2. Section 3 is devoted to the detailed methodology of the adapted face detection technique. The development of the proposed smart stick is briefly explained in Sect. 4. Further, the experimental results are discussed in Sect. 5 and finally concluding remarks are given in Sect. 6.

2 Problem Definition

The problem is to capture the region of single/multiple human faces in an arbitrary image or video. Given an image as a frame taken from the video stream, automatic face detection is considered as complex task due to the presence of multiple objects and cluttered background. We need to extract facial features by ignoring background and anything else in a captured image.

Face detection is a common problem that determines the location, shape, and size of a face. Particularly, human face detection is considered more challenging because while its anatomy is same, there are a lot of environmental and personal factors affect facial features. Face detection becomes more difficult when interpersonal variability (pose, illumination conditions, facial expressions, use of cosmetics, hairstyle, beard, the presence of glasses, etc.) and extra personal variability (age, gender, race, etc.) involved.

The above-mentioned objective at reasonable accuracy rate is achieved using Haar-cascade classifier whose detailed description is provided in the subsequent section.

3 Face Detection

Haar-like features, developed by Viola and Jones in [19], have been used in this paper to detect the human face(s) because of its computation speed and reliability. This feature makes it suitable for real-time applications and described below.

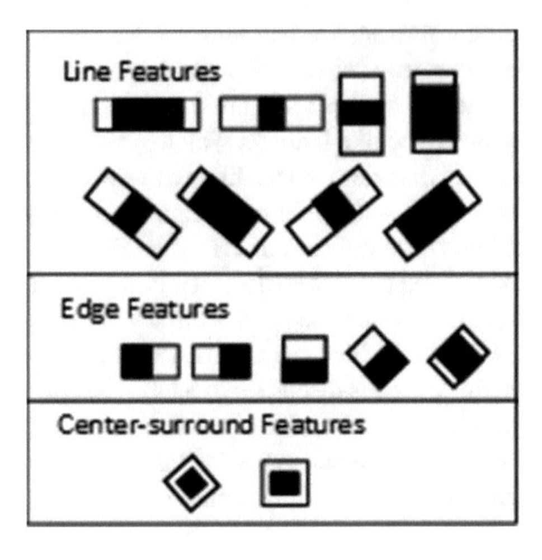

Fig. 1 Sample Haar features which have been used for rapid face detection

3.1 Haar-Like Features

A Haar-like feature is computed by considering neighboring regions at a particular position in a sub-window in an image. The feature is computed by summing up the intensity values of all pixels in each region and then calculate the difference of these sums to compute the integral portion of the image as shown in Fig. 2 and expressed in (1). The difference is then used to categorize subsections of an image. Figure 1 explains how to calculate each feature. A feature is calculated as a value and this value is the difference of sum of white pixels and sum of black pixels in every rectangle. For instance, eyes subsection is commonly darker than the cheeks subsection in a human face image. A Haar-like feature, in this case, is a set of two neighboring rectangular areas above the eye and cheek regions. The references [19, 21] can be made for more details.

Haar features start scanning the image from left to right towards the bottom to extract facial features. These Haar features scan the image several times using different Haar features in order to extract maximum facial features. The features are computed rapidly using the concept of the integral image which requires only four values at the corners of the rectangular window to calculate the sum of all pixels inside any rectangular section as shown in Fig. 2. In Fig. 2, the sum of all pixels within rectangle A is represented by G1. G2 represents the sum of all pixels within the rectangle A + B. Similarly, G3 and G4 represent the sum of all pixels within A + C and A + B + C + D, respectively.

The sum of the pixels within the rectangle D can be computed as

$$\begin{aligned} &G1 = A,\ G2 = A + B,\ G3 = A + C,\ G4 = A + B + C + D \\ &D = G4 - G2 - G3 + G1 \\ &D = (A + B + C + D) - (A + B) - (A + C) + (A) \end{aligned} \tag{1}$$

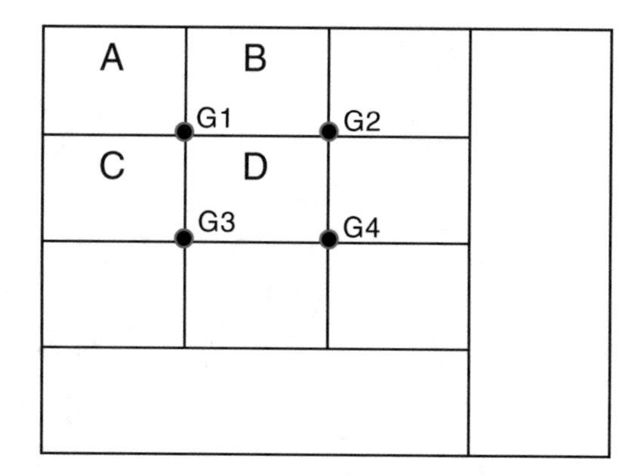

Fig. 2 Computation of integral image. A sample integral image (rectangular window D) is computed here

The integral image at the position (x, y) can be computed as

$$ii(x, y) = \sum_{x' \le x, y' \le y} i(x', y'), \tag{2}$$

where, ii(x, y) is the integral image and i(x, y) is the original image. The integral image ii(x, y) over a single pass is computed as follows:

$$s(x, y) = i(x, y) + s(x, y - 1)$$
$$ii(x, y) = s(x, y) + ii(x - 1, y),$$

where, s(x, y) is the cumulative sum of rows, $ii(-1, y) = 0$ and $s(x, -1) = 0$.

3.2 *Classification Function*

Viola and Jones in [19] suggested the window size 24 × 24 (i.e., size of D window in Fig. 2) as the base window to start scanning and evaluating Haar features in the image. If we consider a number of parameters such as horizontal, vertical, diagonal, etc. (see Fig. 1) then we have to calculate millions of features (weak classifiers) which is practically infeasible. One solution to this problem is to refer the AdaBoost learning algorithm which extracts the best features among all possible features. The AdaBoost algorithm constructs a strong classifier using a linear combination of weak classifiers. A learning algorithm is designed to select a single rectangle feature which best classifies the positive and negative data. The weak learner determines the optimal threshold θ_i classification function as

$$h_i(x) = \begin{cases} 1 & p_i f_i(x) < p_i \theta_i \\ 0 & \text{otherwise} \end{cases}, \tag{3}$$

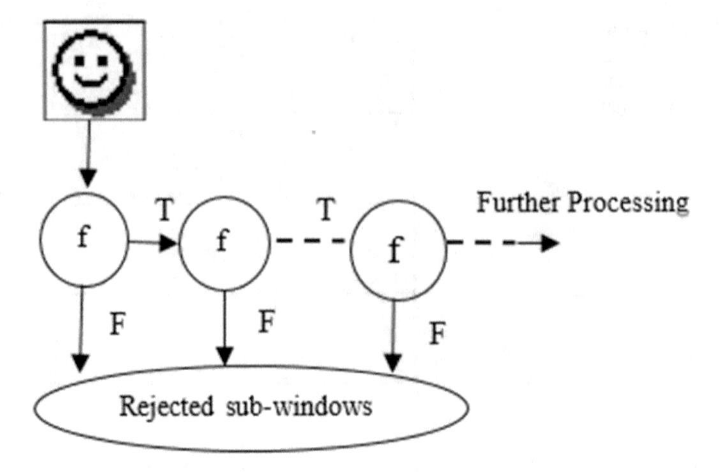

Fig. 3 Cascade of classifiers

where, $h_i(x)$ is the weak classification function, f_i is the feature, θ_i is the threshold and p_i is parity. Here, x is the size of the sub-window, i.e., 24×24.

3.3 Cascade of Classifiers

Finally, a series of weak classifiers used in cascade as shown in Fig. 3. The reference for more detailed description can be made to [20]. The system detects objects based on extracted features by moving the specified window over the image many times. Each time, the system calculates different feature and labels the specific region as positive or negative by the present location of the moving window. A negative result indicates that the object and/or portion of the object is not found in this specific region. Therefore, the moving window is moved to the next position. On the other hand, a positive result indicates the object and the system move to the next stage of classification.

The method for constructing a cascade of classifiers, shown in Fig. 3, increased the face detection performance in terms of computation time. The method becomes speedy because it rejects a number of negative sub-windows in the process of detecting positive instances. The rejection of a large number of sub-windows before proceeding to a more complex classifier helps to achieve fast detection and low false positive rates as well.

The described methodology in this section has been implemented in the developed smart stick for face detection and so human detection in front of the visually impaired person. The overall development features available with the stick and the working flow chart is briefly explained in the subsequent section.

4 Development of Smart Stick

In this paper, a cheap but durable smart stick is developed which is an improved version of the smart stick introduced in [1] and having the following characteristics:

- The designed stick is able to detect the front as well as sideways obstacles using the camera and a set of ultrasonic sensors.
- The stick is able to detect an obstacle of any height and can alert the person by telling the distance using ultrasonic sensors.
- The stick is able to tell whether the obstacle is human or other using face detection feature.
- The stick is able to detect pit, downstairs, and other dangerous hurdles using another sonar sensor which is equipped at the bottom of the stick.
- The stick attentive the person using vibrations. The vibration is generated by the vibration motor which is equipped with the stick. The vibration and the sound send to the ear of the user via wireless communication.
- The user gets the accurate distance of the obstacle on the basis of ultrasonic sensors.
- The proposed system achieved better response time before colliding the obstacles. Results are shown in the subsequent section.
- The stick is eased of use and fabricated on the plastic material to make it light weight, durable, and easy to carry.
- The training of the product is not expensive or time-taking.

The developed smart stick, shown in Fig. 4, is an embedded device which is integrated with a microcontroller, camera, vibration motor, distance sensors, Bluetooth modules, and other necessary components to make it function. Figure 5 shows the working flow diagram of the stick.

The proposed stick is able to detect obstacles of any kind which are present in front of the user. The accurate location with orientation of any obstacle can be achieved using a series of ultrasonic sensors. The stick provides alert facilities in two forms (i) vibration in the stick and, (ii) sound in the ear of the person using earphone. The strength of the vibration and audio track depend on how far the obstacle is situated.

5 Experimental Results and Discussion

5.1 Experimental Setup

The described face detection technique has been implemented in hardware using a set of Raspberry Pi board and Pi camera (see Fig. 4b). This setup of the Pi camera and Raspberry Pi microcontroller is fitted in the smart stick, which is developed by us and shown in Fig. 4c. The stick is an improvement of the smart stick introduced in [1]. The Pi camera captures the video and transmits it to Raspberry Pi microcontroller. The developed smart stick helps the visually impaired persons for navigation in

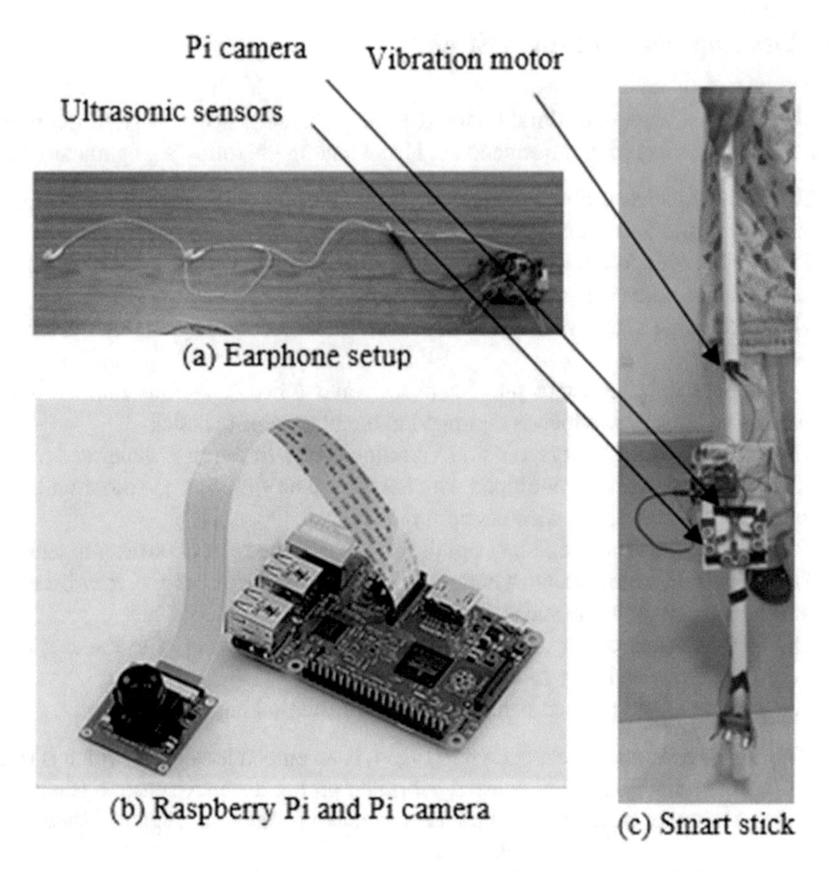

Fig. 4 (a) An earphone setup is connected with the smart stick using Bluetooth (b) the system of the Raspberry PI connected with PI camera, and (c) the developed smart stick

indoor/outdoor environments. The smart stick consists a series of ultrasonic distance sensors to calculate the distance of in-front obstacle. Along with vibration in the stick to alert the person, an audio description of the environment is also facilitated through the earphone (see Fig. 4a). The earphone set is connected with the smart stick through wireless communication.

5.2 Implementation

OpenCV-Python library has been used to implement Haar-cascade classifier in Python programming environment. This library binding with Python is designed specifically for computer vision and machine learning problems. The operating system used to run python scripts in Raspberry Pi board is Raspbian. Interfaces for high-speed GPU operations based on CUDA and OpenCL are also under active

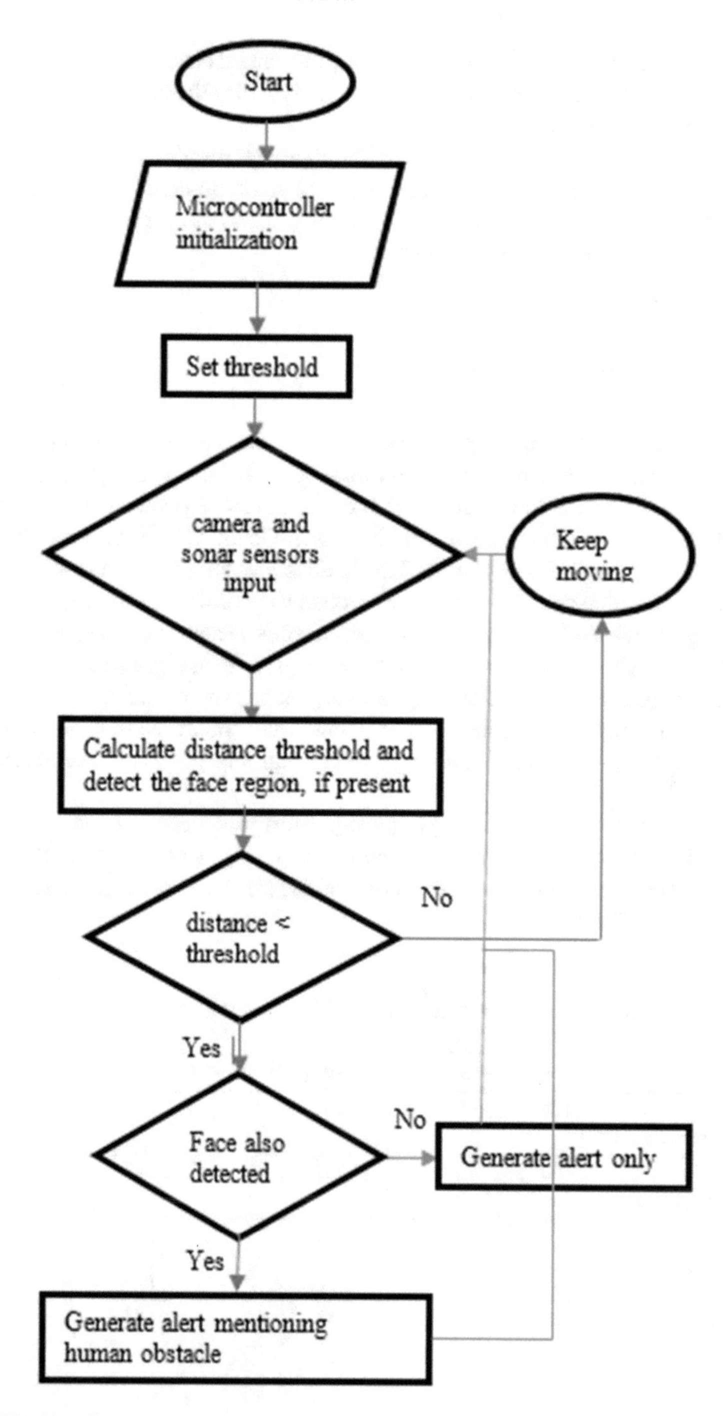

Fig. 5 Working flow diagram of the developed smart stick

development. OpenCV is available with many pretrained classifiers such as the face, eyes, smile, etc. We have used these libraries and modify the source code according to Haar-cascade classifier.

We have implemented the face detection approach on a Raspberry Pi Model B RASP-PI-3 @ 1.2 GHz microcontroller with 64-bit quad-core ARMv8 CPU with 1 GB RAM. All source code is written in Python 3 and the experiments were conducted in indoor and outdoor environments.

5.3 *Results and Discussion*

The Haar-cascade algorithm detects the face in an image by undergoing aforesaid steps. The Pi camera captures the video continuously and one image per second is set as frame rate due to the low configuration of the developed embedded system, i.e., smart stick. With the help of available libraries in OpenCV, a set of image database has been trained for real-time testing. The database contains more than 4000 images of more than 500 persons. The database is prepared according to our requirements by combining the selected images of BioID face detection database, FaceScrub, VALID multimodal database and AR-Face database. A more than eight images have been taken of each person with simple/complex backgrounds and variation in illumination.

In the testing phase, a number of experiments were conducted in various environments. A video frame may contain single and/or multiple faces. Some representative results are shown in Fig. 6.

In terms of speed, the used light weight algorithm produced reasonable fast results with face detection in $\approx$0.7 s which is good enough to generate alert message while walking with a smart stick. We have examined the accuracy of the proposed system and is shown in Table 1.

Table 1 Face detection accuracy with average time taken in real-time experiments with different environment settings

Test images	Image tested	Average time taken	Accuracy rate (%)
Set 1 Indoor direct image	100	$\approx$0.5	98
Set 2 Outdoor direct image	100	$\approx$0.7	97
Set 3 Indoor image on screen	100	$\approx$0.6	89
Set 4 Outdoor image on screen	100	$\approx$0.8	91

Fig. 6 Some representative snapshots of the hardware results detecting single and multiple faces of human in real and on screen. The video is capturing by the Pi camera which is embedded in the smart stick. The method is successfully detecting faces in various conditions and effects. The lowest accuracy, while detecting faces from real and screened object simultaneously, is close to 90%

6 Conclusion and Future Work

After conducting hardware experiments on a large number of images containing simple/complex backgrounds and human face(s), it is observed that Haar-cascade classifier is able to perform satisfactorily in embedded systems also in real-time environments with different settings. The accuracy does not significantly affect by illumination, backgrounds, wearing glasses, a number of faces, etc., as shown in the results section. This face detection approach requires very less time in computing and so one of the best suitable techniques in terms of speed and reliability to implement in hardware systems. In near future, we shall compare the accuracy performance and implement some face recognition algorithms in the same hardware system.

References

1. Sharma S, Gupta M, Kumar A, Tripathi M, Gaur MS (2017) Multiple distance sensors based smart stick for visually impaired persons. In: The 7th IEEE annual computing and communication workshop and conference. Las Vegas, pp 1–5
2. Zafeiriou S, Zhang C, Zhang Z (2015) A survey on face detection in the wild: past, present and future. Comp Vis Image Underst 138:1–24
3. Jun B, Choi I, Kim D (2013) Local transform features and hybridization for accurate face and human detection. IEEE Trans Pattern Anal Mach Intel 35(6), 1423–1436
4. Felzenszwalb PF, Girshick RB, McAllester D, Ramanan D (2010) Object detection with discriminatively trained part-based models. IEEE Trans Pattern Anal Mach Intel 32(9), 1627–1645
5. Li H, Lin Z, Brandt J, Shen X, Hua G (2014) Efficient boosted exemplar-based face detection. In: IEEE conference on computer vision and pattern recognition. Columbus, pp 1843–1850
6. Yang B, Yan J, Lei Z, Li SZ (2014) Aggregate channel features for multi-view face detection. In: IEEE international joint conference on biometrics. Clearwater, pp 1–8
7. Bellakhdhar F, Loukil K, Abid M (2013) Face recognition approach Using gabor wavelets, PCA and SVM. IJCSI Int J Comp Sci Issues 10(2):201–207
8. Abhishree TM, Latha J, Manikantan K, Ramachandran S (2015) Face recognition using gabor filter based feature extraction with anisopropic diffusion as a pre-processing technique. Proc Comput Sci 45:312–321
9. Pujol FA, Pujol M, Morenilla AJ, Pujol MJ (2017) Face detection based on skin color segmentation using fuzzy entropy. Entropy 19(1):1–22
10. Jiang H, Miller EL (2017) Face detection with the faster R-CNN. In: 12th IEEE international conference on automatic face & gesture recognition. Washington, DC, pp 650–657
11. Gross R, Matthews I, Baker S (2004) Appearance-based face recognition and light-fields. IEEE Trans. Pattern Anal Mach Intel 26(4), 449–465
12. Wolf L (2009) Face recognition, geometric vs. appearance-based. Encycl Biom. pp 347–352
13. Delac K, Grgic M, Liatsis P (2005) Appearance-based statistical methods for face recognition. In: 47th international symposium ELMAR-2005. Zadar, pp 151–158
14. Rabbani MA, Chellappan C (2007) A different approach to appearance–based statistical method for face recognition using median. IJCSNS Int J Comput Sci Netw Secur 7(4), 262–267
15. Barnouti NH, Al-dabbagh SSM, Matti WE (2016) Face recognition: a literature review. Int J Appl Info Syst 11(4):21–31
16. Dwivedi S, Gupta N (2017) A new hybrid approach on face detection and recognition. Int J Adv Res Ideas Innov Tech 3(3), 485–492
17. Chihaoui M, Elkefi A, Bellil W, Amar CB (2016) A survey of 2D face recognition techniques. Computers 5:1–28
18. Muqeet MA, Holambe RS (2017) Local appearance-based face recognition using adaptive directional wavelet transform. J King Saud Univ Comput Info Sci (Article in Press) https://doi.org/10.1016/j.jksuci.2016.12.008
19. Viola P, Jones M (2004) Robust real-time object detection. Int J Comput Vision 57(2):137–154
20. Menezes P, Barreto JC, Dias J (2004) Face tracking based on haar-like features and eigenfaces. IFAC Proc Vol 37(8):304–309

21. Wang YQ (2014) An analysis of the Viola-Jones face detection algorithm. Image Process Line 4:128–148
22. Zhang X, Gonnot T, Saniie J (2017) Real-time face detection and recognition in complex background. J Sig Info Process 8:99–112

Web-Based Service Recommendation System by Considering User Requirements

Neha Malviya and Sarika Jain

Abstract In this age of Internet and service delivery almost all the kinds of services and products are available online for selection and use. In addition of that for a single kind of product or service a number of different vendors and service providers are exist. Additionally all the providers are claimed to provide most valuable services. In this context to compare and find the appropriate service according to the end client a service recommendation system is required. The aim of this recommendation system design is to understand the client current requirements and explore the database for recovering the most likely services. In order to demonstrate the issues and solution of this domain a real-world problem namely hotel booking service is used. On the problem of this recommendation system design is treated as a search system on structured data source. Thus to find the suitable outcomes from the proposed working model quantum genetic technique is used. That technique first accepts the dataset information and the user requirements, after that the encoding of information is performed in binary values. Additionally the query sequence is treated as binary string with all 1s. Finally the genetic algorithm is implemented for finding the fit solution among all the available binary sequences. The generated seeds from the genetic algorithm are treated as final recommendation of search system. Additionally the fitness values are used to rank the solutions. The implementation and result evaluation is performed on JAVA technology. After that the performance using time and space complexity notified. Both the performance parameters demonstrate the acceptability of the work.

Keywords Data mining · Service recommendation · Genetic algorithm
Searching · Datasets

N. Malviya (✉) · S. Jain
Department of Computer Science and Engineering,
Sri Aurobindo Institute of Technology, Indore, India
e-mail: nehamalviya2415@gmail.com

D. K. Mishra et al. (eds.), *Data Science and Big Data Analytics*,
Lecture Notes on Data Engineering and Communications Technologies 16,
https://doi.org/10.1007/978-981-10-7641-1_25

1 Introduction

A rich amount of services and service providers are available now in these days online. The QoS of services are only visible on the basis of the feedbacks and online user's review. But which services are appropriate for the different user's need is a complicated issue. In this presented work a web service recommendation model is proposed for investigation and design to find most suitable and appropriate web service according to the end client's need. The recommendation systems are basically the techniques of prediction or searching the user behaviour relevant information according to available items. In order to find a suitable and efficient recommendation system for online services the data mining technique based recommendation system is proposed.

The key problem is taken from the real world issues for instance for a single kinds of service a number of service providers available. Additionally all the service provider's clime that their services are most suitable for the users but the client's needs is different from offered services. Thus a system is required to which analyse the service provider's offered service and the end client's requirements and suggest most fit service among available services. In this context the proposed work focused on online hotel service providers for investigation and solution design. Additionally it is recognized that for analysing both the sides of available inputs the heuristic based search solution is a suitable technique. Therefore the proposed technique utilizes the quantum genetic technique for providing solutions for the end client needs.

2 Proposed Work

This chapter explains the proposed system designed for recommending the online services. Therefore the detailed methodology of the system design and proposed algorithm is associated in this chapter.

2.1 System Overview

The soft computing techniques are enabling us to find much appropriate and likely data among a large amount of patterns. A rich domain of applications where high accurate results are required during search of patterns the soft computing approaches are employed such as accurators, temperature regulators, feature selection, inventory and production management and other. In this presented work a new application of soft computing approach is introduced for making recommendation system. Recommendation systems are basically a kind of suggestion system which understand the requirements and find best match among available items or products. The proposed recommendation system is intended to find the appropriate web-based service which is offered by different service providers.

In this context a most popular industry which is frequently serving their clients online is taken as initial problem namely the hotel industry. A significant amount of online users search for this service and sometimes they are not satisfied with the actual requirements. Therefore in this presented work the solution is targeted for designing a solution or recommendation system for online hotel service. The system consist of two modules first the user search space and second the solution space which is available on dataset input. In order to obtain the best match solution according to the user's need the quantum genetic algorithm is used as search algorithm for generating the solution according to the user need. The genetic algorithm is one of the popular and effective search technique in soft computing that climes to provide at least on solution at any point of time during the problem solving. In addition of that the concept of quantum is used to encode the solution according to the problem space for finding optimal solutions. This section provides the overview of proposed web-based service recommendation system. The next section provides the detailed design of the solution.

2.2 *Methodology*

The proposed system architecture for finding recommendations according to user requirements is described using Fig. 1. The figure contains the used steps for processing of data and obtaining required solution.

Load dataset: data mining is a technique of data analysis thus in any data mining application the dataset is an essential part of mining. In this presented work the online hotel booking provider's dataset is used. This dataset contains the different kinds of services offered by the service providers. Additionally the user feedback basis list of service values is also incorporated. This data is available online for download. That is the first phase of system where user selects the dataset from the local machine directory. The proposed system read the dataset and holds it into a data structure for performing analysis.

Pre-process data: data preprocessing applied on data for improving the dataset quality. Therefore the cleaning and transformation of data is performed in this phase. In this work to improve the quality of data the null values from dataset is removed. To remove the null value based instances all the dataset instances are evaluated individually and the instance with null value is replaced with not applied.

Input user requirements: that is an additional provision created for finding the user requirements. Therefore all the attributes individual values are treated as individual service and for each service all the dataset is evaluated. During this for each service the available unique values from the dataset is computed and appended into a dropdown list. By making selection from these available service and values user is performing search. Therefore the problem space is to search the user's selected values from the dataset contents.

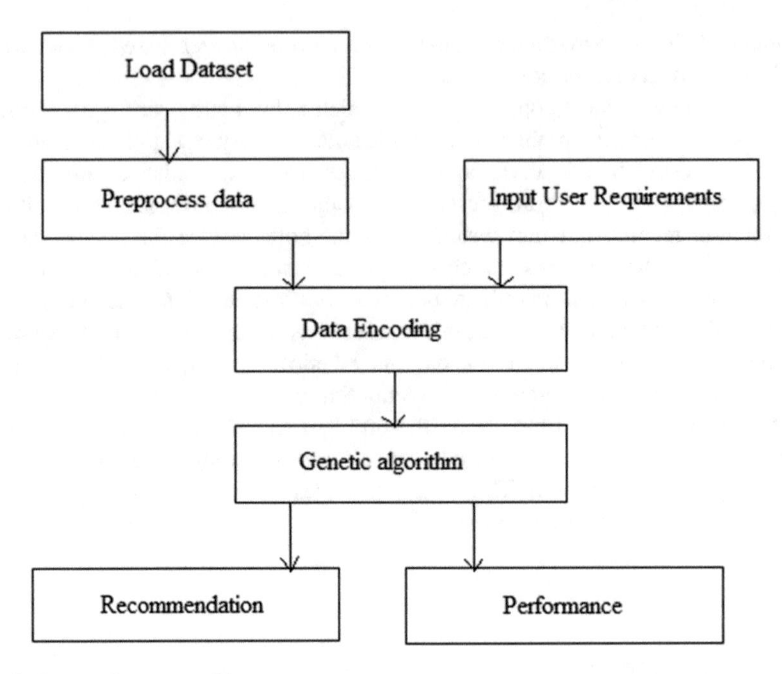

Fig. 1 Proposed system architecture

Data encoding: this phase accept the values of both parts of the system first from the user input interface and second the available dataset. After that all the dataset is compared with the user input query and the encoding of each solution sequence is performed. In this context when the dataset instance value matched with the user required values the outcome is noticed as 1 otherwise it is kept as 0. The process of dataset encoding is given using Table 1.

Genetic algorithm: after generating the binary encoded strings the genetic algorithm is applied on binary dataset. Here the entire dataset is treated as population for the genetic algorithm. After performing the search using genetic algorithm the system generates fit solutions according to their fitness values. The working of genetic algorithm is described as:

Genetic algorithm is genetically inspired search process that finds the optimum solution in huge search space. The available resources are genetically treated to find the fittest response among a number of solutions, which is basically an iterative process for discovering more appropriate solution. This search technique guarantees to find the best solution, but intermediate solution is also produces in each progressive steps. Therefore before use of this algorithm the primary functioning of the genetic algorithm is required to learn. The genetic algorithms make use of three key thought for answer discovery: reproduction, natural selection and diversity of the genes [1].

Genetic Algorithms procedures a couple of entities these entities are the system of symbols that are contributing in solution space. The new generation is formed using

Table 1 Encoding of data

Input: Dataset D, User input U Output: binary encoded strings B
Process: 1. $[row, col] = readDataset(D)$ 2. $[col, values] = readUserInput(U)$ 3. $for\ i = 1; i \leq row; i++$ a. $temp = null;$ b. $for\ j = 1; j \leq col; j++$ i. $if(D_{i,j} = U_j)$ 1. $temp.add(1)$ ii. $else$ 1. $temp.add(0)$ iii. End if c. End for d. $B.add(temp)$ 4. End for 5. Return B

the selection process and genetically encouraged operators. The fleeting description of the total pursuit process is given as.

Generate Initial Population—Primarily the genetic algorithms are started with the arbitrarily generated orders, with the permitted alphabets for the genes. For shortening the computational procedure all the generated population orders have the similar number of symbols in separately order.

Check for Termination of the Algorithm—for stopping the genetic algorithm a stopping principles is compulsory to shot for discovering the finest solution. It is likely to stop the genetic optimization procedure by means of

1. value of the fitness function,
2. Highest number of iterations
3. and fixing the number of generation.

Selection—This is a process of choosing the finest symbols amongst all entities, in this situation for determining the novel population two operators are used viz. crossover and mutation. In this state the ascending of sequences is accomplished and using these best entities is transferred to the fresh generation. The exclusivity assurances, that the value of the optimization function cannot produce the vilest results.

Crossover—The crossover is fundamentally the process of recombination the entities are selected by selection and recombined with each other. Using this new order is achieved. The purpose is to get new population entities, which get the finest likely characteristics (genes) of their parent's entities.

Mutation—The arbitrary variation on some of the genes assurances that even if none of the entities comprise the essential solution genes, it is still likely to generate them by means of mutation process by randomizing the exploration.

New generation—The selected entities from the selection procedure combined with those genes that are managed with the crossover and mutation for subsequent generation development.

According to the description in [2] the traditional genetic algorithm can defined using the below specified genetic pseudo code (Table 2).

Recommendation: the generated solutions using the genetic algorithm are used as the recommendation of the system. In addition of that the fitness values of the genetic search process is used here as the ranking base.

Performance: the performance of the proposed search system is also figured in terms of time requirements and memory usages and reported at the same time.

2.3 Proposed Algorithm

This section provides the steps of the system working therefore the entire processes of the system described above are summarized in small steps using Table 3.

3 Results Analysis

This chapter provides the details about the performance evaluation of the proposed service recommendation system. Thus, this section includes the computed result parameters by which the performance of the system is demonstrated.

3.1 Memory Consumption

Processes which are required some amount of main memory for execution of the current task. Additionally that is assigned dynamically according to the requirements of processes. The memory usages of the process or algorithm also termed as the memory consumption or the space complexity of algorithms. The memory requirements of the algorithm are computed using the following formula: Fig. 2 shows the Memory Usages.

$$MemoryUsed = Total\,Assigned\,Memory - Total\,Free\,Space$$

Table 2 Genetic algorithm

Input: instance Π, size α of population, rate β of elitism, rate γ of mutation, number δ of iterations Output: solution X

Process
// initialization
1. generate α feasible solutions randomly;
2. save them in the population POP;
//Loop until the terminal condition
3. for $i = 1$ to δ do
//Elitism based selection
4. number of elitism $ne = \alpha \cdot \beta$;
5. choose the best ne solutions in PoP and save them in PoP_1;
//Crossover
6. number of crossover $nc = (\alpha - ne)/2$;
7. for $j = 1$ to nc do
a. randomly select two solutions X_A and X_B from PoP ;
b. generate X_C and X_D by one-point crossover to X_A and X_B;
c. save X_C and X_D to PoP_2 ;
8. end for
//Mutation
9. for $j = 1$ to nc do
a. choose a solution X_j from PoP_2;
b. mutate each bit of X_j under the rate γ and generate a new solution X_j' ;
c. if X_j' is unfeasible
i. update X_j' with a feasible solution by repairing X_j' ;
d. end if
e. update X_j with X_j' in PoP_2;
10. end for
//Updating
11. update $PoP = PoP_1 + PoP_2$;
12. end for
13. Returning the best solution
14. return the best solution X in PoP ;

Table 3 Proposed algorithm

Input: dataset D, user requirements R Output: recommended services S
Process: 1. $T = readDataset(D)$ 2. $P_N = preprocessData(T)$ 3. $U = readUserinput(R)$ 4. $i = 1; i \leq N; i++$ for a. $B = encodeData(P_i, U)$ 5. End for 6. $S = geneticAlgo.search(U, B)$ 7. Return S

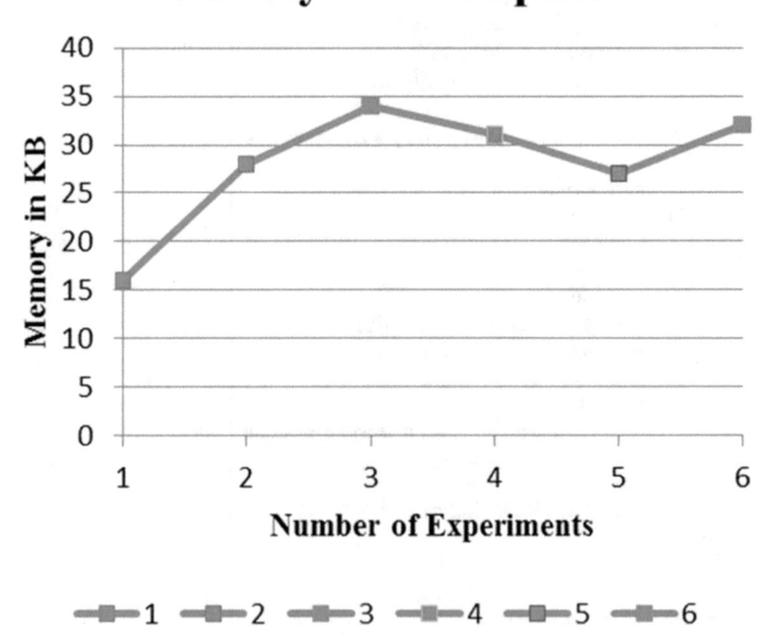

Fig. 2 Memory usages

Table 4 shows the Memory Usages in searching Data sets.

The amount of space required for proposed service recommendation model is demonstrated using Table 4 and Fig. 2. The table provides the values of memory

Table 4 Consumed memories

Number of experiments	Proposed service recommendation system
1	16
2	28
3	34
4	31
5	27
6	32

requirements in KB (kilobytes) and figure visualize the pattern of requirements on the basis of values. In this figure X-axis includes the number of experiments performed and Y-axis shows the corresponding memory consumption. According to the obtained performance the requirements of the system is increases as the amount of data increases for processing. But according to the variations observed the amount of requirements are considerable according to the data size.

3.2 Time Consumption

Processes take an amount of time for processing the input data according to the algorithms evaluation. This time requirement of the algorithm is termed as the time consumption of algorithm processing or the time complexity of the system. The time of the search processing is computed using the following formula:

$$TimeConsumption = Algorithm\,Start\,Time - Algorithm\,Endtime$$

Table 5 shows the Time Consumption in searching results. The time requirement for the proposed service recommendation model is provided in Fig. 3 and Table 5. Table includes the two attributes first the size of data for processing and second is the amount of time required for performing the evaluation. The time is measured herewith in terms of seconds The X-axis of figure includes the experimentation and the Y-axis delivers the time requirements to visualize figure. According to the found performance of the proposed recommendation system is varying in time for the same dataset. Therefore, we included different experiment for the same graph and observed resulting values. Time consumption depends on the data size which we taken for processing of algorithm.

Table 5 Time consumption

Number of experiments	Proposed service recommendation system
1	22.98
2	26.11
3	24.26
4	28.88
5	32.16
6	28.01

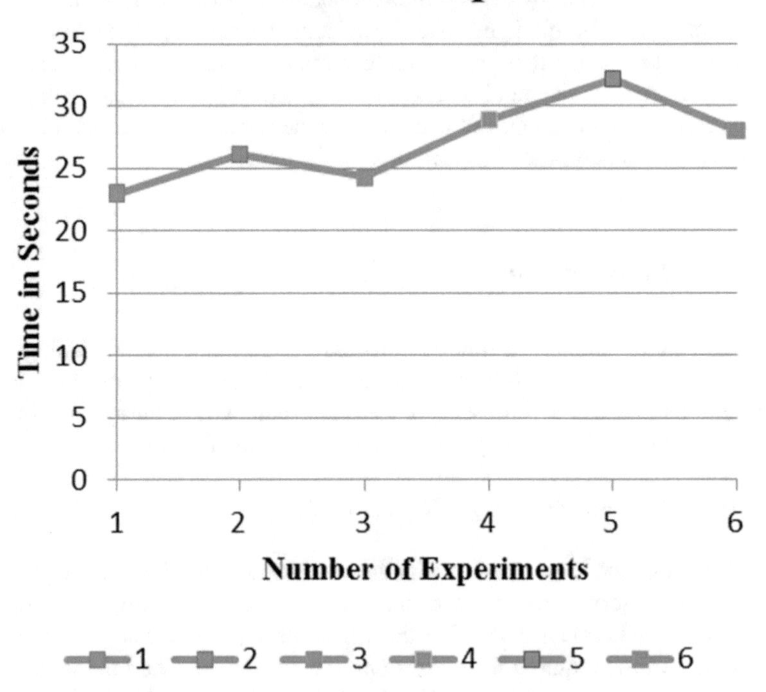

Fig. 3 Time consumption

4 Conclusion and Future Work

The aim of the proposed work is to design a recommendation system based on web-based service offering service providers according to their QoS (quality of service). The key design and implementation of the proposed technique is accomplished and this chapter presents the conclusion of work performed with future extension possibility.

Table 6 Performance summary

S. no.	Parameters	Remark
1	Time consumption	Less time consumed during the search process it increases with the amount or number of evaluation cycles
2	Memory usages	The acceptable amount of memory resource is consumed for making search over increasing amount of data. Most of the time it demonstrate the consistent outcomes

4.1 Conclusion

The term recommendation system belongs from the ecommerce where for suggesting good products this system is implemented. In addition to that in such recommendation systems the user behaviour is also incorporated. The presented work is provides a new model of recommendation system for web-based service selection. Now in these days a number of service providers and product vendors are offering their products and services only. On the other hand the online users getting confused which services are much appropriate for them. Therefore in order to find optimal selection of online services a new recommendation model is proposed.

The proposed recommendation system works on the basis of available or offered services by the service providers and their relevant feedback values. According to the needs or requirements of the end clients the services are explored and optimum services are selected and suggested to clients. In order to demonstrate the key issues of designing the service recommendation model for online offered services a real-world problem of hotel booking service is selected. Most of the time the end client is worried about the offers and the ground reality in this context a solution is required. The proposed model uses the concept of quantum genetics for finding the suitable recommendations according to client's requirements. The model is works in two phases first it processes the dataset for refinement and data encoding using the quantum concept. Here for encoding the user requirement is also required. After encoding the solution space is converted into binary strings. In next module the encoded data is used as the solution space to search the user requirements. The final outcome of the search algorithm is accepted as recommended service through the proposed system.

The implementation of the proposed technique is performed using JAVA technology and JAVA-based data structure. After implementation the evaluation of performance is performed for both space complexity and time complexity. The Table 6 holds the summary of the performance.

According to the obtained results the performance of the proposed web-based service recommendation system is acceptable for both time as well as space consumption. In near future the extensions are described in next section.

4.2 Future Work

The main aim of the work to implement a web-based service recommendation model for improving user acceptability is completed successfully. In near future the following extensions are feasible for work.

1. Currently the system is implemented only with genetic algorithm in near future more soft computing techniques are explored and optimum technique is implemented with the system.
2. Current system is not evaluated on the real-world data it is only designed and tested for the predefined dataset and their attributes in near future the efforts are made to involve real world attributes.
3. Current system does not include the social media reviews and feedbacks; it is suggested to involve both with sentiment analysis technique to enhance more the current recommendation engine.

References

1. Genetic algorithms for optimization, programs for MATLAB ® Version 1.0 User Manual
2. Xuan J, Jiang H, Ren Z Pseudo code of genetic algorithm and multi-start strategy based simulated annealing algorithm for large scale next release problem. Dalian University of Technology
3. Hu Rong, Dou Wanchun, Liu Jianxun (2014) ClubCF: A clustering-based collaborative filtering approach for big data application. IEEE Trans Emerg Topics Comput 2(3):302–313
4. Chuchra R (2012) Use of data mining techniques for the evaluation of student performance: a case study. Int J Comput Sci Manag Res 1(3)
5. Padhy N, Dr Mishra P (2012) The survey of data mining applications and feature scope. Int J Comput Sci Eng Inf Technol (IJCSEIT) 2(3), 43–58
6. Dunham MH, Sridhar S (2006) Data mining: introductory and advanced topics, 1st edn. Pearson Education, New Delhi. ISBN: 81-7758-785-4
7. Delmater R, Hancock M (2002) Data mining explained-a manager's guide to customer-centric business intelligence. Digit Press, Boston
8. Bhatnagar A, Jadye SP, Nagar MM (2012) Data mining techniques & distinct applications: a literature review. Int J Eng Res Technol (IJERT) 1(9)
9. Industry application of data mining. http://www.pearsonhighered.com/samplechapter/0130862711.pdf

Unsupervised Machine Learning for Clustering the Infected Leaves Based on the Leaf-Colors

K. Ashok Kumar, B. Muthu Kumar, A. Veeramuthu and V. S. Mynavathi

Abstract In data mining, the clustering is one of the important processes for categorizing the elements into groups whose associated members are similar in their features. In this paper, the plant leaves are grouped based on the colors in the leaves. Totally, three categories are specified to represent the leaf with more green, leaf with yellowish shades and leaf with reddish shades. The task is performed using image processing. The leaf images are processed in the sequence such as image preprocessing, segmentation, feature extraction, and clustering. Preprocessing is done to denoize, enhance, and background color fixing for betterment of result. Then, the color-based segmentation is done on the preprocessed image for generating the sub-images by clustering the pixels based on the colors. Next, the basic features such as entropy, mean, and standard deviation are extracted from each sub-images. The extracted features are used for clustering the images based on the colors. The image clustering is done by the Neural Network architecture, self-organizing map (SOM), and K-Means algorithm. They are evaluated with various distance measuring functions. Finally, the city-block in both method produced the clusters with same size. This cluster set can be used as a training set for the leaf classification in future.

Keywords Infected leaves · Clustering · Image processing · Unsupervised machine learning · SOM · K-Means algorithm

---

K. Ashok Kumar (✉) · B. Muthu Kumar · A. Veeramuthu
Sathyabama University, Chennai, India
e-mail: ashokkumar.cse@sathyabamauniversity.ac.in

V. S. Mynavathi
Institute of Animal Nutrition, Tamil Nadu Veterinary and Animal Sciences University, Chennai, India

D. K. Mishra et al. (eds.), *Data Science and Big Data Analytics*,
Lecture Notes on Data Engineering and Communications Technologies 16,
https://doi.org/10.1007/978-981-10-7641-1_26

1 Introduction

Image processing is playing a vital role towards soil status, lighting level, infection and defects, quality of product, etc., in agricultural field. It is possible to monitor and control the agricultural process by image processing technique along with the advanced communication techniques. The current status of the agricultural field will be captured as a photo by using the unmanned robotic technique. Then, the captured images can be processed and forwarded to the server for getting the required service for the field.

In general, the plant image is processed either as a gray scale or as a colored image. The decision-making is based on the features extracted from the processed image. Commonly the features can be extracted globally or locally. Global features are nothing but which are considered for the entire image. The local features which are considered within the segmented or connected components from the original image. In general, the common features to be considered for the process are edges, transferred images, shapes, colors, textures, etc. Data Mining techniques are applied on the extracted features for decision-making process.

Data Mining is nothing but extracting or mining the new knowledge from the huge volume of existing data. The knowledge discovery is done by applying the following processing steps. In this paper, the same set of processes is carried out on images for knowledge discovery from the available images. The processing steps are

- Data cleaning [Image preprocessing]
- Data Integration [Images collected from various environments, with various colored images]
- Data Selection [Leaves which are green in nature are selected for the process]
- Data transformation [Performing segmentation and feature extraction from the segmented images for the mining process]
- Data Mining [SOM and K-Means are applied for grouping or clustering the leaves based on the colors]
- Pattern evaluation [Evaluated using precision, recall and F1 score values]
- Knowledge Presentation [Charts and SOM maps are used to visualize the clusters]

The paper is organized as related works, Infected leaves categorization, results and discussion and ended with conclusion.

2 Related Work

Hanson et al. [1] applied the cropping, resizing and filtering as preprocessing tasks on the leaf images. The image was split into various region based on K-Means clustering technique. The features such as shape, color, and texture were used to classify the leaves using SVM. The classification is done to classify the leaf based on the disease affected. GLCM was used for texture feature extraction. In general, GLCM is used to find the pixel information in a horizontal, vertical, or diagonal way. GLCM can be applied only on gray scaled images.

Naik and Sivappagari [2] Leaf disease classification was done using segmentation technique. Various leafs were tested with the classification algorithm. Only 5% of images from each category were considered for training. SVM and Neural Networks are used for the classification process. Their performances were compared with the classification gain. It was suggested at the end that, NN performed better than SVM. The preprocessed tasks were not specified clearly. The Neural Network selection criteria were also not specified clearly.

Tigadi and Sharma [3] proposed an automated system for identifying the Banana leaf diseases. Totally, the diseases are classified into five types. Images were captured using digital camera with size of 35. Out of which, 25 leaf images were used for training and 10 images were used for testing the classification system. Image cropping, resizing, and color conversion were applied on the images as preprocessing tasks, mainly for image enhancement. Actual images were converted into gray and HSV color model for easy processing. Two different features were considered such as Histogram of Template features and color features which included the mean and standard deviation. As in the description, Multi-layered feed-forward architecture was used for the leaf disease classification. The disease infection level was assigned as grade based on the percentage of infected area.

Wanjare et al. [4] used color and GLCM texture features for their process. Euclidean distance measure was applied between the input image and the images in the database. The matching is like a direct matching. It was developed as a web based process.

Gutte and Gitte [5] used segmentation, feature extraction and classification steps as a sequence of steps for plant disease recognition. Apples and grapes plant diseases were considered. Segmentation was used to cluster the similar featured pixels together as a group. This segmentation helped to identify the infected region on the leaf. The techniques and methods were not defined clearly to specify for what task, what technique or method was applied.

Robotics are used in agricultural field for monitoring, harvesting, etc. [6, 7]. Robotic-based web cam connected with PC was used to capture the images. It was proposed to use MATLAB for performing image processing on the captured data. More information were declared about the microchip and various sensors with their tasks. But, the image processing and image classification were specified only in the abstract level, but not specified in a clearly anywhere inside the paper [6].

Wable et al. [7] Sig-Bee based personalized network for data communication, IR sensor for measuring the light level, LM35 sensor for temperature measuring, sensor for soil moisture measuring, camera for capturing the leaf as image. ARM 7 controller for making the decision based on the input image, and a microcontroller for auto switch on/off the light and fan were used for image capturing task. Image processing task was done based on thresholding method. Here, images were converted into gray scale and then to binary scale. The features selected for process, the techniques applied for classification, training size, etc., were not specified in a detailed way.

Agrawal and Mungona [8] Cotton plant leaves were used for identifying the infected region. Training directory was created initially. From the training directory, the required data were fetched as training set. Image preprocessing was done with contrast enhancement algorithm. The infected region was detected with ROI. Eigen features were used with Multi-Atlas algorithm for classification.

Zhang et al. [9] applied RGB to HSI color model conversion for smoothing the progress of the colors in the satisfied manner. The texture features are combined with the color features for improving the similarity identification among the objects. The object is converted into gray image for developing the spatial gray level dependence matrices. These matrices are used for texture analysis.

Karale et al. [10] applied color conversion from RGB to HSI model. Color and texture features were used for image characterization using content-based image retrieval.

Chaudhary et al. [11] applied color transformation before segmentation. Color spot segmentation was processed using Otsu threshold method.

Ashokkumar and Thamizharasi [12] texture extraction was done using segmentation-based Fractal Texture Analysis algorithm (SFTA). The given image is converted into gray scale image for segmentation. SFTA is applied to decompose the gray image into a set of binary images. The fractal dimensions of the split images are calculated to define the segmented texture patterns.

3 Infected Leaves Categorization

3.1 Image Processing Task

The infected leaves are categorized based on the color of the leaves. The infection in the leaves will be reflected by turning the green color into yellowish or brownish or white shaded colors, with or without dots on the leaves. Since the process is based on the colors, different infected leaves are collected from web and framed as a data repository. From the repository, the infected green-in-nature leaves are selected manually for the clustering process. The selected images are passed as input for the image preprocessing task. Here, the noise is removed by 2D Median filter for generating the de-noised image (DI), enhanced de-noised image (EDI) is produced by Histogram equalizer and unique background color is fixed as black (BEDI). The outcome of an image is shown in Fig. 1.

The preprocessed image is taken for segmentation. Here, color-based segmentation is done by K-Means clustering method. Three sub-images are generated from the original image by segmentation. The outcome of the segmented sub-images is shown in Fig. 2.

The basic features such as entropy, mean, and standard deviation of the three sub-images are extracted. For an image, three features from three sub-images, totally, nine are extracted and framed as a vector for the clustering process.

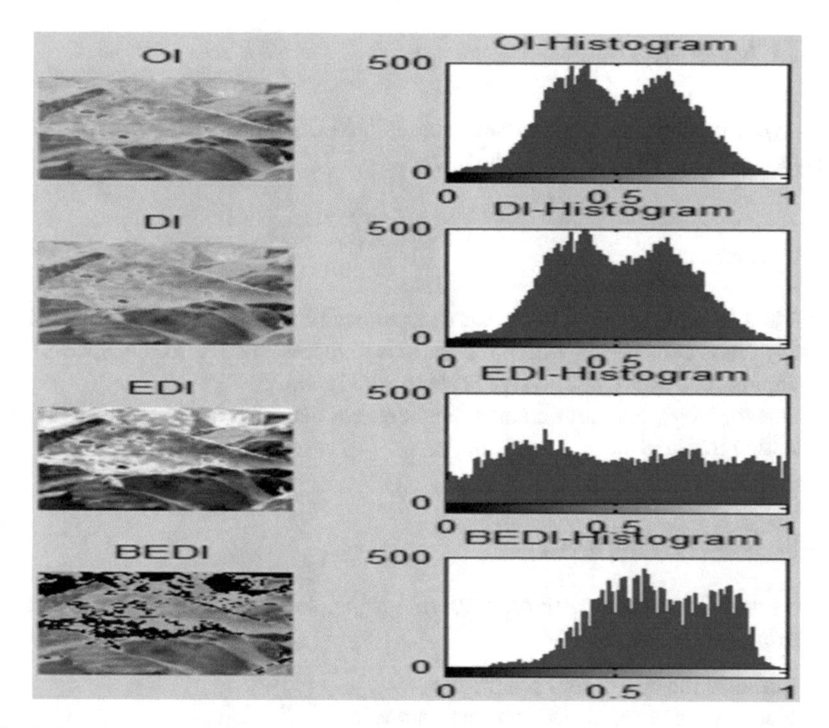

Fig. 1 Preprocessed leaf image with histogram representation

Fig. 2 Segmented sub-images

3.2 Clustering

Two types of clustering techniques such as Self-Organizing Map (SOM) and K-Means clustering are applied on the images.

3.2.1 Som

SOM is a Neural Network works on unsupervised learning concept. In this paper, the network is constructed with three neurons to represent the three classes with the nine inputs. The SOM architecture is shown in Fig. 3.

SOM topology, neighbor connection, neighbor distances are given as SOM representation in Fig. 4.

3.2.2 K-Means Clustering

This is working on the partition clustering principle. The algorithm for K-Means clustering is

i. Initialize the center for each cluster.
ii. Notify the nearest cluster to every data value.
iii. Set the location of each cluster to the mean of all data points fitting into that cluster.
iv. Repeat steps ii and iii until all data elements are converged to one of the cluster center.

The performance of the clustering technique is measured using precision, recall and F1 values.

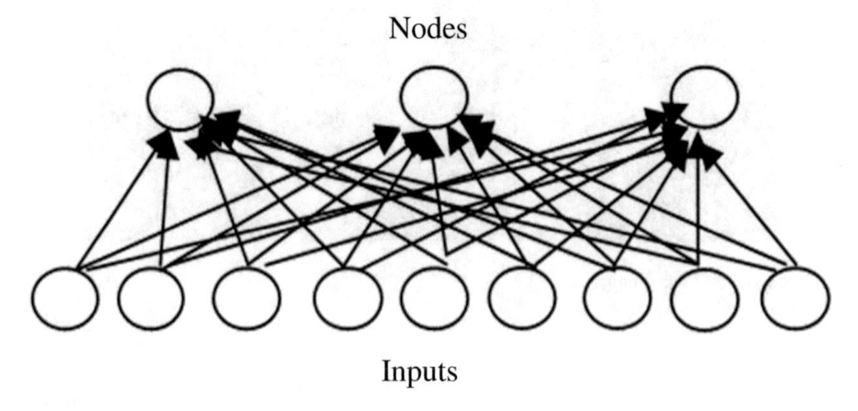

Fig. 3 SOM architecture

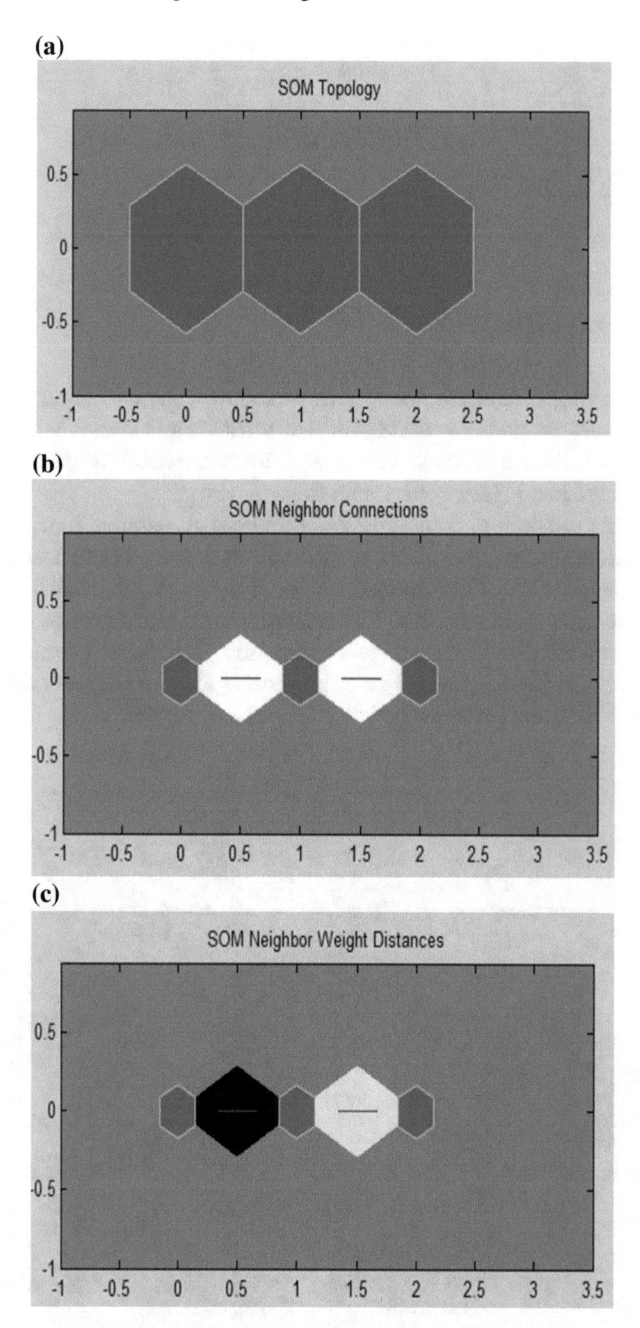

Fig. 4 SOM representation **a** SOM topology, **b** SOM neighborhood connections, **c** SOM neighborhood weight distances

$$Precision = \frac{TP}{TP + FP} \tag{1}$$

$$Recall = \frac{TP}{TP + FN} \tag{2}$$

$$F_1 = 2 \times \frac{Recall \times Precision}{Recall + Precision} \tag{3}$$

4 Results and Discussion

The two clustering techniques are applied to evaluate the clustering performance. The entire process is done in MATLAB. The clustering is started with 16 images. Then, 46 images, 70 images and 120 images. The clustering of SOM is generated as a hit values, Fig. 5 and cluster chart, Fig. 6.

In K-Means clustering, four different distances measuring functions such as squared-Euclidian, City-block, cosine, and correlation are applied and compared with the SOM hit values. User-defined routine is developed to evaluate the number of elements in each clusters with K-Means clustering. This routine is executed after clustering the elements with each distance measuring functions. The clustered count is given in Table 1. C1, C2, and C3 are representing the three clusters. SOM hits for the same set of data are given as the last row in the same table.

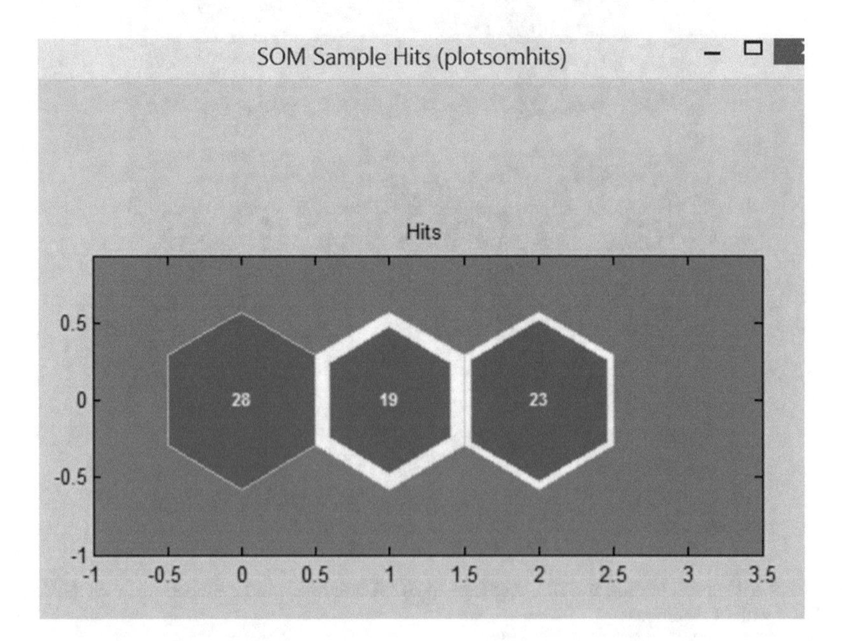

Fig. 5 SOM hit representation

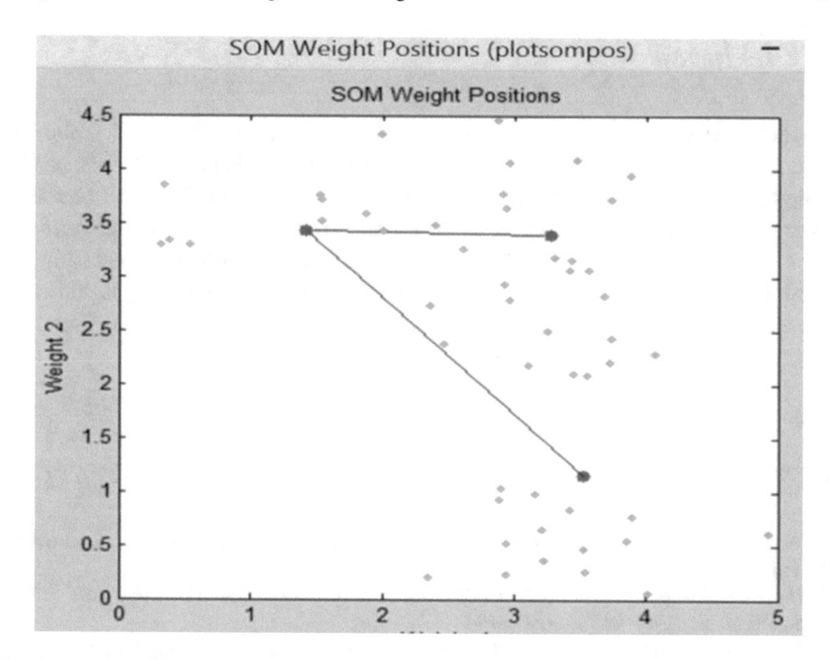

Fig. 6 Cluster chart representation

Table 1 Clusters with size

Clustering techniques		C1	C2	C3
K-Means	City block	14	17	15
	Squared euclidean	13	18	15
	Cosine	13	18	15
	Correlation	15	16	15
SOM		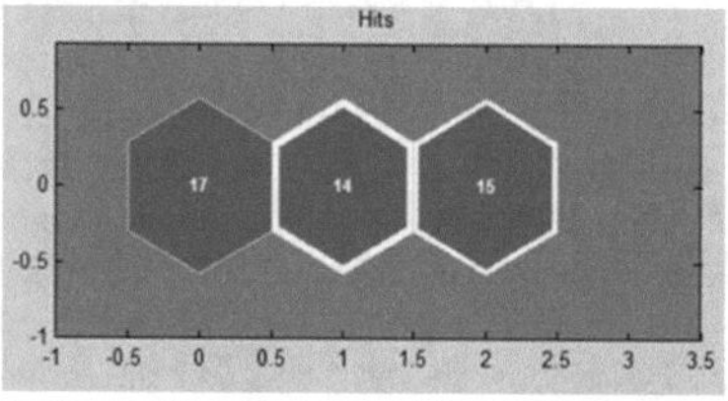		

The City-Block outcome is similar as SOM. Not only with this size, with all different sizes, these two have generated the same set of cluster size. This makes to fix K-Means with City-Block distance measuring function for clustering.

The performance of the cluster is tested with precision, recall and F1 values. The outcome of precision is 1 and recall 0.97 with F_1 as 0.98. The cluster is better when F_1 is close to 1. Both, SOM and K-Means with City-Block function performed well in grouping the leaves based on the colors.

5 Conclusion

The infected leaves are grouped into three clusters based on the colors. Yellowish, brownish, and greenish colored leaves are cluster groups. SOM and K-Means clustering are applied for clustering. In K-Means, four different distances measuring functions are applied for selecting the best performed function for the process. It is observed that the cluster size of SOM is matching with City-Block distance function in K-Means clustering. From the F_1 value, it is clear that the clustering is performed with 97%. These clusters can be used as a training set data for the infected leaf classification process.

References

1. Hanson J, Joy A, Francis, J (2016) Survey on image processing based plant leaf disease detection. Int J Eng Sci 2653–2655
2. Naik MR, Sivappagari CMR (2016) Plant leaf and disease detection by using HSV features and SVM classifier. Int J Eng Sci 3794
3. Tigadi B, Sharma B (2016) Banana plant disease detection and grading using image processing. Int J Eng Sci 6512
4. Wanjare M, Kondekar A, Mane S, Kamble S (2016) Disease detection on cotton leaves and its possible solutions. Int J Eng Sci 3559
5. Gutte VS, Gitte MA (2016) A survey on recognition of plant disease with help of algorithm. Int J Eng Sci 7100
6. More RB, Bhosale DS, Scholar, ME, JSPM's Bsiotr W (2016) Agrobot-a robot for leaf diseases detection. Int J Eng Sci 7352
7. Wable AA, Khapre GP, Mulajkar RM (2016) Intelligent farming robot for plant health detection using image processing and sensing device. Int J Eng Sci 8320
8. Agrawal MNJ, Mungona SS (2016) Application of multi-atlas segmentation in image processing. Int J Eng Sci 6409
9. Zhang SW, Shang YJ, Wang L (2015) Plant disease recognition based on Plant leaf image. J Anim Plant Sci 25:42–45
10. Karale A, Bhoir A, Pawar N, Vyavhare R (2014) Review on detection of plant leaf diseases using color transformation. Int J Comput Trends Technol 15(3):114–116
11. Chaudhary P, Chaudhari AK, Cheeran AN, Godara S (2015) Color transform based approach for disease spot detection on plant leaf. Int J Comput Sci Telecommun 3(6):65–70
12. Ashok Kumar K, Thamizharasi K (2015) Gesture controlled robot using MEMS accelerometer for eradication of weeds. Indian J Sci Technol 8(5). https://doi.org/10.17485/ijst/2015/v8i5/60481

Real-Time Big Data Analysis Architecture and Application

Nandani Sharma and Manisha Agarwal

Abstract Real-Time Big Data Analysis systems are those systems that process big data in given deadline or time limit. These types of systems are used to analysis a big data that is using data from some real world environment to analysis, predicate the solution to real-world problem. In this paper, we deal with architecture of this type of system what is basic structure of this type of system and their application in different area. We also categories theses type of in two main categories real-time system and near real-time system.

Keywords Real-time big data analysis · Type of real-time analysis · Near real-time system · Application of RTBDA

1 Introduction

"Real-time big data isn't just a process for storing pet bytes or Exabyte's of data in a data warehouse, it's about the ability to make better decisions and take meaningful actions at the right time "says Michael Minelli, co-author of Big Data, Big Analytics [1].

Real-Time Big Data analysis is basically processing of stream of data in motion and analysis that data to come to a conclusion or decision and that decision is use in different application. Real-Time Big Data Analysis is applied in much area as earth observatory system, DRDO NETRA project. And many more we here discusses

N. Sharma (✉) · M. Agarwal
Banasthali University, Vanasthali, India
e-mail: jamnandanii@gmail.com

M. Agarwal
e-mail: mani1811@gmail.com

D. K. Mishra et al. (eds.), *Data Science and Big Data Analytics*,
Lecture Notes on Data Engineering and Communications Technologies 16,
https://doi.org/10.1007/978-981-10-7641-1_27

some of them after a general introduction of type of real-time analysis, big real-time analysis stack [2] and a five phases of process model [2].

2 Big Data

Data refer to raw set of information. But if we read, hear, or see word big data we simply interpreted as the large clusters of data but this is not the case. Big data does not mean data large in volume. Big data is basically unstructured huge data set which gives high speed analysis report on that data. Unstructured data means data set is heterogeneous data can be collected from various sources and in various format. Source for data can be sensor, websites, telecommunications data, business, and financial transactions data and many more. These gathered data are processed to give future prediction.

Big data acquire poly-structure data and analysis that data at high speed just to make good future decision.

Volume is one of the important constraints which big data hold there are other Vs which big data hold and these Vs are following:

1. Volume: As its name suggest "big data" is collection of a very large amount of data and we have to handle this large data.
2. Variety: This character makes big data different from traditional data analysis system. As big data has capacity to handle a combination structural, semi-structural, and unstructured data from different sources and analysis.
3. Velocity: It may be assumed that analysis data in large volume will decrease the velocity of computation. But big data maintain its velocity perfectly to handle to handle real-time data within given constraints.
4. Value: Value is the important v of all five Vs. This v addresses the requirement of valuation. Value is main buzz for big data because it is important for IT infrastructure system businesses to store large amount of value in database.
5. Veracity: Veracity deal with authenticity, trustworthiness, origin, etc. In such a large volume of data give feature veracity of complete data is difficult to achieve. There may be some dirty data also.

3 Real-Time System

Real-time systems are those system that work in bound of deadline. Real-time system has time limit to complete the task. The limit to complete task known as deadline can be soft deadline or hard dead line. Soft deadline can be miss deadline will not harm environment as well as system. In the hard real-time system if we miss the deadline some catastrophic loss happens to the system or environment of the system. Real-time system takes decision according to current situation of system and environment.

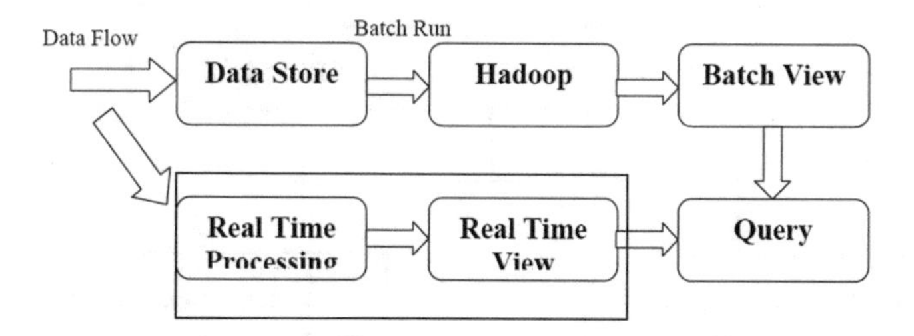

Fig. 1 Big data and real-times system [3]

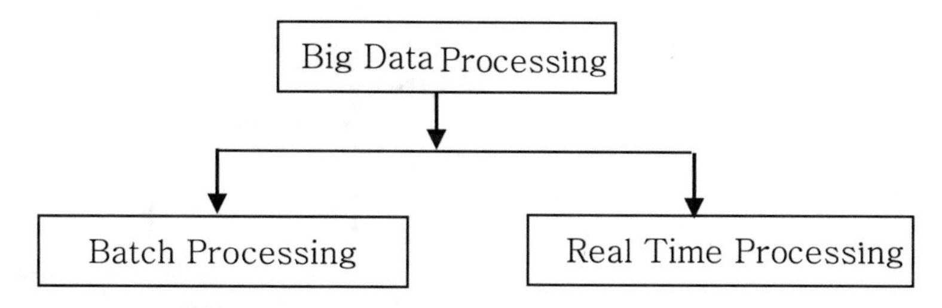

Fig. 2 Big data processing

4 Real-Time Big Data Analysis

Big data processing can be done in two ways Batch and real-time processing. In Batch processing us analysis historically or previously stored information to take future decision. In real-time processing we take real-time stream of data and make decisions at the current instance of time (Fig. 1).

In real-time processing time is an important constrain for completion of work. Real-time big data analysis is a combination of big data decision making power to be meeting with the time constraints of real-time system. Real-time big data analysis collected data in from different real-time source as sensor, business transaction, etc. And process them in parallel and predicted the decision for betterment (Fig. 2).

5 Type Real-Time Processing

Real-time analysis can be of two type real time and near real time (Fig. 3).

Real-time analysis is one in which analysis of data or decision-making from data is necessary to be completed within a specified time limit. A very large data should be analysis in blink of eye and decision is taken. In real-time analysis system data is analysis in deadline its main and important feature of system (Fig. 4).

Near real-time analysis is which in data may be analysis in the given time limit. In these type of analysis system deadline is only one point to complete the analysis it may miss the deadline. So in near real-time system if deadline missed will not affect

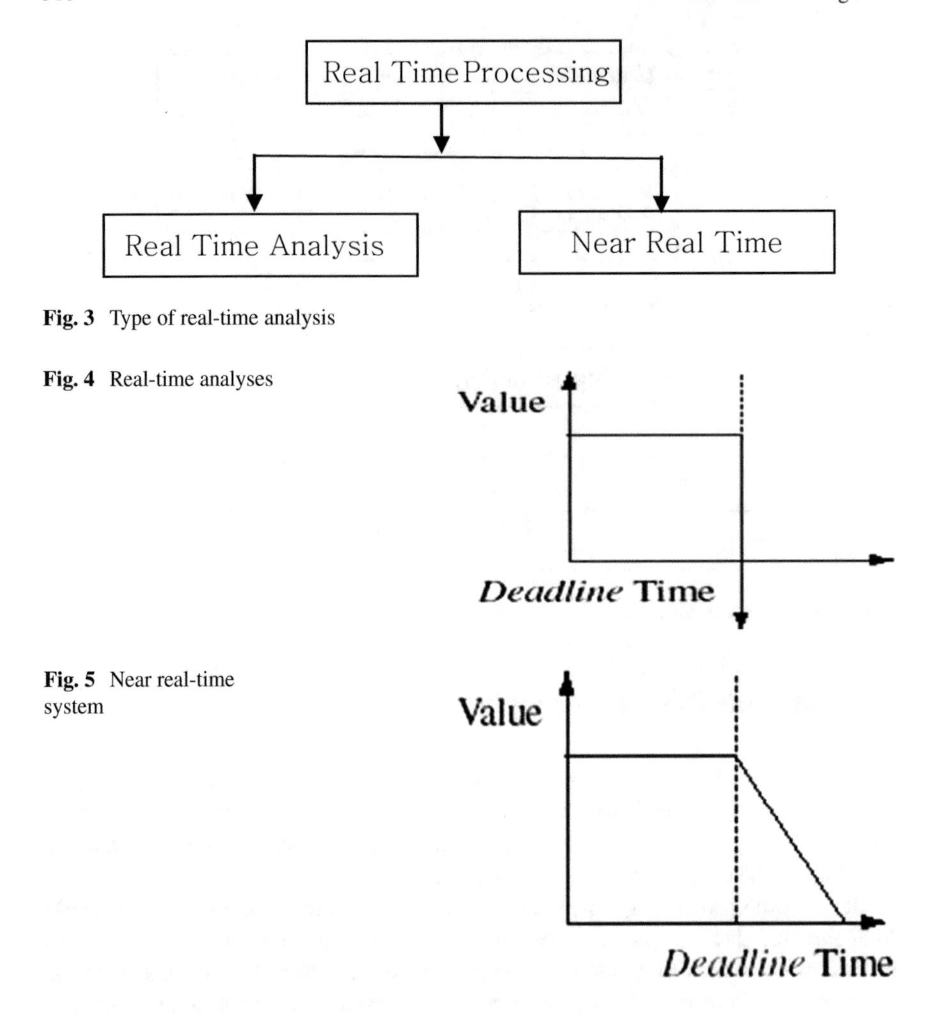

Fig. 3 Type of real-time analysis

Fig. 4 Real-time analyses

Fig. 5 Near real-time system

other decision analysis. As hadoop is not suitable for real-time t analysis but it may be used for near real-time system (Fig. 5).

6 Real-Time Big Analysis Data Stack

We need to generalize working of RTBA, so a model is proposed by David smith. David Smith writes a blog on it. According to David smith this model contains four layers and each layer has its own many tools to handle working of particular layer.

- Data layer: At this layer the structured, unstructured and semi structured data from different source are stored and processed with different tools. Example of

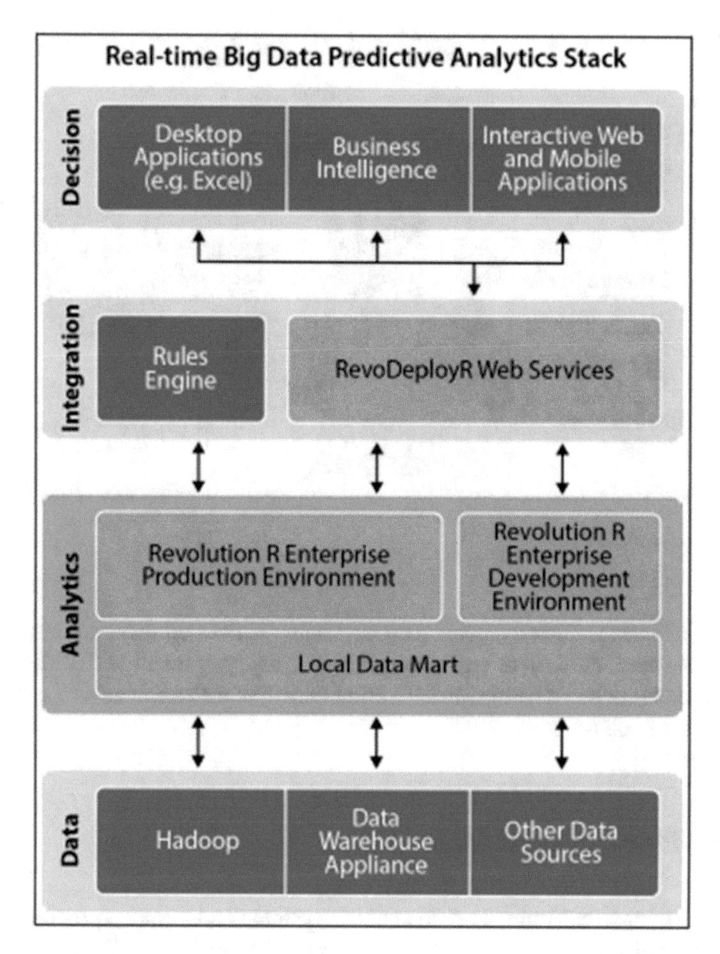

Fig. 6 RTBDA stack [2]

structured data is RDBMS, NoSQ, etc. Example of unstructured data is hadoop reduced map, streaming data from web, sensor, etc.

- Analytics layer: The second layer is analytics layer. This layer takes periodically data and updates from data layer. This transfer of data and updates to other tools of analytics is done by local mart. The analytics layer includes a production environment for deploying real-time scoring and dynamic analytics; a development environment for building models; situated near the analytics engine to improve performance (Fig. 6).
- Integration layer: Integration Layer is an interface to end-user application and analytics engines together. This layer perform dynamic analytics which brokers communication between app developers and scientists with the help of main contain of this layer rules engine or CEP engine and API.

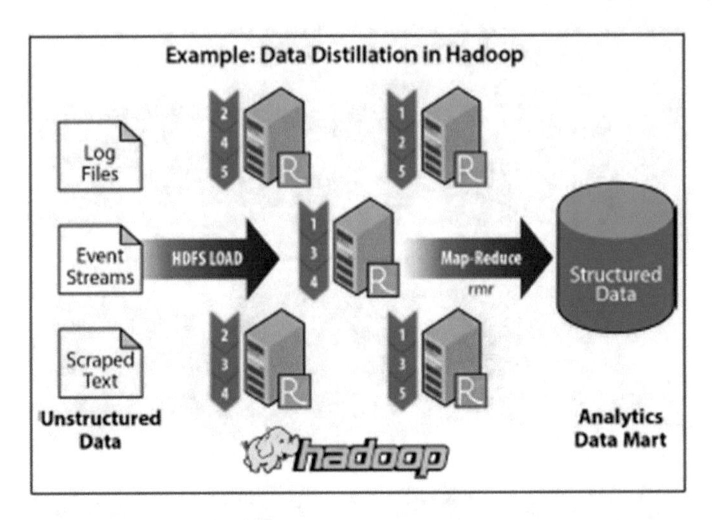

Fig. 7 Data distillation [2]

- Decision layer: Decision layer is the one where result of all layer come visualized to the end-user. And these end-user can access analysis result with different-user applications such as desktop, mobile, and interactive web apps, as well as business intelligence software.

7 Five-Phase Process Model [2]

Development of a real-time big data analysis system is an iterative process and that development of system is explained by Smith's five phase process model. This model has five phase data distillation, model development, validation and deployment, real-time scoring, and model refreshment.

Data Distillation—Input to his phase is unstructured data from different source as sensor, website, social media, and video, audio. First of all we extract feature from individual unstructured data and then combine disparate data. Then from that combined data we filtered data for our own interest. This data become input to the data modeling development. This data is also exporting sets of distilled data to a local data mart (Fig. 7).

Model development—Different stage of this phase is feature selection, sampling, and aggregation; variable transformation; model estimation; model refinement; and model benchmarking. In this phase data is of distillation phase is used of constructing refine predication model after compare dozen of model for a powerful and robust real-time algorithm [3] (Fig. 8).

Validation and development—In this phase model prepared and checked with real-time stream of data. If model works it can be deployed into real-time environment.

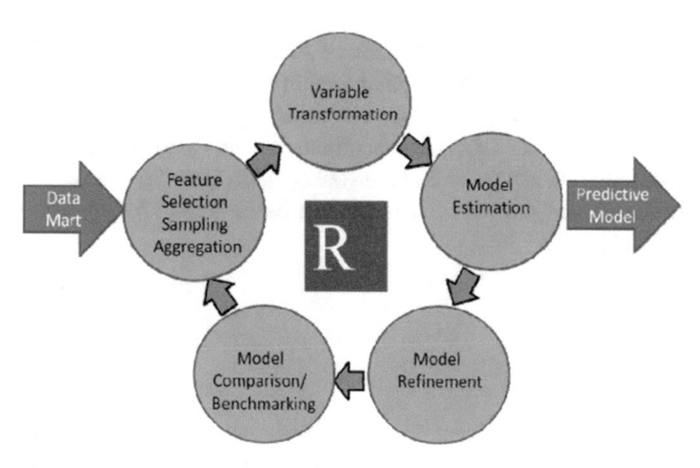

Fig. 8 Modeling distillation [3]

Real-time scoring—Real-time scoring is the phase where the model separated from data in data mart. Real-time data is given to the model with somewhat different hardware.

Model Refresh—Change is a part of life cycle development so the model of real-time model changes periodically according to the need of real-time environment. But we use the capability of reuse in old model after validation and deployment.

8 Application of RTBDA

With the emerging of Internet, hand handled devices and digitalization of data in management the digital data grow exponential daily. That led to a real-time big data analysis system. These types of system help in making decision in many areas for human life. Some of applications of real-time big data analyzer are following:

Google analytics: With the increase website in internet the requirement of a tool to analysis website traffic and client behavior come in senior. This requirement is fullfilled by free available web analytics Google analytics. Google analytics is use to analysis the behavior of people accessing particular web site.

DRDO NETRA: DRDO NETRA is an Internet traffic surveillance and network motoring system developed by DEFENCE RESEARCH AND DEVELOPMENT ORGANISATION of India.

Amazon kinesis: Amazon kinesis is an application use to create real-time dashboard, generate alert and implement dynamic pricing and advertising and more.

Real-Time Business Intelligence—Real-time business intelligence (RTPI) is an approach to data analytics that enables business users to get up-to-the-minute data by directly accessing operational systems or feeding business transactions into a real-time data warehouse and business intelligence (BI) system. This type of system is used in many different fields as detection of ATM fraud, taking strategic action of business management and specious passenger in airlines, railways, or bus.

Earth Observatory System—Earth observatory system generates a huge amount of data every second. That data is first acquired at Data Acquisition Unit (DAU). At second phase that data is filtered and load balanced in Data Processing Unit. At last that filtered data is analysis and some decision is made at Data Analysis and Decision Unit (DADU) [4].

9 Conclusion

In this article, an overview of Real-Time Big Data Analysis system, stack of RTBDA, Model of RTBDTA and also some of application area of RTBDA The contain of article have given regardless of the fact that accessible information, tools, and techniques available in the literature. Although this paper have not given detailed and all application of real-time big data analysis.

References

1. Venkata Krishna kumar S, Ravishankar KS (2015) A study of real time big data analysis. Int J Innov Res Comput Commun Eng (An ISO 3927 2007 certificate organisation) 3(8). ISSN(Online): 2320-9801
2. Barlow M (2013) Real time big data analytics: emerging architecture. O'Reilly Media
3. Meshram AD, Kulkurn AS, Hippargi SS (2016) Big data analystic using real time architecture. IJLTET 6(4). ISSN: 2278-621X
4. Rathore MM, Ahmad A, Paul A, Daniel A (School of Computer Science and Engineering, Kyungpook National University, Daegu, Korea) Hadoop based real-time big data architecture for remote sensing earth observatory system
5. Munagapati K, Usha Nandhini D (2015) Real time data analytics. Int J Appl Eng Res 10(3):7209–7214. ISSN 0973-4562
6. Rajasekar D, Dhanamani C, Sandhya SK (2015) A survey on big data concepts and tools. IJETAE, ISO 9001:2008 Certified Journal 5(2). ISSN 2250-2459

Missing Value Imputation in Medical Records for Remote Health Care

Sayan Das and Jaya Sil

Abstract In remote area where scarcity of doctors is evident, health kiosks are deployed for collecting primary health records of patients like blood pressure, pulse rate, etc. However, the symptoms in the records are often imprecise due to measurement error and contain missing value for various reasons. Moreover, the medical records contain multivariate symptoms with different data types and a particular symptom may be the cause of more than one diseases. The records collected in health kiosks are not adequate so, imputing missing value by analyzing such dataset is a challenging task. In the paper the imprecise medical datasets are fuzzified and fuzzy c-mean clustering algorithm has been applied to group the symptoms into different disease classes. In the paper missing symptom values are imputed using linear regression models corresponding to each disease using fuzzified input of 1000 patients' health-related data obtained from the kiosk. With the imputed symptom values new patients are diagnosed into appropriate disease classes achieving 97% accuracy. The results are verified with ground truth provided by the experts.

Keywords Rural health care · Missing value · Regression model · Fuzzification Classification

1 Introduction

Experimental or observed datasets are often incomplete due to missing attribute values [1], a common phenomenon for erroneous acquisition systems and adverse

S. Das (✉) · J. Sil
Indian Institute of Engineering Science and Technology, Shibpur, Howrah, India
e-mail: sayan.das57@gmail.com

J. Sil
e-mail: js@cs.iiests.ac.in

D. K. Mishra et al. (eds.), *Data Science and Big Data Analytics*,
Lecture Notes on Data Engineering and Communications Technologies 16,
https://doi.org/10.1007/978-981-10-7641-1_28

environmental conditions. Missing data items hide important information, essential to decision-making in critical situations. In medical domain missing value imputation is a necessary step to extract accurate knowledge, especially in developing an automatic disease diagnosis system.

In rural India due to shortage of expert manpower, healthcare services face a real challenge which could be handled to some extent by developing intelligent systems for providing services at primary level [2]. In remote villages health kiosks are set up for obtaining basic health related attribute values of the patients like blood pressure, pulse rate, height, weight, etc. On medical reports some attributes or symptoms are left blank, because of measurement error or absence of response from the patients. Knowledge extraction by ignoring missing value or fixing it without analyzing the records may generate wrong inferences. In case of health care a patient is either treated wrongly or not avail any treatment resulting potential health hazard in the system. The paper aims at developing an autonomous disease diagnosis system by imputing missing value(s) of symptoms using computational method.

Several challenges are to be dealt while predicting missing value in medical records compared to the conventional datasets. Many techniques have been developed so far to deal with the problem of missing value. In [3], authors discuss top ten algorithms of data mining which are most promising in imputing missing value. An approach has been proposed [4, 5] for estimating the missing values in continuous domain. The easiest way to impute the missing value is to replace each missing value with the mean of the observed values for that variable according to [6] but it does not produce good classification and correlation is negatively biased [6]. The imputed values provide incorrect representation of the population because the shape of the distribution is distorted by adding the values equal to the mean [7]. The missing value is imputed with an observed value, closer in terms of distance in Hot-Deck [8] imputation method. Hot-Deck (HD) randomly selects a value from a pool of observations that matches based on the selected covariates. The HD method is not efficient for small size of data set due to lack of chance of close relation in all aspects of the data set [9]. The authors employ non-parametric discretization technique [10] and try to impute the missing values in the medical records. In the work [11, 12] researcher handles the missing values in dengue fever data set which may be numeric or categorical. Various approaches to deal with the missing values using benchmark data sets are described in [13]. The performance of classification of medical records are presented [14, 15] after predicting the missing values. However, the existing methods consider precise data set for estimating missing value unlike the scenario of rural India. In rural area due to lack of domain knowledge, response of the patients are often imprecise. Moreover, chance of measurement error is evident in recording the symptom value due to inadequate skilled manpower. Other than uncertainty in medical records [2], predicting missing value from inadequate medical data is also not addressed in the literature.

This paper aims at imputing missing symptom values of the medical records collected from health kiosks situated in rural India. First the imprecise symptom data sets are fuzzified [2] with proper semantics in consultation with the experts and medical literature [16–19]. Gaussian membership functions with different means and standard deviations are used to calculate the degree of membership value of the fuzzy variables, representing the symptoms. In medical record, a particular symptom may be the cause of multiple diseases and so fuzzy clustering algorithm [20] has been applied to group the symptoms into different disease classes. In the paper we employ regression method [21] to impute missing value of symptoms with respect to each disease, known after clustering the data. Regression model has been developed using relationship between different symptoms obtained by analyzing the medical records of patients. We obtain 97% accuracy in diagnosing diseases with the imputed missing symptom value while applied on the original and simulated data sets. The results are verified with ground truths provided by the experts.

The rest of the paper is organized as follows: Sect. 2 describes methodology applied to compute missing value. Section 3 provides the experimental results and discussion. Finally, Sect. 4 concludes the paper with future work.

2 Methodology

In remote villages health workers collect health records (pulse, high blood pressure, low blood pressure, height, weight, BMI, temperature and SpO_2) of people using different sensors and analyzed the data for diagnosing primary level of the diseases of the patients. However, medical database are multivariate, imprecise and often contains missing value. Ignoring missing value in the medical records may result erroneous diagnosis due to absence of important information. Imputation of missing value is problematic, particularly in remote area where users' input is often vague, data collection is not perfect due to lack of skill workers and same symptom may be the cause of multiple diseases. As a first step symptoms are fuzzified and data are clustered using fuzzy c-means clustering algorithm to group the symptoms with respect to the disease.

2.1 Fuzzification of Records

Imprecise symptoms are represented using fuzzy sets with proper semantic based on the standard medical science data [16–19], given in Table 1. For example, a sample patient may have blood pressure 156/96 mmHg and pulse rate 87 bpm. With reference to the respective standard value the symptoms of the patients are represented using fuzzy sets "High", "Low" and "Normal". Membership functions of symptoms like "blood pressure", "pulse rate", etc. are shown in Fig. 1, used to fuzzify the symptom values. Table 2 shows primary symptoms of diseases, obtained in consultation with the experts.

Table 1 Standard health attributes

Input variable	Range	Fuzzy set
Blood pressure (Systolic/Diastolic) in mm HG Standard: (120/80) mm HG	(90–120)/(60–80)	Normal
	(121–139)/(81–89)	Pre-High
	(140–159)/(90–99)	High-Stage 1
	More than 160/More than 100	High-Stage 2
	Less than 90/60	Low
Pulse in bpm Standard: 72 bpm	(60–100)	Normal
	More than 100	High
	Less than 60	Low
SpO_2 in mm Hg	Below 90	Low
	90–94	Moderate
	Above 94–99	Normal
BMI	Below 18.5	Under weight
	18.5–24.9	Healthy
	25–29.9	Over weight
	Above 30	Obesity
Temperature in fahrenheit	Below 95	Low
	97.7–99.5	Normal
	99.6–Above 100	High

Table 2 Disease related primary symptoms

Disease	Related symptoms
Cold/Cough	High blood pressure, low blood pressure, pulse, temperature
Breathlessness	High blood pressure, low blood pressure, pulse, SpO_2
Fever	Temperature, high blood pressure, low blood pressure, pulse
Abdominal pain	High blood pressure, low blood pressure, pulse, BMI
Insomnia	High blood pressure, low blood pressure, pulse, SpO_2
Acidity	High blood pressure, pulse, SpO_2, BMI
Vomiting	High blood pressure, low blood pressure, pulse, SpO_2
Knee pain	High blood pressure, low blood pressure, pulse, temperature
Headache	High blood pressure, low blood pressure, pulse, SpO_2
Diarrhea	High blood pressure, low blood pressure, pulse, temperature

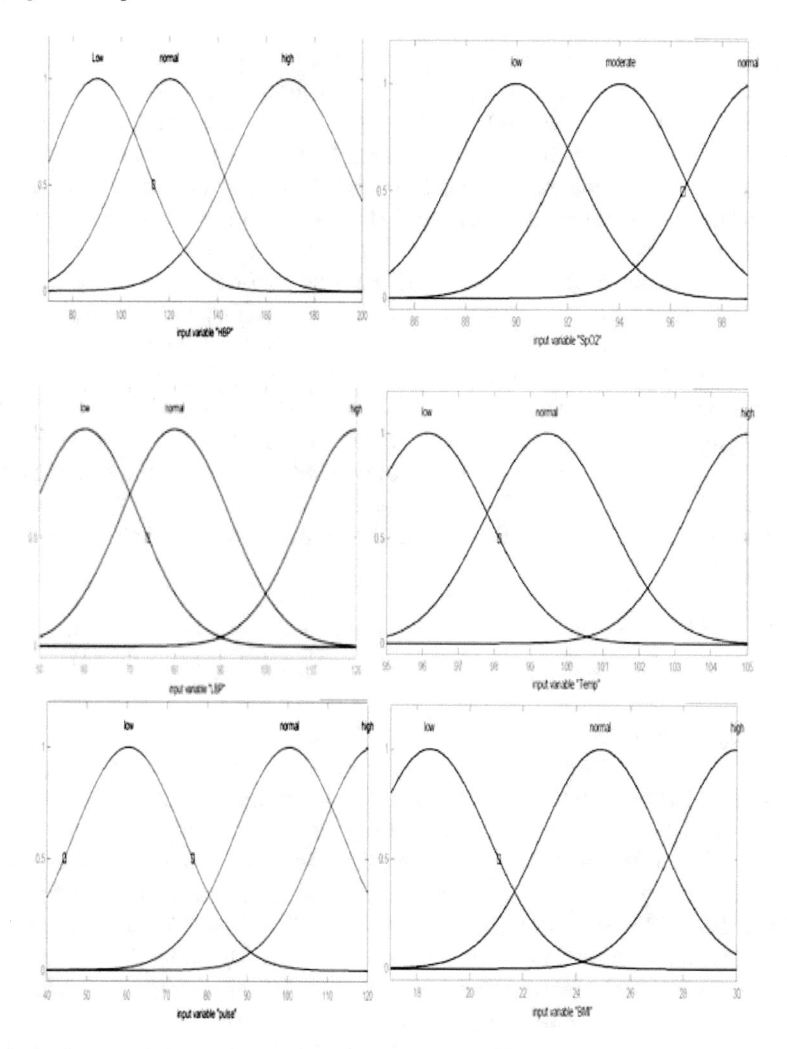

Fig. 1 Membership curves of "systolic (HBP)", "diastolic (LBP)", "pulse (PUL)", SpO_2", "Temp (T)" and "BMI"

2.2 Missing Value Estimation

Medical records contain missing value at random and in the paper missing value has been estimated using linear regression model where relation between the symptoms are framed by analyzing the patient dataset with respect to each disease (each row of Table 2). The regression models are built using the symptoms without missing value and mean square error (MSE) cost function. The aim of this approach is to impute the missing symptom values so that the predicted values best fit along the regression line without any residual variance.

In the paper "blood pressure" symptom is dealt using a novel approach since it is associated with two other symptoms, i.e., "systolic" (High Blood pressure) and "diastolic" (Low blood pressure). "Systolic" (S) symptom is represented using fuzzy sets *systolic High* (S_H) with membership $\mu_x S_H$, *Low* (S_L) with membership $\mu_x S_L$ and *Normal* (S_N) with membership $\mu_x S_N$. Similarly, "diastolic" (D) value is represented as *Low* (D_L) with membership $\mu_y D_L$, *High* (D_H) with membership $\mu_y D_H$ and *Normal* (D_N) with membership $\mu_y D_N$, for $\forall x \in X$ and $\forall y \in Y$, where X and Y represent universe of discourse of "systolic" and "diastolic" values, respectively. Cartesian product ($\times$) between two fuzzy sets $S = \{\mu_x S_H, \mu_x S_L, \mu_x S_N\}$ and $D = \{\mu_y D_H, \mu_y D_L, \mu_y D_N\}$ are evaluated using Mamdani rule [22] to obtain activation of a particular valid combination of "systolic" and "diastolic" symptoms of each patient.

For example, say a patient has "breathlessness" disease and assume "pulse" (PUL) symptom, an important input for diagnosing the disease is missing from the patient's record. Other related symptoms associated to "breathlessness" diseases are "high blood pressure" (HBP), "low blood pressure" (LBP) and "SpO_2". The relation between symptom PUL and other symptoms for that particular disease are defined using Eq. (1).

$$\begin{aligned}\mu_r \text{PUL} = {} & a_1 \cdot \max\left[\{\mu_x S_H, \mu_x S_N, \mu_x S_L\} \times \{\mu_y D_H, \mu_y D_N, \mu_y D_L\}\right] \\ & + a_2 \cdot \left[\max\{\mu_z(SpO_2)_L, \mu_z(SpO_2)_M, \mu_z(SpO_2)_N\}\right]^{-1}\end{aligned} \quad (1)$$

$z \in Z$ representing universe of discourse of symptom "SpO_2" and $r \in R$ representing universe of discourse of symptom "PUL". a_1 and a_2 are regression coefficients for disease "breathlessness" evaluated by minimizing the MSE cost function. $\mu_z(SpO_2)_L$, $\mu_z(SpO_2)_M$ and $\mu_z(SpO_2)_N$ are membership values of "Low", "Moderate" and "Normal" fuzzy sets, respectively representing oxygen saturation level (SPO_2). Equation (1) is framed based on the observation that either $\mu_x S_H$ is "High"/"Low"/"Normal" or $\mu_x S_L$ is "High"/"Low"/"Normal" of each patient. The same rule follows for the set D with elements $\mu_y D_H$, $\mu_y D_N$ and $\mu_y D_L$.

After predicting the missing value of symptom PUL of a patient, it is defuzzified [23] to know the crisp value and used for disease diagnosis.

Say, a patient has symptoms such as HBP, LBP and SpO_2 with values 152, 95 and 93 unit, respectively. The missing symptom value of "PUL" is evaluated using Eq. (1) as

$$\mu_r \text{PUL} = a_1 0.75 + a_2 1.4,$$

where a_1 (0.63) and a_2 (0.262) are regression coefficients and the membership value of "PUL" is calculated as 0.84. After defuzzification the missing value of pulse is 107 unit.

For HBP as missing symptom in "breathlessness" disease, the relationship between "blood pressure" and other related symptoms (PUL, SpO_2) are framed using Eq. (2). Equation (3) represents the same for LBP symptom as missing value.

Table 3 Disease related to basic symptoms

Disease	Relationship between related symptoms
Cold/Cough	$\mu_r PUL = a_1 \cdot \max [\{\mu_x S_H, \mu_x S_N, \mu_x S_L\} \times \{\mu_y D_H, \mu_y D_N, \mu_y D_L\}] + a_2 \cdot [\max \{\mu_z(SpO_2)_L, \mu_z(SpO_2)_M, \mu_z(SpO_2)_N\}]^{-1} + a_3 \cdot [\max \{\mu_t(T)_H, \mu_t(T)_N, \mu_t(T)_L\}^{-1}]$ (Missing value pulse)
Fever	$\mu_t T = a_1 \cdot [\max [\{\mu_x S_H, \mu_x S_N, \mu_x S_L\} \times \{\mu_y D_H, \mu_y D_N, \mu_y D_L\}]]^{-1} + a_2 \cdot [\max \{\mu_z (PUL)_H, \mu_z (PUL)_N, \mu_z (PUL)_L\}]^{-1}$ (Missing value temperature)
Abdominal pain	$\mu_m BMI = a_1 \cdot [\max \{\mu_r (PUL)_H, \mu_r (PUL)_N, \mu_r (PUL)_L\}]^{-1} + a_2 \cdot \max [\{\mu_x S_H, \mu_x S_N, \mu_x S_L\} \times \{\mu_y D_H, \mu_y D_N, \mu_y D_L\}]$ (Missing value BMI)
Insomnia	$\mu_r PUL = a_1 \cdot \max [\{\mu_x S_H, \mu_x S_N, \mu_x S_L\} \times \{\mu_y D_H, \mu_y D_N, \mu_y D_L\}] + a_2 \cdot \max [\{ \mu_z(SpO_2)_L, \mu_z(SpO_2)_M, \mu_z(SpO_2)_N\}]^{-1}$ (Missing value PUL)
Acidity	$\mu_z SpO_2 = a_1 \cdot \max [\{\mu_x S_N, \mu_x S_L\} \times \{\mu_y D_H, \mu_y D_N, \mu_y D_L\}]^{-1} + a_2 \cdot \max [\mu_r (PUL)_H, \mu_r (PUL)_N, \mu_r (PUL)_L]^{-1} + a_3 \cdot \max [\mu_m(BMI)_H, \mu_m(BMI)_N, \mu_m(BMI)_L]^{-1}$ (Missing value SpO_2)
Vomiting	$\mu_x S = a_1 \cdot \max [\mu_r(PUL)_H, \mu_z (PUL)_N, \mu_r(PUL)_L] + a_2 \cdot [\max\{\mu_z(SpO_2)_L, \mu_z(SpO_2)_M, \mu_z(SpO_2)_N\}]^{-1}$ (Missing value HBP) $\mu_x D = a_1 \cdot \max [\mu_r(PUL)_H, \mu_z (PUL)_N, \mu_r(PUL)_L] + a_2 \cdot [\max \{ \mu_z(SpO_2)_L, \mu_z(SpO_2)_M, \mu_z(SpO_2)_N\}]^{-1}$ (Missing value LBP)

$$\mu_x S = a_1 \cdot \max\left[\mu_r(PUL)_H, \mu_r(PUL)_N, \mu_r(PUL)_L\right] + a_2 \cdot \left[\max\{\mu_z\left(SpO_2\right)_L, \mu_z\left(SpO_2\right)_M, \mu_z\left(SpO_2\right)_N\}\right]^{-1} \quad (2)$$

and

$$\mu_x D = a_1 \cdot \max\left[\mu_r(PUL)_H, \mu_r(PUL)_N, \mu_r(PUL)_L\right] + a_2 \cdot \left[\max\{\mu_z\left(SpO_2\right)_L, \mu_z\left(SpO_2\right)_M \mu_z\left(SpO_2\right)_N\}\right]^{-1} \quad (3)$$

Table 3 shows relationship between symptoms for different disease like cold/cough, fever, abdominal pain, insomnia, acidity, and vomiting.

3 Results and Discussion

We demonstrate the method using 1000 patients' health data acquired from the health kiosks situated in remote areas. We consider general diseases like cold/cough, fever, abdominal pain, insomnia, acidity, vomiting, knee pain, headache and diarrhea. The primary symptoms which are involved for these diseases are pulse, high blood pressure, low blood pressure, BMI, SpO_2 and temperature. The missing values in the paper are selected at random. Since the patient may have overlapping diseases, Fuzzy c-mean clustering algorithm has been applied to cluster the patient records with respect to diseases. In the paper, we use Davies-Bouldin (DB) index [24] to point out the

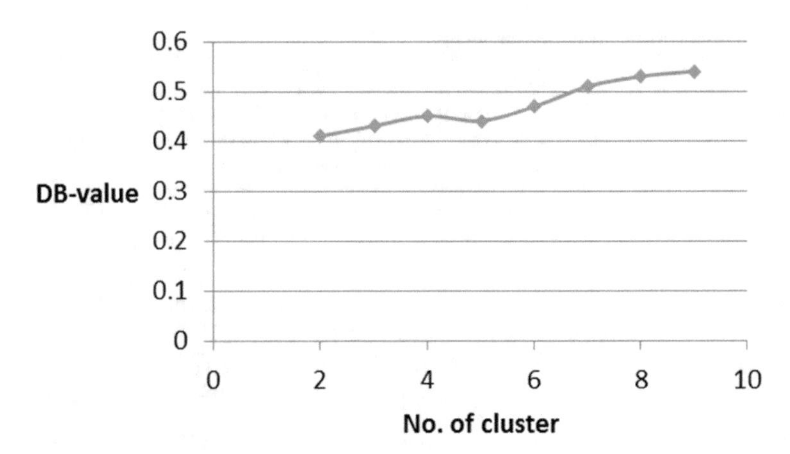

Fig. 2 DB value versus number of cluster

Table 4 Error of missing symptom value for respective diseases

Disease	Average error of missing symptoms					
	Pulse	HBP	LBP	BMI	SpO_2	Temp
Breathlessness	0.82	0.72	0.57	0.32	**0.39**	0.12
Cold/Cough	0.75	0.87	0.79	0.81	1.08	**0.27**
Fever	1.05	0.59	0.65	0.59	0.75	**0.2**
Abdominal pain	0.67	0.68	0.76	**0.42**	0.53	0.63
Insomnia	0.55	**0.35**	**0.43**	0.98	0.75	0.31
Acidity	0.62	0.86	0.94	0.65	0.82	**0.48**

correct number of clusters. DB index takes into account both the error caused by representing the data vectors with their cluster centroids and the distance between clusters [25]. Figure 2 shows DB versus number of cluster plot using patients' database and it shows that the datasets are clearly separable as the number of clusters or disease class labels increases for the given data set.

Table 4 shows average error in predicting missing value of different symptoms with respect to each disease considering 100 patients per disease and we assume that each symptom has 40% missing value. Finally, using 10-fold cross-validation technique 97% accuracy is achieved after imputing missing symptom value. Figure 3 shows ROC plot as the performance of proposed method. Table 5 gives comparative study of missing symptom imputation methods.

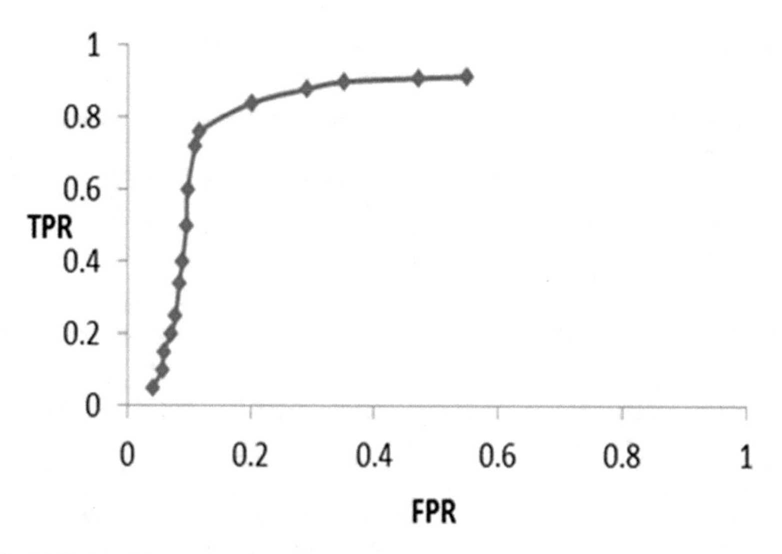

Fig. 3 ROC plot of the proposed method

Table 5 Comparative study of the proposed method with others

Techniques	Error (%)	Accuracy (%)	Misclassification rate	Sensitivity (%)	Specificity (%)	True positive rate (TPR)	False positive rate (FPR)	Precision
Hot deck	3.0–5.0	91.03	0.0897	85	91	0.91	0.342	0.934
Mean substitution	2.0–4.0	92.19	0.0781	88	95	0.93	0.211	0.952
Proposed method	0.3–1.5	97.38	0.0262	90.9	98.09	0.98	0.191	0.987

4 Conclusions

Medical data from remote area are usually found to be incomplete as in many cases on medical reports some attributes can be left blank, because they are inappropriate for some class of illness. In this work we examined the performance of proposed missing value imputation method using linear regression model. 3-D plot between the symptoms shows that the relations between the symptoms are not linear in some cases. Thus the error may increase to find the missing symptom value. The results are compared with the value already known in the record. It is worth to mention that the system is for primary health care and in case of any emergency patients are referred to the experts.

Acknowledgements This work is supported by Information Technology Research Academy (ITRA), Government of India under, ITRA-Mobile grant [ITRA/15(59)/Mobile/Remote Health/01].

References

1. Tian J et al (2012) A fuzzy clustering approach for missing value imputation with non-parameter outlier test
2. Das S, Sil J (2017) Uncertainity management of health attributes for primary diagnosis. In: International conference on big data analytics and computational intelligence (ICBDACI). https://doi.org/10.1109/ICBDACI.2017.8070864
3. Wu X, Kumar V, Quinlan JR et al (2007) Top 10 algorithms in data mining. Knowl Inf Syst 14(1):1–37. https://doi.org/10.1007/s1011500701142
4. Jabbar MA et al An evolutionary algorithm for heart disease prediction. In: Communications in computer and information science, vol 292. Springer, 378–389. http://dx.doi.org/10.1007/978-3-642-31686-9_44
5. Jabbar MA, Deekshatulu BL, Chandra P Graph based approach for heart disease prediction. In: Proceedings of the third international conference on trends in information, telecommunication and computing, Volume 150 of the series lecture notes in electrical engineering, pp 465–474
6. Roddick JF, Fule P, Graco WJ (2003) Exploratory medical knowledge discovery: experiences and issues. SIGKDD Explor Newsl 5(1), 94–99. http://doi.acm.org/10.1145/959242.959243
7. Schneider T (2001) Analysis of incomplete climate data: Estimation of mean values and covariance matrices and imputation of missing values. J Clim 14:853–871
8. Thirukumaran S, Sumathi A (2012) Missing value imputation techniques depth survey and an imputation Algorithm to improve the efficiency of imputation. In: 2012 fourth international conference on advanced computing (ICoAC), Chennai
9. Andridge R, Little R (2010) A review of hot deck imputation for survey non-response. Int Stat Rev 78(1):40–64
10. Sebag M, Aze J, Lucas N ROC-based evolutionary learning: application to medical data mining. In: Artificial evolution volume 2936 of the series lecture notes in computer science, pp 384–396
11. Krishnaiah V, Narsimha G, Subhash Chandra N Heart disease prediction system using data mining technique by fuzzy K-NN approach. In: Emerging ICT for bridging the future—Proceedings of the 49th annual convention of the computer society of india (CSI). Series advances in intelligent systems and computing, vol 337, pp 371–384
12. Joshi S, Nair MK Prediction of heart disease using classification based data mining techniques. In: Computational intelligence in data mining—volume 2, volume 32 of the series smart innovation, systems and technologies, pp 503–511
13. Khaleel MA, Dash GN, Choudhury KS, Khan MA Medical data mining for discovering periodically frequent diseases from transactional databases. In: Computational intelligence in data mining—volume 1, volume 31 of the series smart innovation, systems and technologies, pp 87–96
14. Madhu G et al (2012) A novel index measure imputation algorithm for missing data values: a machine learning approach. In: IEEE international conference on computational intelligence & computing research
15. A novel discretization method for continuous attributes: a machine learning approach. Int J Data Min Emerg Technol 4(1), 34–43
16. Fauci AS, Kasper DL, Harrison R (1950) Harrison's principles of internal medicine
17. https://healthfinder.gov. Accessed 22 Oct 2017
18. http://www.mayoclinic.org. Accessed 21 Oct 2017
19. Glynn M, Drake WM (2012) Hutchison's clinical methods
20. Bezdek JC, Ehrlich R, Full W (1984) FCM: the fuzzy c-means clustering algorithm. Comput Geosci (Elsevier)
21. Ryan TP (2008) Modern regression methods. John Wiley & Sons
22. Mamdani EH, Assilian S (1975) An experiment in linguistic synthesis with a fuzzy logic controller. Int J Man Mach Stud 7(1):1–13
23. Fortemps P, Roubens M (1996) Ranking and defuzzification methods based on area compensation. Fuzzy Sets Syst (Elsevier)

24. Davies DL, Bouldin DW (1979) A cluster separation measure. IEEE Trans Pattern Anal Mach Intell 1(2):224–227
25. Kärkkäinen I, Fränti P (2000) Minimization of the value of Davies-Bouldin index. In: Proceedings of the LASTED international conference signal processing and communications, pp 426–432

Recommendation Framework for Diet and Exercise Based on Clinical Data: A Systematic Review

Vaishali S. Vairale and Samiksha Shukla

Abstract Nowadays, diet and exercise recommender frameworks have gaining expanding consideration because of their importance for living healthy lifestyle. Due of the expanded utilization of the web, people obtain the applicable wellbeing data with respect to their medicinal problem and available medications. Since diseases have a strong relationship with food and exercise, it is especially essential for the patients to focus on adopting good food habits and normal exercise routine. Most existing systems on the diet concentrate on proposals that recommend legitimate food items by considering their food choices or medical issues. These frameworks provide functionalities to monitor nutritional requirement and additionally suggest the clients to change their eating conduct in an interactive way. We present a review of diet and physical activity recommendation frameworks for people suffering from specific diseases in this paper. We demonstrate the advancement made towards recommendation frameworks helping clients to find customized, complex medical facilities or make them available some preventive services measures. We recognize few challenges for diet and exercise recommendation frameworks which are required to be addressed in sensitive areas like health care.

Keywords Health care · Food · Nutrition · Diet and exercise recommendation frameworks

V. S. Vairale (✉) · S. Shukla
Faculty of Engineering, CHRIST (Deemed to be University), Bangalore, Karnataka, India
e-mail: vairale.sheshrao@res.christuniversity.in

S. Shukla
e-mail: samiksha.shukla@christuniversity.in

D. K. Mishra et al. (eds.), *Data Science and Big Data Analytics*,
Lecture Notes on Data Engineering and Communications Technologies 16,
https://doi.org/10.1007/978-981-10-7641-1_29

1 Introduction

One of the main considerations in today's life is right diet and adequate exercise, particularly, for the individuals experiencing some minor or significant health issues. Generally, people stay ignorant of significant causes behind inadequacy or overabundance of different key nutrients in their diet, for example, calcium, proteins, and vitamins, and how to correct such nutrients through appropriate diet. Presently, for different sorts of data, individuals rely on the web. They utilize web crawlers like Google to look data over the web. The query must be correct with the goal that it will give the data identified with patients' medicinal issues. But, it is hard to get the imperative data adequately as there is part of data accessible on the web. People are constrained for the hunt of right diet and exercise suggestions along with their own preferences.

Recommendation frameworks are information gathering structures which help people to settle on decisions by suggesting items based on verification and guide them to choose specific items wisely. Recommendation frameworks look at the clients' personal information (content-based approach) with some important characteristics matches with other clients' information (Collaborative Filtering), learning-based (Knowledge-based recommender structure) and together all (Hybrid). Food suggestion applications made with the objective to assist the clients step by step by proposing best food items and set of activities according to patient's health condition and inclinations.

Diet and physical activity recommendations in earlier systems are made up in typical frame which is not fitting for a broad assortment of clients, who have variation in age groups, sex, weight, height, lifestyle, eating regimen and exercise slants et cetera. Further, diet regimen and fitness activity are correlated. Also, diet and fitness activity suggestions incorporate data from various regions like personal profile, nutrition, medicinal information and physical movement. A modified eating regime and activity proposal system considers customers' health profile, food and fitness inclinations and empowers the customer to cure the therapeutic issue with some level by adopting healthy lifestyle.

Several systems proposed distinctive food suggestion frameworks. These frameworks can be classified as: (i) Food proposal frameworks [1, 2], (ii) Menu suggestions [3], (iii) Diet suggestions [4], (iv) Health suggestions for various diseases like diabetes and cardiovascular [5, 6], and (v) Recipe suggestions [7]. Most of the systems for diet recommendation proposals exist for patients with diabetes, yet still there are less structures which grouped together all domains and incorporated them into one framework, for eating regimen and physical action suggestion. Despite the way that all the learning bases have their own limitations, specific complexities and working measures; and these all information bases are interrelated for eating standard and physical movement recommendations for patients. Thus, an efficient diet and exercise recommendation framework is the need of society which will facilitate customer to evaluate their own nourishment essentials, provides up to date data about sustenance and exercise as for their sickness and upgrade client's wellbeing by interrelating

Table 1 List of recommendation frameworks in food domain

Recommendation techniques	Papers	Food domain approaches
Collaborative filtering	Svensson et al. [8]	Online grocery store
	Elahi et al. [9]	Considering user preferences
	Berkovsky and Freyne [10]	Recipe generation
Content-based	Elsweiler et al. [11]	Health issues
Knowledge-based	El-Dosuky et al. [12]	Personalized nutrition service
	Aberg [13]	Handling malnutrition issue
Hybrid	Berkovsky and Freyne [10]	Food items rating

distinctive information base like food and its nutrients, work out, client profile and level of disease.

This paper summarizes existing research work done with diet and exercise recommendation frameworks which give proposals by considering the clients' inclinations and their nutritional requirements. A review of some existing work identified with the utilization of recommender frameworks in the diet and exercise domain is given in Table 1.

This paper is composed in following sections. In Sect. 2, we provide a broad idea of recommendation methods used in general for users. In Sect. 3, we examine healthcare recommendation frameworks and how effectively these frameworks can assist people to pick healthy foods that suits to their inclinations and medicinal condition. We present a concise review on diet and exercise recommendation based on clinical data. In Sect. 4, we identify some challenges for diet and exercise recommendation frameworks with respect to user profile, algorithms for recommendations, availability of data sets and the level of disease is considered as topic for future work. Lastly, in Sect. 5, we conclude our review on diet and food recommendation frameworks based on clinical data.

2 Recommendation Frameworks

Because of substantial data over-burdens activated by the Internet, separating/discovering useful data turns out to be difficult to process. In this specific circumstance, recommender frameworks turned into an effective tool to separate valuable data and convey it in a productive way. A recommender framework predicts the inclinations of clients for non-rated products and suggests new things to clients. With these advantages of recommender frameworks, coming up with new proposal approaches and incorporating them in various fields rise to a great degree of extension. Recommender frameworks are used to recommend services and items (e.g., films, books, electronic gadgets, and travel) which best address to the clients' requirement and inclinations [14]. However, in the healthcare domain, the recommender frameworks have been found as a powerful solution to help clients to adapt to the immense measure of accessible information identified with foods/exercises. Numerous frameworks have been proposed for creating customized recommendations. Following are the most widely used recommendation techniques.

2.1 Collaborative Filtering Recommendation Systems

Collaborative filtering is widely used and researchable method of recommender frameworks. The essential thought of this method is to utilize the opinions of the group for generating proposals. First, customers ought to give rating to some given things either implicitly or explicitly. After that, the recommender system recognizes the closest neighbors whose likes are like the given client and suggests items that the closest neighbors have preferred [15]. Collaborative filtering is usually developed by utilizing following techniques: user-based [16], item-based [17], model-based [18], and matrix factorization [19].

2.2 Content-Based Recommendation Systems

This framework makes customized recommendations by accessing the data regarding available product information (e.g., disease and restricted food items) and client data depicting what the clients like [20]. The fundamental idea of a content-based framework is to break down the data with respect to client inclinations and product data utilized by the client, and after that suggest the product based on available data. The research around this method basically depends on suggesting things with printed material, for example, website pages, books and reports.

2.3 Knowledge-Based Recommendation Systems

These systems are considered as an answer for handling a few issues created by traditional methodologies (e.g., ramp-up issues [21]. Additionally, these frameworks are particularly helpful in areas in which the quantity of present product's inputs is relatively less (e.g., food domain) or when clients need to characterize their requirements explicitly (e.g., "vegetarian or non-vegetarian"). There are two fundamental methodologies for creating these recommender frameworks: case-based- and constraint-based suggestion [14]. This approach utilizes clients' inclinations to suggest specific products, and afterward considers clients' feedback to improve the accuracy of recommendations [21].

2.4 *Hybrid Recommendation Frameworks*

These systems depend on the blend of the previously mentioned methods. For example, collaborative filtering techniques have cold-start problem, i.e., new user or new item. While, CB methodologies can handle this issue with the available information about these kinds of items. The paper [22] summarize some hybrid methods which join both collaborative and content-based approaches, containing weighted, switching, blended, and feature combination.

3 Healthcare Recommendation Frameworks

Individuals increasingly utilize the Internet for getting data with regards to diseases and their relevant medications. Presently, numerous online wellness portals as of now give non-customized health data as articles. However, it is difficult to get applicable information for individual's condition and translate this medical terms and connections. Recommender Systems (RS) already enable these frameworks to perform exact data filtering. In this paper, we demonstrate the development made up with recommendation systems helping clients to find customized, complex medicinal information and help them with preventive human services measures.

The advancement and awareness among the patients regarding their health issues lead to move towards healthcare recommendation systems [23]. This transition challenges researchers with new inquiries both in conventional recommender issues and in domain related issues. The key ideas in HRS that were handled inside the most recent years are the personalization of recommendation frameworks to individual users and their present health status, and the impact of these strategies on the clients trust and the assessment techniques (e.g., interrelated domain) and measures (e.g., client satisfaction) in healthcare recommendation systems [24]. We discuss food, diet and exercise recommendation frameworks in following subsections.

3.1 *Food Recommendation Frameworks*

With consideration of dietary intake of nowadays, it has been seen that there is a more prominent measure of development in the wellbeing related issues due to changing and busty lifestyles. For example, diabetes, cancer and hypertension which are the reason for some unhealthy food habits. The problem can be resolved by giving appropriate eating regimen proposals. In this situation food recommendation frameworks are additionally researched as a potential intends to help individuals to make themselves more fit and healthy [11]. It is beneficial to use food recommendation frameworks as a piece of a procedure for correcting dietary intake of clients. For this situation, food recommendation frameworks not just take in clients' inclinations for food items and nutrients, yet additionally select balanced diet by considering medical issues, nutritious necessities, and past eating practices.

Several research works have been proposed for various recommendation frameworks identified with food and diet. These frameworks are utilized for food suggestions, menu proposals, diet plan suggestions, health suggestions for specific diseases, and recipe proposals. Dominant part of these suggestion frameworks separate clients' inclinations from various sources like clients' ratings [25, 26], choice of recipes [27, 28], and browsing and medical history [29–31]. For example, in [27], a recipe suggestion framework is proposed utilizing social routing framework. The social routing framework separates clients' selections of recipes and consequently suggests the recipes. In [29], a recipe suggestion framework is recommended that is fit for learning similitude measure of recipes utilizing swarm card-sorting. The previously mentioned proposal frameworks need in taking care of a typical issue known as cold start issue. All these frameworks must wait for the other clients to enter enough information for the successful proposals. A portion of the business applications like [22, 32] offer clients for a survey study to get clients inclinations in a short timeframe. For example, the review utilized by [22] is particularly intended to coordinate the way of life of the client, i.e., healthy persons, sportsman, pregnant, and so on. The overview additionally endeavors to disallow different food items which do not coordinate the client's way of life. The study [32] designed questionnaire through which a client answers diverse inquiry regarding his/her way of life, food inclinations, supplement intake, and food habits. The framework once extricates all the essential data is then ready to suggest distinctive meal plan for every day and week-by-week premise.

3.2 Diet and Exercise Recommendation Frameworks

A diet recommendation framework gives specific food items that assists the individuals to balance the nutritional requirement as per their health conditions. The current diet suggestion frameworks can be categorized into three types.

1. This technique considers the common limitation in food intake by some health issues, it endorses the food items which ought to kept away from in the eating routine suggested by specialists. It just gives the requirement identified with ailments and that may display a few inconveniences in which this strategy does not consider the dynamic participation of clients since it is a latent eating regimen solution for clients that does not consider the level of diseases.
2. This technique endorses the eating regimen through dietician's counselling, it gets the eating behavior of patients based on the data acquired from different inquiries and recommends the eating routine by looking at the recommended table of nutrients. It is a technique that utilizes the recommended table of nourishment with regards to the inclination of patients utilizing surveys. Although, this strategy can give personalized diet recommendation framework to user contrasted with

that of the ordinary framework, it isn't sufficient to apply the personalized diet recommendation for clients since it does not consider the different conditions in clients' eating regimen, consumption level, eating routine, and exercise in a short timeframe.

3. This technique that endorses a diet set provided by the database and introduced with calorie table, it gives the eating regimen list suggested by specialists or dietician using domain information. However, this strategy has a restriction that cannot give customized administrations since it endorses the eating regimen that will be executed by clients without considering the diversity among the clients.

Lee et al. [33] examines and arranges individual data, diabetes, and sustenance regions. Fuzzy induction is used to make suggestions. On time glucose level and physical movement space are totally dealt with by manual judgments of area specialists, which can be managed by a few methods for framework mechanization.

Kovasznai [34] concentrates on recommending allowed, not allowed, proposed and remarkably embraced plans for sustenance from client's close to home and well-being related information. It needs in considering customer's slants and proposing sustenance things with the fitting sum.

Lin et al. [35] build up a structure which empowers nutritionists to give better organizations, and help customers to track physiological data, supper and exercise information on week by week premise. Culture, client inclinations, and illness point of view are not thought about.

Faiz et al. [36] make an application named SHADE which delivered eat less with sustenance records and its amount in view of the clients' inclinations and it proposes customer's favored activities with intervals and intensity.

Agapito et al. [37] developed a structure that can build the customer's prosperity profile, and provides the personalized dietary recommendation according to the wellbeing status. The profile is nurtured by utilizing dynamic continuous overviews masterminded by restorative authorities and requested by the customers. Need to incorporate new customary nourishment or/and formulas and association among customers and medicinal specialists has made more intuitive.

Kljusurić et al. [38] displayed a nutrition-based framework utilizing fuzzy logic demonstrating technique for schools in Croatia. Every day menu value, meal inclination, nutrition and vitality estimation of foods considered as input values. They consider meal choices as contribution since eating habits are distinctive in different spots of Croatia. Their product is appropriate for arranging of menu which has ideal cost and satisfactory supplement intake.

Kurozumi et al. [39] proposed a Japanese eating regimen assessment framework that utilizes Fuzzy Markup Language and foods list distributed by "Japan Diabetes Society". Dietary level is assessed by computing the supplement rates of foods chosen for meal and proposed technique as indicated by the "Japanese dietician database".

Another work considers the illnesses were exhibited by Chen et al. [40]. An eating routine proposal was offered by utilizing Fuzzy Rules and Knapsack method that uses clients' height, weight, exercise level, renal capacity, hypertension, elevated cholesterol and inclination information.

Mamat et al. [41] proposed "Fuzzy Multi-Objective Linear Programming" application that gives an ideal eating routine incorporates with enough supplements at a day with sensible cost. The expert system utilized as a part of this work to perceive the kind of ailment as indicated by the side effects listed by clients. Primary strategy utilized as a part of the eating routine evaluation frameworks is Fuzzy Logic and its variations.

Mamat et al. [42] made an adjusted eating routine arranging utilizing Fuzzy Linear Programing approach. In this arranging, evaluation of starch and fat sums in the nourishment that taken each day, foods costs and the measure of supplements that required to take daily. This approach is considered for women aged of 30 years.

Mák et al. [43] have exhibited a condition that characterizes dietary and physical movement consultative issue. Proposed condition communicated as "Various leveled Multi-Objective Dietary Menu Planning Problem with Harmony" permits the formalization of any eating regimen and physical movement arranging issue.

Al-Nazer et al. [32], presented semantic web concept and ontology method to assess the client's food choices and health status related customer's data and use this data to get the relevant information with the objective that customers can influence to enquire top-ranked foods and exercises. A semantic model which utilizes the personalization strategies in perspective of composed domain ontologies, designed by the domain specialists, to endorse the imperative support that is appropriate with individual's requirements.

Tao et al. [44], presented a customized ontology structure for learning base and considering customer profiles. This system captured users' profiles from both a global learning base and customer close-by case storage facilities. The ontology structure is assessed by separating it with other models in the web data collection methods.

Hsiao et al. [45], presented a customized eating regime structure which is not simply makes a translation of supplement recommendation into sensible dish choices, yet moreover recognizes reactions from customers to change their meal plans. The outcomes demonstrated on food requirements and can be satisfied by the regular eating routine plan which empowers a customer to change the course of action easily. The rules generated by SmartDiet are effective to enhance the health status of a user.

Chiang et al. [46], presented customized activity planning method is proposed to exercise daily some physical activities and balanced the calorie intake. With the help of individual's daily updates and expert's input, the recommendation outcomes were adjusted.

Villarreal et al. [47] proposed an ontology system for clients. As a legitimate examination, health, and personal profile, user's needs, food choices are joined together to suggest the meal plan. Physical activity is an important part in diabetes administration was disregarded.

Cantais et al. [48] presented food ontology with Protégé and its blends and data offering to "PIPS" structure. The advancement and issues have been discussed. "Pallet Reasoner" provides ontology structure while dismissing its rule value and acceptances.

Kim et al. [49] conveyed the likelihood of a modified food proposal model for clients. Predefined XML design is provided for data exchange and compromise

Table 2 Manually reviewed papers on diet and exercise recommender framework based on clinical data with their top concepts

Year	#Papers reviewed	Concepts
2003	2	Diet recommendation, without consideration of exercise domain and expert's decisions
2005	3	Health information systems
2006	2	Semantic proposal tool for health data
2008	1	Provides diet and exercise suggestion list
2009	3	Fuzzy ontology model for diet plans
2010	3	Healthy meal planning tool
2011	6	Rule based methods, Semantic web ontology are used to generate diet plans
2012	4	FML, meal plans suggestions, health issues
2013	6	Physical activity suggestions, ontology and semantic web, diet and physical activity recommendations
2014	6	SWRL rules, integration of food and exercise domain
2015	6	Diet and exercise suggestions, experts in loop
2016	6	Fuzzy sets, food and nutrition, exercise recommendations
2017	2	Dietician, fuzzy logic, meal and exercise plans

instrument among various areas. User's food interests have not inspected for recommending balanced meal plan.

Kovasznai [34] and Khan et al. [50] presented case-based technique for meal suggestions. A constraint generated procedure is provided where a clause is verified at every center point and based on the estimation of the quality, built-in resulting rule is endorsed. Customer choices for foods are not considered in [34]. An automated framework for the proposition is proposed in [50]. When an expert describes actions for a specific case, the framework keeps these data and produces a result when similar case appears in future. Various domains, for instance, individual's data, food and health databases are taken as a one structure without creating any data exchange framework between other domains.

Using clustering method [1] food items are assembled in various groupings in perspective of their different supplement estimations. Each gathering itself contains standard, compelled and avoidable supports from diabetes point of view.

Izumi et al. [51, 52] used OWL structure to indicate individual and physical activity areas using Protégé to provide wellness and exercise suggestions. Both individual and exercise spaces data are clubbed together and portrayed with an ontology structure. Eating regimen suggestions were not considered. Table 2 presents methods that designed for diet regimen and fitness recommendations.

4 Research Challenges

There are issues to resolve in building the system itself and difficulties with regards to the customer and their collaboration along with the structure. Considering the reliance of the framework on the client, we first distinguish the difficulties regarding the client. At that point we decide the issues concerning the calculations utilized as a part of recommender frameworks.

4.1 Availability of Data Sets

The fate of diet and exercise recommender frameworks firmly depends on interdisciplinary joint effort and coordinated over organizations. Recommender frameworks have begun to nurture at the point when data collections are turned out to be open and quality measurements become accessible. We would like to see more public datasets for health recommender framework which are useful in modeling techniques, testing client inputs and designing new measurements for the area of health recommendation frameworks. Beside offline assessments conceivable through these information collections, it is important that online assessments still assume a broad part in assessing a recommender framework. Client real responses may differ from expectations produced using offline information.

4.2 Food/Nutritional Data

Providing suggestions to balance the nutritional needs, may be identified with lack of healthy sustenance, weight reduction or to anticipate nourishment-based diseases [11, 53]. Suggestions may be nourishment substitution items [54, 55], meal (Breakfast, Lunch, Dinner) [56], or specific to balanced diet. The eating habits [57] and its unpredictable and social dependencies [10] must be considered.

4.3 Physical Activity Data

Providing suggestions on what exercises to perform may be related to find exercises that are mostly liked and inspiring and furthermore match the client's necessities and requirements. A recommender system could incorporate location information and climate information to find exercises that are ideal for the clients setting.

4.4 Dietician-in-Loop

This approach is an essential requirement for generating accurate recommendations [58]. The medical experts in these kinds of recommendation frameworks are not only consumer of computerized data, as well as a somebody who can intelligently control recommendations and tools. The specialist as an expert inside the circle of a recommendation framework can ensure that the generated information is correlated with patients' health information on basis implicit knowledge while the recommender framework can incorporate patient information and treatment results and conceivable (side-)effects identified with past choices [59–61].

4.5 Recommendation Algorithms

Enhancing the nature of recommendations by incorporating different requirements (e.g., health conditions, nutritional needs, the preferences for foods and exercises) into the recommendation procedure. Considering more principles and constraints in the proposal procedure will improve the idea of suggestions [62]. The challenge is the way to recommend foods and exercise which meets health circumstances and nutritional needs of clients, and taking care of preferences of clients. In this situation, recommender frameworks require extra efforts from clients since clients need to report the daily food intake consistently and this can keep clients demotivating from utilizing the framework for longer period.

5 Conclusion

We give an outline of recommendation frameworks in the diet and exercise domain with respective of three distinct sorts of food/nutrient recommendation frameworks. This paper introduces some current investigations in the diet and exercise domain, which fundamentally concentrates on customized recommendations to people, by considering their choices of food and physical activity and nutritional requirement based on clinical data. Next, we consider the investigations which focus at counseling of healthy nourishment items in different health conditions. Recommendation strategies are utilized as a part of many diet recommendation frameworks. Moreover, crossover approaches are additionally utilized to enhance the recommender's execution. Although, in various settings, food and exercise recommendation frameworks consider as an important part in giving foods, meeting inclinations and sufficient nutritional requirements of clients and in addition convincing them to follow healthy eating practices. A few difficulties with respect to client data, recommendation approaches, changing eating practices, availability of data set are discussed as challenges for future.

References

1. Phanich M, Pholkul P, Phimoltares S (2010) Food recommendation system using clustering analysis for diabetic patients. In: International conference on information science and applications
2. Ge M, Elahi M, Fernaández-Tobías I, Ricci F, Massimo D (2015) Using tags and latent factors in a food recommender system. In: Proceedings of the 5th international conference on digital health 2015—DH'15
3. Runo M (2011) FooDroid: a food recommendation app for university canteens. Swiss Federal Institute of Theology, Zurich
4. Su CJ, Chen YA, Chih CW (2013) Personalized ubiquitous diet plan service based on ontology and web services. Int J Inf Educ Technol 3(5):522
5. Evert AB, Boucher JL, Cypress M, Dunbar SA, Franz MJ, Mayer-Davis EJ, Yancy WS (2014) Nutrition therapy recommendations for the management of adults with diabetes. Diabetes Care 37(Supplement 1):S120–S143
6. LeFevre ML (2014) Behavioral counseling to promote a healthful diet and physical activity for cardiovascular disease prevention in adults with cardiovascular risk factors: US preventive services task force recommendation statement. Ann Intern Med 161(8):587–593
7. Freyne J, Berkovsky S (2013) Evaluating recommender systems for supportive technologies. Hum–Comput Interact Ser 195–217
8. Svensson M, Laaksolahti J, Höök K, Waern A (2000) A recipe based on-line food store. In: Proceedings of the 5th international conference on intelligent user interfaces IUI'00. ACM, New York, NY, USA, pp 260–263
9. Elahi M, Ge M, Ricci F, Fern´andez-Tob´ıas I, Berkovsky S, Massimo D (2015) Interaction design in a mobile food recommender system. In: IntRS@recsys, CEUR-WS.org, CEUR workshop proceedings, vol 1438, pp 49–52
10. Berkovsky S, Freyne J (2010) Group-based recipe recommendations: analysis of data aggregation strategies. In: Proceedings of the fourth ACM conference on recommender systems. ACM, pp 111–118
11. Elsweiler D, Harvey M, Ludwig B, Said A (2015) Bringing the "healthy" into food recommenders. CEUR Workshop Proc 1533:33–36
12. El-Dosuky MA, Rashad MZ, Hamza TT, El-Bassiouny AH (2012) Food recommendation using ontology and heuristics. AMLTA, Springer, Commun Comput Inf Sci 322:423–429
13. Aberg J (2006) Dealing with malnutrition: a meal planning system for elderly. AAAI spring symposium: argumentation for consumers of healthcare
14. Burke R, Felfernig A, Göker M (2011) Recommender systems: an overview. AI Mag. 32:13
15. Ekstrand M (2011) Collaborative filtering recommender systems. Found Trends® Hum-Comput Interact 4:81–173
16. Asanov D (2011) Algorithms and methods in recommender systems. Berlin Institute of Technology, Germany, Berlin
17. Sarwar B, Karypis G, Konstan J, Reidl J (2001) Item-based collaborative filtering recommendation algorithms. In: Proceedings of the tenth international conference on World Wide Web—WWW'01
18. Koren Y, Bell R, Volinsky C (2009) Matrix factorization techniques for recommender systems. Computer 42:30–37
19. Bokde D, Girase S, Mukhopadhyay D (2015) Matrix factorization model in collaborative filtering algorithms: a survey. Proced Comput Sci 49:136–146
20. Pazzani MJ, Muramatsu J, Billsus D (1996) Syskill and Webert: identifying interesting web sites. In: Proceedings of the thirteen national conference on artificial intelligent, vol 1, pp 54–61
21. Burke R (2000) Knowledge-based recommender systems. In: Encyclopedia of library and information systems, vol 69. Marcel Dekker, pp 180–200
22. Burke R (2002) Hybrid recommender systems: survey and experiments. User Model User-Adap Inter 12(4):331–370

23. Schäfer H, Hors-Fraile S, Karumur R, Calero Valdez A, Said A, Torkamaan H, Ulmer T, Trattner C (2017) Towards health (aware) recommender systems. In: Proceedings of the 2017 international conference on digital health—DH'17
24. Genitdaridi I, Kondylakis H, Koumakis L, Marias K, Tsiknakis M (2013) Towards intelligent personal health record system: review, criteria and extensions. Proced Comput Sci (Elsevier)
25. Freyne J, Berkovsky S (2010) Intelligent food planning. In: Proceedings of the 15th international conference on intelligent user interfaces—IUI'10
26. Forbes P, Zhu M (2011) Content-boosted matrix factorization for recommender systems. In: Proceedings of the fifth ACM conference on recommender systems—RecSys'11
27. Svensson M, Höök K, Cöster R (2005) Designing and evaluating kalas. ACM Trans Comput-Hum Interact 12:374–400
28. Geleijnse G, Nachtigall P, van Kaam P, Wijgergangs L (2011) A personalized recipe advice system to promote healthful choices. In: Proceedings of the 15th international conference on intelligent user interfaces—IUI'11
29. Van Pinxteren Y, Geleijnse G, Kamsteeg P (2011) Deriving a recipe similarity measure for recommending healthful meals. In: Proceedings of the 15th international conference on intelligent user interfaces—IUI'11
30. Ueda M, Asanuma S, Miyawaki Y, Nakajima S (2014) Recipe recommendation method by considering the user's preference and ingredient quantity of target recipe. In: Proceedings of the international multi conference of engineers and computer scientists, vol 1
31. Rehman et al (2017) Diet-right: a smart food recommendation system. KSII Trans Internet Inf Syst 11
32. Al-Nazer A, Helmy T, Al-Mulhem M (2014) User's profile ontology-based semantic framework for personalized food and nutrition recommendation. Proced Comput Sci 32:101–108
33. Lee C, Wang M-H, Hagras H (2010) A type-2 fuzzy ontology and its application to personal diabetic diet recommendation. IEEE Trans Fuzzy Syst
34. Kovasznai G (2011) Developing an expert system for diet recommendation. In: 6th IEEE international symposium on applied computational intelligence and informatics (SACI)
35. Lin E, Yang D, Hung M (2012) System design of an intelligent nutrition consultation and recommendation model. In: 9th international conference on ubiquitous intelligence and computing and 9th international conference on autonomic and trusted computing
36. Faiz I, Mukhtar H, Qamar A, Khan S (2014) A semantic rules & reasoning based approach for diet and exercise management for diabetics. In: IEEE international conference on emerging technologies (ICET)
37. Agapito G, Calabrese B, Guzzi P, Cannataro M, Simeoni M, Care I, Lamprinoudi T, Fuiano G, Pujia A (2016) DIETOS: a recommender system for adaptive diet monitoring and personalized food suggestion. In: IEEE 12th international conference on wireless and mobile computing, networking and communications (WiMob)
38. Kljusurić JG, Kurtanjek Ž (2003) Fuzzy logic modelling in nutrition planning-application on meals in boarding schools. In: Current studies of biotechnology, Vol. III-Food
39. Kurozumi K et al (2013) FML-based Japanese diet assessment system. In: IEEE international conference on fuzzy systems (FUZZ)
40. Chen R-C et al (2013) Constructing a diet recommendation system based on fuzzy rules and knapsack method. In: International conference on industrial, engineering and other applications of applied intelligent systems. Springer
41. Mamat M et al (2013) Fuzzy multi-objective linear programming method applied in decision support system to control chronic disease. Appl Math Sci 7(2):61–72
42. Mamat M et al (2012) Fuzzy linear programming approach in balance diet planning for eating disorder and disease-related lifestyle. Appl Math Sci 6(103):5109–5118
43. Mák E et al (2010) A formal domain model for dietary and physical activity counseling. In: International conference on knowledge based and intelligent information and engineering systems. Springer
44. Tao X, Li Y, Zhong N (2011) A personalized ontology model for web information gathering. IEEE Trans Knowl Data Eng 23:496–511

45. Hsiao J, Chang H (2010) SmartDiet: a personal diet consultant for healthy meal planning. In: IEEE 23rd international symposium on computer-based medical systems (CBMS)
46. Chiang J, Yang P, Tu H (2014) Pattern analysis in daily physical activity data for personal health management. Pervasive Mob Comput 13:13–25
47. Villarreal V, Hervás R, Fdez AD, Bravo J (2009) Applying ontologies in the development of patient mobile monitoring framework. In: 2nd international conference on ehealth and bioengineering—EHB 2009, Romania
48. Cantais J, Dominguez D, Gigante V, Laera L, Tamma V (2005) An example of food ontology for diabetes control. Working notes of the ISWC 2005 workshop on ontology patterns for the semantic web. Galway, Ireland
49. Kim J-H, Lee J-H, Park J-S, Lee Y-H, Rim K (2009) Design of diet recommendation system for healthcare service based on user information. In: Fourth international conference on computer sciences and convergence information technology
50. Khan AS, Hoffmann A (2003) Building a case-based diet recommendation system without a knowledge engineer. Artif Intell Med 27:155–179
51. Izumi S, Kuriyama D, Itabashi G, Togashi A, Kato Y, Takahashi K (2006) An ontology-based advice system for health and exercise. In: Proceedings of the 10th IASTED international conference on internet and multimedia systems and applications 535-029, pp 95–100
52. Izumi S, Kuriyama D, Miura Y, Yasuda N, Yotsukura R, Kato Y, Takahashi K (2007) Design and implementation of an ontology-based health support system. Technical report of IEICE SS2006-82, pp 19–24
53. Rokicki M, Herder E, Demidova E (2015) Whats on my plate: towards recommending recipe variations for diabetes patients. In: Proceedings of UMAP15
54. Freyne J, Berkovsky S (2010) Recommending food: reasoning on recipes and ingredients. In: International conference on user modeling, adaptation, and personalization. Springer, pp 381–386
55. Achananuparp P, Weber I (2016) Extracting food substitutes from food diary via distributional similarity. arXiv:1607.08807
56. Ge M, Ricci F, Massimo D (2015) Health-aware food recommender system. In: Proceedings of the 9th ACM conference on recommender systems, pp 333–334
57. Harvey M, Ludwig B, Elsweiler D (2012) Learning user tastes: a first step to generating healthy meal plans. In: First international workshop on recommendation technologies for lifestyle change
58. Kieseberg P, Malle B, Fru¨hwirt P, Weippl E, Holzinger A (2016) A tamper-proof audit and control system for the doctor in the loop. Brain Inf 3(4):269–279
59. Kieseberg P, Weippl E, Holzinger A (2016) Trust for the doctor-in-the-loop. In: European research consortium for informatics and mathematics (ERCIM) news: tackling big data in the life sciences, vol 104, issue 1, pp 32–33
60. Malle B, Kieseberg P, Weippl E, Holzinger A (2016) The right to be forgotten: towards machine learning on perturbed knowledge bases. In: Proceedings of IFIP WG 8.4, 8.9, TC 5 international cross-domain conference on availability, reliability, and security in information systems, CD-ARES 2016 and workshop on privacy aware machine learning for health data science, PAML 2016, Salzburg, Austria, August 31–September 2. Springer, pp 251–266
61. Rossetti M, Stella F, Zanker M (2016) Contrasting offline and online results when evaluating recommendation algorithms. In: Proceedings of the 10th ACM conference on recommender systems, pp 31–34
62. Mika S (2011) Challenges for nutrition recommender systems. In: CEUR-WS.org, workshop proceedings on context aware intelligent assistance, pp 25–33

Security Assessment of SAODV Protocols in Mobile Ad hoc Networks

Megha Soni and Brijendra Kumar Joshi

Abstract The basic requirement in mobile ad hoc network (MANET) is to achieve Secure routing. Dynamic characteristics of MANET offers many challenges to achieve security parameters such as availability, integrity, confidentiality authentication, and non-repudiation. To hinder the normal routing operation malicious nodes make use of the vulnerable routing protocols. The real challenge to achieve secure routing as secure versions of the routing protocols is also vulnerable for routing attack. In this paper, We assess and compare the security of Ad hoc On demand Distance Vector routing protocol and Secure Ad hoc On demand Distance Vector routing protocol under different types of routing attack like blackhole and Replay attack.

Keywords MANET · Security parameters · AODV · SAODV blackhole attack Replay attack

1 Introduction

In MANET wireless medium used by randomly moving nodes to forward data packets for others nodes which are not within the range of direct broadcast. Ad hoc routing protocols adapted by network topology which undergoes quick and dynamically changing. This features is lacking in routing by wired networks. The salient features of MANETs are in auto-configure mode and the capability to work in an infrastructure less network. Above features of MANETs makes it advantageous in field like military

M. Soni (✉) · B. K. Joshi
MCTE, Mhow, India
e-mail: meghasoni@svceindore.ac.in

B. K. Joshi
e-mail: brijendrajoshi@yahoo.co.in

D. K. Mishra et al. (eds.), *Data Science and Big Data Analytics*,
Lecture Notes on Data Engineering and Communications Technologies 16,
https://doi.org/10.1007/978-981-10-7641-1_30

operations, emergency rescue and disaster relief, etc., by providing cost effective fast installable and simply reusable solution.

The various routing protocols designed for MANETs [1] are focused at optimizing network routing performance. A protocol finds a route only when needed is named as reactive protocol. This is one of the MANET routing protocol which has achieved more attention as compared to other type of routing protocols. This feature allows the adhoc on demand protocols to perform better than the table driven routing protocols, which find and keep record of all feasible paths in the MANET even for those may not be utilized [2].

Emphasis on security must be given in real world MANETs [3]. Various attacks on routing in MANETs is interrupt the normal rote finding and set up process. These attacks can be brought by one or more malicious nodes. In recent literature several protocols on secure routing techniques have been proposed to protect from attacks. Cryptographic techniques based schemes are used to provide features like, authentication, message integrity and non repudiation.

The main focus of this paper is to analyze the security of a popular MANET on demand routing protocol, Ad hoc On demand Distance Vector (AODV) [2] and its secure version Secure Ad hoc On demand Distance Vector (SAODV) [4]. We describe how an attacker can be disrupt AODV routing by launching different routing attacks and find that the similar type of routing attacks are unsuccessful in SAODV because it uses asymmetric cryptography scheme. We showed that SAODV is also susceptible to certain kinds of replay attacks. A brief overview of AODV and SAODV routing protocols are discussed to assess their security by simulations of attack.

2 AODV and SAODV Working

AODV is an on demand reactive routing protocol that finds routes only when required. Sequence numbers are used to ensure that routes are fresh. A route request (RREQ) packet are broadcasted by a sending node to find a root to destination node. The RREQ contains the broadcast ID, current sequence number and IP address of node's. RREQ is received by the destination node and sends a route reply packet (RREP) on the same path which is setup during the process of route discovery. When failure in link occur a route error packet (RERR) is sent by intermediate node to the source and destination nodes [4].

SAODV added a feature of asymmetric or public key encryption to secure routing messages of AODV against different routing attacks. Hash chains and digital signatures are used by SAODV to secure both the mutable header field and non-mutable fields. These fields are hop count, source and destination addresses details, sequence number of all nodes. Calculation of hash chain by a source node are as follows: To sets the Max Hop Count to the Time To Live (IP header) first it creates a seed or random number. After that it calculates hash field by Max. Hop Count and seed and finally generates a digital signatures. This techniques makes it tough for a malicious node to change the routing messages since every node has a unique digital signature and it can only be generated by itself. When routing message processed for verification any decrement in the hop count field of RREQ packets or RREP packets by an attacker can be noticed, since hash field is checked by all the intermediate nodes.

To broadcast RERR, only signature of the packet is required to send. Before forwarding the RREQ, RREP and RERR packets, the destination nodes or intermediate nod authenticate the hash chain field and digital signatures of node.

3 Security Flaws in AODV and SAODV

The reasons of attacks on AODV protocol are:

- AODV is on demand reactive protocol.
- Message broadcasting is used in AODV Protocol.
- It is lacking in mobility management/flat routing.
- Shortest path algorithms is used AODV.
- No authentication process for non mutable field.
- It only maintain the record of neighborhood node.
- AODV is vulnerable to real time attack.
- AODV is not able to observe the next hop activities [5].
- The major factors which makes SAODV vulnerable to different attacks are:
- Partial authentication.
- No discrimination between old and new routing massage packets.
- A malicious node can easily spoof IP address of a node because IP header part is not secure.

4 Security Attacks on MANET

MANETs are not fully secured from various attacks. Attackers can drop of traffic network, modify control message or forward routing message to other node.

Goal of Attacker are:

- To decrease the overall throughput of network.
- To increase packets latency.
- To collapse link between two nodes.
- To change the packets root for increase link bandwidth.

The purpose of malicious node to lunch attacks in MANETs is to disrupt the normal operation of network or to take the routing information. Attacks can be generally defined into two category, Active attacks and Passive attacks. The purpose of launching an active attack in the network is to damage information and operation by inserting data or information in the network. Spoofing, and impersonation are examples of active attack. The aim of passive attack is to exchange data without

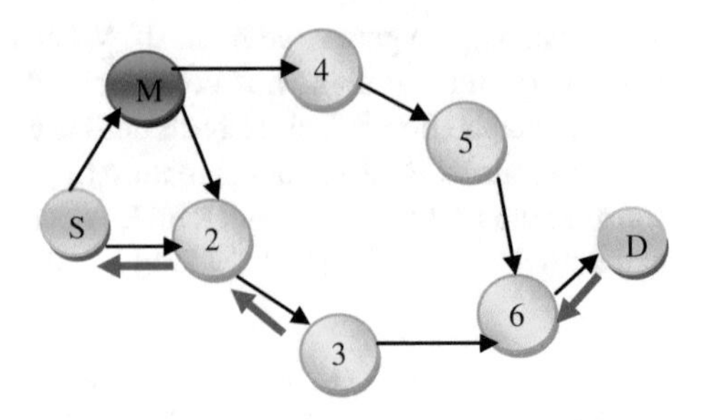

Fig. 1 Network topology

affecting the normal operation. It is hard to identify passive attack as compare to active attack.

Some of the most common attacks on MANET are:

Figure 1 shows the MANET topology in which S Node wants to send a data packets to D node. Node M is a malicious node.

A. *Route Disruption Attak*

An attacker initiates the route disruption attack in the MANET network by sending false or modified RREQ, RREP and RERR packets. AODV routing is interrupted while SAODV is able to defend the normal routing operation under attack condition [6].

B. *Route Invasion Attack*

To launch route invasion attack an attacker modifying and forging false RREQ and RREP packets in MANET. This attack can be successful against AODV but in SAODV a node is able to oppose this type of attack by identifying such packets and drop them.

C. *Blackhole Attack*

In blackhole attack, an attacker advertized that it has minimum path to the destination node and first sends false information of root to the source node, when root is established it can be receives the packets from source node [7].

The malicious Node A spoof IP address of destination nodes D from RREQ packets and launched blackhole attack. It sends fake RREPs which contents highest value of destination sequence numbers. AODV can be targeted by blackhole attack in two modes i.e., External blackhole attack in which a malicious node is not a part of network and Internal blackhole attack in which a malicious node is a part of network [8] but it was not successful in SAODV.

D. *Replay Attack*

A malicious node implements replay attack, by buffering the routing packets of one of the communicating parties and after some time replays those messages. By spoofing IP address and sending old RREQ packets, the attacker tries to misguided the destination node that the sending node wants to communicate again. In replay mode, the destination node will send RREP packets to the malicious node. In this an attacker is able to established the connection with the destination node and can send its own interest data packets.

5 Security Threats in AODV and SAODV

AODV is more vulnerable to different routing attacks as it has not strong security features and it can be effortlessly targeted by a attacker. The purpose of design secure routing protocols are to achieve security parameters like authentication, availability, confidentiality, Integrity and non-repudiation and it is AODV can effortlessly be targeted by a attacker to interrupt its routing. To disrupt routing in AODV an malicious node can be adapt following techniques [8]:

- To degrade network performance and increase routing delay an attacker generate fake RERR packets.
- RREP and RREQ packets forge or modify by attacker.
- In blackhole attack malicious node sends fake RREPs of highest sequence numbers.
- To disrupt the normal routing operation attackers make a tunnel/wormhole.
- To receive or drop data packets it Spoof source or destination IP address and block legitimate network node.
- Attacker form routing loops and initiate sleep deprivation attack.
- A malicious node lunch resource consumption attack to exhaust node batteries.

SAODV protocol is designed to deal with only specific attack and it do not offer a complete secure routing solution, Likewise other secure routing protocols SAODV has some security limitations; Prevention from attacks like replay, wormhole or tunneling is not easy in SAODV.

6 Performance Parameter

To evaluate the MANET performance in normal and in presence of attack various parameter such as Energy per Data, Packet Delivery Ratio, throughput and end to end delay is usee [9].

A. *Throughput*

The relative amount of total number of packets delivered to total simulation time is defined as throughput.

B. *Packet Delivery Ratio* (*PDR*)

PDR is obtained in terms of total received packets at destination node and the total of transmitted packets by source node.

C. *Delay*

To pass a packet through the network the average time acquired by it, this time can be defined as delay.

D. *Energy per Data* (*EDR*)

Amount of energy consumed by the node to delivered per byte data is defined EDR.

7 Simulation

Performance of AODV and SAODV are analyzed in normal and presence of blackhole attacks condition.

Simulation shows that when the number of nodes was increases throughput and PDR is decreases in network AODV performs batter than SAODV protocol in throughput and PDR. Because control overhead is increases due to cryptographic security schemes.

Energy per Data and Packet delay are high if we increased the number of nodes in MANET. EPD and Packet delay is higher for SAODV as compare to AODV protocol because of the extra processing and verification of cryptographic schemes (Figs. 2, 3, 4 and Tables 1, 2, 3).

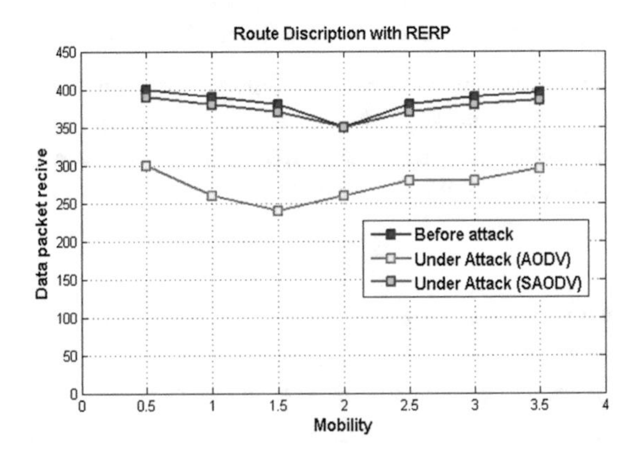

Fig. 2 Route discription with RREP on AODV and SAODV

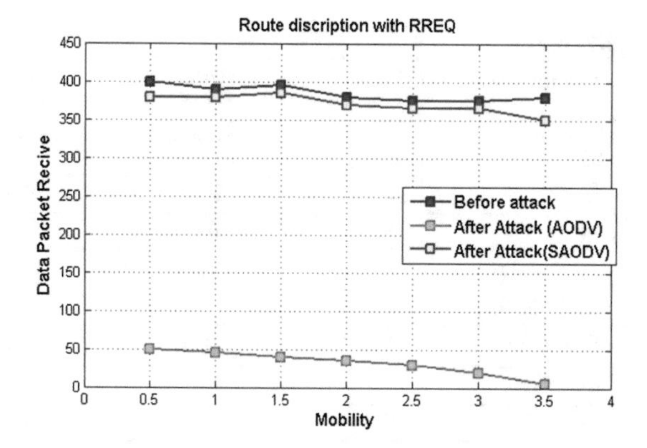

Fig. 3 Route discription with RREQ on AODV and SAODV

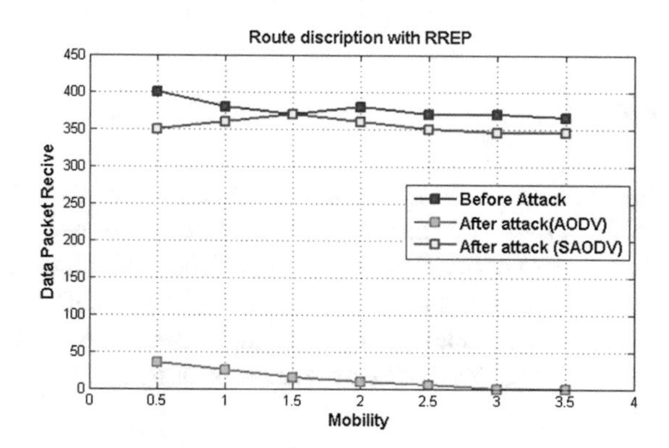

Fig. 4 Route discription with RERR on AODV and SAODV

Table 1 Performance parameter

S. no	Parameter	Value
1	Time of simulation	100 s
2	Name of protocol	AODV, SAODV
3	No. of malicious nodes	one
4	Number of nodes	20–100
5	Number of Max. connections	8, 16, 24, 32, 40
6	Pause time	25 s
7	Terrain area	700 × 700

Table 2 AODV performance

No. of nodes	Throughput	Delay	PDR	EDR
20	82	0.4	95	4
40	82	0.6	90	5.5
70	78	1.5	80	9.5
100	70	1.8	76	12

Table 3 SAODV performance

No. of nodes	Throughput	Delay	PDR	EDR
20	80	0.6	90	7.5
40	78	0.9	80	8
70	70	1.9	70	14
100	60	2.4	60	16

8 Conclusion

This paper paying attention on effects of attacks on security of SAODV and AODV protocols in MANETs based on the comparison on all mentioned parameters. We have observed the behavior of SAODV in presence of blackhole attack up to a defined level and it is found that blackhole attack is not successful in SAODV. AODV is more vulnerable for attacks due to modification of RREQ, RREP and RERR. To provide better security and stability in MANETs more secure routing and security mechanism is required to build up.

References

1. Perkins CE (2008) Ad hoc networking. Pearson Publication, India, pp 175–179
2. Zapata MG, Asokan N (2002) Securing ad hoc routing protocols. In: WISE, Sept 2002
3. Sharma M, Joshi BK (2016) A mitigation technique for high transmission power based wormhole attack in wireless sensor networks. In: ICTIBIG 2016. IEEE, Nov 2016
4. Abusalah L, Khokhar A, Guizani M (2008) A survey of secure mobile ad hoc routing protocols. IEEE Commun Surv Tutor 10(4):78–93
5. Arshad J, Azad MA (2006) Performance evaluation of secure on-demand routing protocol for Mobile ad hoc networks. IEEE
6. Maan F, Abbas Y (2011) Vulnerability assessment of AODV and SAODV routing protocols against network routing attacks and performance comparison. In: IEEE Wireless Advanced, pp 36–41
7. Kannhavong B, Nakayama H, Nemoto Y (2007) A survey of routing attacks in mobile ad hoc networks. IEEE Wirel Commun 14(5):85–91
8. Joshi BK, Soni M (2016) Security assessment of AODV protocol under wormhole and DOS attacks. In: IC3I2016. IEEE, Dec 2016
9. Soni M, Joshi BK (2016) Security assessment of routing protocols in mobile ad hoc networks. In: ICTIBIG2016. IEEE, Nov 2016

10. Ning P, Sun K (2003) How to misuse AODV: a case study of insider attacks against mobile ad hoc routing protocols. In: IEEE systems, man and cybernetics society information assurance workshop, pp 60–67, June 2003
11. Soni M, Joshi BK (2017) Security assessment of DSDV protocol in manet. Int. J Adv Comput Eng Netw (IJACEN) 5
12. Ramanthan S, Steenstrup M (1996) A survey of routing techniques for mobile communication networks. pp 89–104
13. Boukerche A, Turgut B, Aydin N, Ahmad MZ, Bölöni L, Turgut D (2011) Routing protocols in ad hoc networks: a survey. Comput Netw 3032–3080

Secure Sum Computation Using Homomorphic Encryption

Rashid Sheikh and Durgesh Kumar Mishra

Abstract Secure sum allows cooperating parties to compute sum of their private data without revealing their individual data to one another. Many secure sum protocols exists in the literature. Most of them assume network to be secure. In this paper we drop that assumption and provide a protocol that is applicable to insecure networks as well. We used additive homomorphic encryption technique for secure sum computation.

Keywords Secure sum · Secure multiparty computation · Homomorphic encryption

1 Introduction

Many research works today focuses on the preservation of privacy during joint computation by multiple parties. It has become more relevant because of the prevailing rate of joint data processing many government departments, joint computation by many financial organisations, and joint projects taken by many companies. Researchers have devised many protocols where multiple cooperating parties can jointly compute a function of their private data without revealing individual data to one another and successfully knowing the value of evaluated function. This area of information security is called Secure Multiparty Computation (SMC) [1]. Its objective is to get the correct result while keeping the individual data secret. Formally, SMC allows evaluation of f(x1, x2, x3, …, xn) where xi is the private data of the party Pi. All party must be able to know the correct value of f but no party should be able to learn other's private data.

R. Sheikh (✉)
Mewar University, Chittorgarh, India
e-mail: prof.rashidsheikh@gmail.com

D. K. Mishra
Sri Aurobindo Institute of Technology Indore, Indore, India

D. K. Mishra et al. (eds.), *Data Science and Big Data Analytics*,
Lecture Notes on Data Engineering and Communications Technologies 16,
https://doi.org/10.1007/978-981-10-7641-1_31

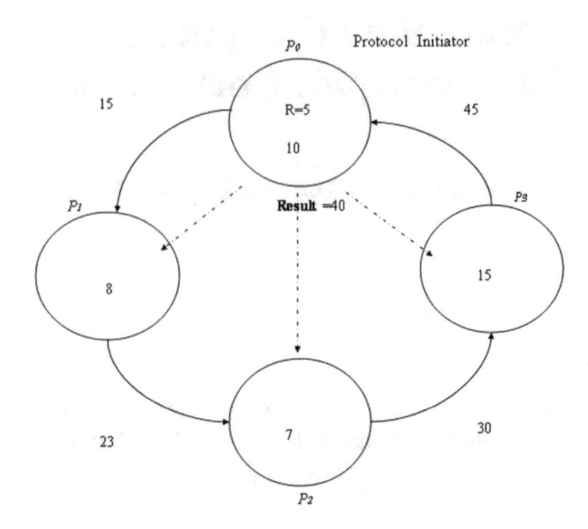

Fig. 1 Secure sum protocol [2]

There are two SMC models in use, Idea model and real model. In ideal model there is a third party called Trusted Third Party (TTP) to which all the participating parties supply data for common function evaluation. The TTP must evaluate the function and distribute the result to all the parties. The TTP is responsible to keep individual data secret from other parties. If the TTP colludes with other parties to leak data the model will see a failure. In actual practice the government organisations and the government approved organisations work as TTP. In real model, all the parties themselves compute the common function without any TTP. They set some protocol to achieve goals of SMC.

On the basis of their behaviour, the parties can be honest, semi-honest, and malicious. An honest party follows the protocol and never tries to learn the secret data of other parties. A semi honest party follows the protocol but curious to know others data. A malicious party neither follows the protocol nor respects privacy of others.

Many mathematical functions can be evaluated securely using SMC techniques. Secure sum allows joint parties to compute sum of their individual data without the private data being revealed to other parties. A secure sum protocol was proposed by Clifton et al. In 2002 [2]. They proposed to arrange all the parties in a ring. One of the parties initiates the protocol by choosing a secret random number and adding its private data to the chosen random number. The initiator now sends the sum to the next party in the ring. The receiving party simply adds its private data to the received sum. It sends the newly computed sum to the next party in the ring. The process is repeated until the sum is received by the initiator. The initiator simply subtracts the random number and sends sum to all the parties. Figure 1 depicts Clifton's secure sum protocol where parties P0 through P3 are arranged in a ring, R is the random number chosen and communication is in clockwise direction. The protocol can work only when the network lines are secured and the parties are semi honest.

We extended the work by dropping the use of random number [3–5]. In all our protocols we allowed all the parties to break their data in shares. The sum of these shares is taken by the parties. Finally the sum of all the shares is the secure sum. This achieves the goal of accuracy of result and the privacy of the individual data. We also assumed the network line to be secure and the parties to be semi-honest.

In this paper we drop the assumption of secure network and propose a secure sum protocol which can work in insecure network as well. We used homomorphic encryption function which allows to get the secure sum over an insecure network. It also eliminates the need of the use of the random number.

The paper outline is as follows: Sect. 2 describes the literature survey regarding our work. In Sect. 3 informal and formal description of our protocol is presented. Section 4 presents the analysis of the protocol and last section concludes the work with a note on the future research.

2 Related Work

Researchers first time paid attention to SMC in 1982 when Yao proposed solution to millionaires' problem [6]. This problem decides who is richer between two millionaires without disclosing individual wealth to one another. It was a two party problem. Goldreich et al. extended it to multiparty using circuit evaluation [7]. The research expanded in many area like private information retrieval [8, 9], privacy-preserving data mining [10, 11], Privacy-preserving geometric computation [12], privacy-preserving scientific computation [13], privacy-preserving statistical analysis [14].

The secure sum protocol proposed by Clifton et al. Is an important milestone but it suffered from a drawback that when two neighbours in the ring collude together, they can learn the data of the middle party just by taking the difference of what they send and receive. Our segmentation approach in [4] eliminated this drawback. In [3] we proposed to change the position of the party after each round of the computation. In [5] we first distributed the segments among parties and then computed the sum.

All the above protocols presume the network lines to be secure. But in actual practice the network is always insecure. Anyone able to intercept can learn the data by eavesdropping. Sniffing technique can be used to capture the packets. Homomorphic property allows encrypting and computing over the encrypted data to get the result. Many homomorphic encryption algorithms are proposed in which Paillier's cryptosystem [15] is frequently used. We use Paillier's secure additive homomorphic public key cryptosystem in our secure sum computation.

A secure additive homomorphic cryptosystem for two messages m1 and m2 satisfies following properties:

1. EPU(m1 + m2) := EPU(m1) + EPU(m2)
2. EPU(km1) := k × EPU(m1)

where PU refers to the public key and E () denotes the encryption function. A constant k is used in the second property.

3 Proposed Protocol

We propose a protocol for secure sum computation using homomorphic encryption. We use symmetric key cryptography where parties share a secret key.

3.1 Informal Description of the Protocol

All the cooperating parties are arranged in a ring network. We use real model of the SMC where no Trusted Third Party (TTP) exists and the parties run protocol among themselves to evaluate the common function of their individual data. One of the parties is designated as the protocol initiator which will start the protocol. The protocol initiator will choose some random number and add to its private data. The sum is encrypted with the public key generated using Paillier cryptosystem, and the ciphertext is sent to the immediate neighbour in the ring. The neighbour adds encrypted data to the received number, and sends newly computed sum to the next neighbour in the ring. The process continues till the protocol initiator receives sum of encrypted data of all the parties. Because of the homomorphic property this is equivalent to the encryption of data of all the parties plus the random number. The initiator will decrypt this using its private key to get the sum of data plus random number. After subtracting the random number the initiator will compute the sum which will be broadcasted to all the concern parties. The proposed architecture is depicted in Fig. 2.

3.2 Formal Description of the Protocol

Formally, as shown in Fig. 2, consider n parties P0 through Pn-1 having private data m0 to mn-1 respectively. All the parties are interested in getting $\sum$mi such that no party will disclose its private data to other party. The public key is denoted as PU and the private key as PR. The protocol for such computation is shown as protocol 1.

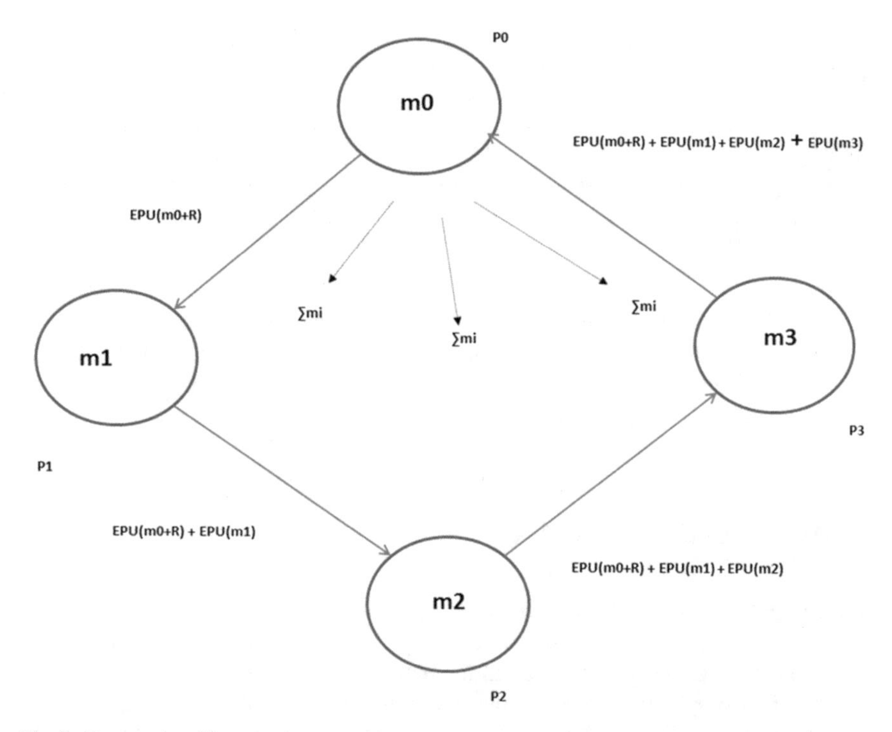

Fig. 2 Proposed architecture for computing secure sum using homomorphic encryption

---

Protocol1: Secure sum computation using homomorphic encryption

---

Require: Parties P0 through Pn-1 having private data m0 to mn-1 respectively. The proptocol initiator party generates and uses a key pair PU and PR.

Ensure: All the parties get ∑mi such that no party will disclose its private data to other party

---

Step 1: Let party P0 is the protocol initiator party. It chooses a non-zero random number R, and adds it to m0. It computes EPU(m0 + R) and sends to P1.

Step 2: For all Pi do, where i = 1 to n−1,
Calculate EPU(m0 + R) + $\sum$EPU(mj) and send to P(i + 1) mod n

Step 3: The protocol initiator computes

$$\mathrm{EPU}(m0 + R) + \mathrm{EPU}(m1) + \mathrm{EPU}(m2) + \cdots + \mathrm{EPU}(mn - 1)$$

Table 1 Complexity comparison

Method	Operations			
	Encryption/decryption	Addition/subtraction	Total	Complexity
Secure sum protocol Clifton et al. [2]	Nil	n+1	n+1	O(n)
Secure sum using homomorphic encryption	n+1	n+1	2(n+1)	O(n)

Step 4: Due to homomorphic property the above expression is equal to EPU(R+m0+m1+m2+ ⋯ +mn−1). After decryption using PR and subtraction of R the initiator gets $\sum$mi which can be broadcasted to all the parties.

4 Analysis of the Protocol

The computation of secure sum using homomorphic cryptosystem protects the data from leakage over the network lines. But this is achieved on the cost of the additional computation of encryption functional at each node of the ring. Table 1 illustrates comparative computations of the proposed protocol with Clifton et al. [2].

Referring to Table 1, its clear that number of operations performed in the proposed method is increased to double as compared to Secure sum protocol of Clifton et al. [2] complexity still remaining the same at O(n).

As we have seen that the parties are sending encrypted data over the network, therefore no intruder can learn the private data of the individual parties. Thus, privacy or confidentiality of the data is protected. Thus, the protocol is suitable for insecure networks. The data is not only secure from cooperating parties but also from the intruders. This is an improvement over previous protocols where the network was assumed to be secure.

5 Conclusion and Future Work

Most of the protocols available in the literature provide protocols for secure sum computation applicable for secure networks only. In this paper we have used additive homomorphic encryption to make it applicable for insecure networks. The protocol is suitable for semi honest parties who follow the steps in the protocol but also curious to learn data of the other parties.

Future work can also be done to make the protocol suitable for malicious parties. Zero knowledge proof protocols can be used for the malicious model. Also, the proposed protocol preserves privacy. Work can be done to protect other security properties like integrity, non-repudiation, etc.

References

1. http://en.wikipedia.org/wiki/Secure_multi-party_computation
2. Clifton C, Kantarcioglu M, Vaidya J, Lin X, Zhu MY (2002) Tools for privacy-preserving distributed data mining. J. SIGKDD Explor Newsl 4(2):28–34. ACM Press
3. Sheikh R, Kumar B, Mishra DK (2010) Changing neighbors k-secure sum protocol for secure multi-party computation. Int J of Comput Sci Inf Secur, USA, 7(1) (Accepted for publication)
4. Sheikh R, Kumar B, Mishra DK (2009) Privacy-preserving k-secure sum protocol. Int J Comput Sci Inf Secur, USA, 6(2):184–188
5. Sheikh R, Kumar B, Mishra DK (2009) A distributed k-secure sum protocol for secure multi-party computation. Submitted to a journal
6. Yao AC (1982) Protocol for secure computations. In: Proceedings of the 23rd annual IEEE symposium on foundation of computer science, pp 160–164
7. Goldreich O, Micali S, Wigderson A (1987) How to play any mental game. In: STOC'87: Proceedings of the nineteenth annual ACM conference on theory of computing, New York, NY, USA: ACM, pp 218–229
8. Chor B, Gilbao N (1997) Computationally private information retrieval (extended abstract). In: Proceedings of 29th annual ACM symposium on theory of computing, El Paso, TX USA, May 1997
9. Chor B, Kushilevitz E, Goldreich O, Sudan M (1995) Private information retrieval. In: Proceedings of the 36th annual IEEE symposium on foundations of computer science, Milwaukee WI, pp 41–50, Oct 1995
10. Lindell Y, Pinkas B (2000) Privacy preserving data mining in advances in cryptography-Crypto2000, lecture notes in computer science, vol 1880
11. Agrawal R, Srikant R (2000) Privacy-preserving data mining. In: Proceedings of the 2000 ACM SIGMOD on management of data, Dallas, TX USA, pp 439–450, 15–18 May 2000
12. Atallah MJ, Du W (2001) Secure multiparty computational geometry. In: Proceedings of seventh international workshop on algorithms and data structures (WADS2001). Providence, Rhode Island, USA, pp 165–179, 8–10 Aug 2001
13. Du W, Atallah MJ (2001) Privacy-preserving cooperative scientific computations. In: 14th IEEE computer security foundations workshop, Nova Scotia, Canada, pp 273–282, 11–13 Jun 2001
14. Du W, Atallah MJ (2001) Privacy-preserving statistical analysis. In: Proceedings of the 17th annual computer security applications conference, New Orleans, Louisiana, USA, pp 102–110, 10–14 Dec 2001
15. Paillier P (1999) Public-key cryptosystems based on composite degree residuosity classes. In: EUROCRYPT'99, Prague, Czech Republic, pp 223–238, 2–6 May 1999

Automated Workload Management Using Machine Learning

K. Deivanai, V. Vijayakumar and Priyanka

Abstract Mainframe System processing includes a "Batch Cycle" that approximately spans 8 pm to 8 am, every week, from Monday night to Saturday morning. The core part of the cycle completes around 2 am, with key client deliverables associated with the end times of certain jobs, tracked by Service Delivery. There are single and multi-client batch streams, a QA stream which includes all clients, and about 2,00,000 batch jobs per day that execute. Despite a sophisticated job scheduling software, and automated system workload management, operator intervention is required, or believed to be required, to reprioritize when and what jobs get available system resources. Our work is to characterize, analyse and visualize the reasons for a manual change in the schedule. The work requires extensive data preprocessing and building machine learning models for the causal relationship between various system variables and the time of manual changes.

1 Introduction

A centralized server is the thing that organizations use to have their business databases, exchange servers, and applications that require a more noteworthy level of security and accessibility than is usually found on littler scale machines. Centralized computer dependably Contain around seventy percent of corporate information from operations (bookkeeping, finance, charging, etc.) Often the "database server" in web-empowered database applications.

Incorporated PCs will be PCs used chiefly by immense relationship for essential applications, conventionally mass data planning. Present-day centralized server PCs have capacities less characterized by their single assignment computational speed (tumbles or clock rate) as by their repetitive interior designing and coming about

K. Deivanai (✉) · V. Vijayakumar · Priyanka
VIT University, Chennai Campus, Chennai 600127, India
e-mail: deivakathir23@gmail.com

V. Vijayakumar
e-mail: vijayakumarv@vit.ac.in

D. K. Mishra et al. (eds.), *Data Science and Big Data Analytics*,
Lecture Notes on Data Engineering and Communications Technologies 16,
https://doi.org/10.1007/978-981-10-7641-1_32

high dependability and security, broad info yield offices, strict in reverse similarity for more seasoned programming, and high use rates to bolster enormous throughput. These machines regularly keep running for a considerable length of time without intrusion, with repairs and even programming and equipment updates occurring amid ordinary operation. For instance, ENIAC stayed in consistent operation from 1947 to 1955. All the more as of late, there are a few IBM centralized server establishments that have conveyed over 10 years of ceaseless business benefit starting 2007, with redesigns not intruding on administration. Centralized servers are characterized by high accessibility, one of the primary purposes behind their life span, as they are utilized as a part of uses where downtime would be expensive or disastrous. The term Reliability, Availability and Serviceability (RAS) is a characterizing normal for centralized computer PCs.

To reprioritize the jobs when it will get available for the system resource, the operator intervention is required or believed to be required. Workloads with operators and the night workloads experience and the expectation of variable in business results in changes to schedule how it executes. These changes are completely captured by the manual operator interventions. Generally the job schedule will execute mostly on the daily basis without the intervention of the operator by leaving the maximum system resources to the workload manager of the system. When the deliverables are met according to the priority for the business based on the job schedule case if that condition did not meet then the operator commands which are captured in the system log and the type of messages signify the changes in the scheduler log.

This approaches enables easily our batch applications while accessing the mainframe data with the early machine learning by using its capability to learn the hidden patterns in the operational data with the help of mathematical modeling algorithms which are readily available. For this approach we are utilizing major have predictive algorithms like K-Means, Decision Tree, Regression tree. When it comes to machine learning approach, the unsupervised learning algorithm like K-Means which used the descriptive statistics to be analyzed the natural patterns and the kind of relationships that are occurring within the operational data on MySQL for Z/OS. This unsupervised learning will be able to identify the clusters of similar records and relationships between different fields. Also supervised learning algorithm is used to train the data to construct decision tree based on the decision tree which is constructed it is used to predict the future values. Based on the classification technique it can be used to dissect it out which kind of group and the new record which is used to insert into Z/OS table. With this kind of approach regression technique is used to predict the future values based on the past history values. Therefore, algorithms are the main part of machine learning and these kind of algorithms lead to aid the executives with more number of evidence based on the decision using the data Z/OS. In particularly IBM machine learning is useful for various dynamic business scenarios.

2 Motivation

To actualize a powerful log analyzer through express programming by individuals a vast scale framework is normally made out of different parts, and those parts might be created with various programming languages. There is no yield structure standard for all log documents, and it is, along these lines, hard to actualize one single log analyzer for the advancement of a huge scale framework through unequivocal programming. Regardless of the possibility that we can execute particular log analyzer for a particular programming, it cannot be regularly utilized as a part of a framework a work in progress without incessant updating. Additionally, as specified in the past passage, even concerning people, it is difficult to identify issues from huge amounts of logs accurately.

3 Problem Statement

To characterize the state of the system using both business exceptions and system workload artifacts to determine if there are patterns in operator's response, capturing this processing knowledge and if the type of manual intervention is predictable [1] and can be automated [2]. Develop a real-time decision support application [3] based on learning how the system state changes and is related to. So that tele-network on operator experience to meet business goals can be reduced while continuing to maximize the use of available resources.

4 Proposed Work

First, this phase consists of extracting the relevant data from the text files and storing them into database on which various analyses would be performed. Generalizing CPU health data for the whole data by linearly extending the data based on fixing the intervals. Also by standardizing the entries into one format and store them as tables in a database. From those files we track the JES commands given by the operator which can be found from the references. Finally in this phase tables are created based on these commands and the relevant information is stored. The database contains all the definitions which have made for planning objects. It likewise holds insights of employment and occupation stream execution and in addition the data as the client Id who has made a protest and with that it indicates when the last question was modified. Upon this we build a machine learning model on top of this so that we can segment the jobs based on the priority and plan it accordingly which has to be executed.

5 Methods

5.1 Mainframe Approach

Centralized server preparing incorporates a "group cycle" that roughly traverses 8 pm to 8 am, every week from Monday night to Saturday morning. The center part of the cycle finishes around 2 AM with key customer deliverables connected with the final days of specific employments, followed by administration conveyance. There are single and multi-customer clump streams, a QA stream which incorporates all customers and around 2,00,000 bunch occupations for every day that executes. In addition to the day-time business transaction there are also clients and confidences information vendor files receive as input into the stream [4]. There is a relative job priority classification scheme. There is a job scheduling application to manage the submission of jobs, based on time and other job or file input delivery dependencies [5].

5.2 Job Scheduling

During the batch cycle, the mainframe system runs at or near 100% of capacity. Despite a sophisticated job scheduling software and automated system workload management, Operation intervention is essential. JES commands by MAVAEN by mainframe centralized servers [6].

5.3 Tivoli Workload Scheduler (OPCA)

TWS1 is a fully automated batch job scheduling system that improves job throughout and greatly reduces operations. TWS helps you arrange and sort out each period of cluster employment execution. Amid the handling day TWS generation control programs deal with the creation control programs deal with the creation environment and computerize most administrator exercises it readies your employments for execution, resolves interdependencies and dispatches and tracks every occupation. Since your occupations start when their conditions are satisfied, idle time is minimized and throughput enhances essentially. Employments never come up short on arrangement and if a vocation falls flat, TWS handler the recuperation procedure with almost no administrator intercession.

5.3.1 Job Entry Subsystem2

MVS (or Z/OC which is the working framework for IBM centralized servers) utilizes a vocation passage subsystem (JES) to get job1 into the working framework, plan occupations for preparing by Maven system and control work yield processing. JES2 is plunged from HASP (Houston programmed spooling needed) which is characterized as a PC program that gives supplementary employment administration capacities, for example, Scheduling, control of employment stream and spooling. JES2 is an utilitarian expansion of the HASP program that gets occupations into the framework and process all yield information created by the occupation. JES2 is the part of Maven that gives the essential capacity to land positions into and yield out of the MAVEN framework. It is intended to give effective spooling, planning and administration offices for the Maven working framework.

MCP commands are for the scheduler (TWS/OPCA) and JES commands are for OS (IBM Z/OS) [7]. MCP commands can get the jobs into the queue including changing its priority and service class. You can even remove the job from the queue using MCP. But one initiators pick up a job from the queue using MCP, if the initiator picks up a job, then MCP commands cannot reach them only JES commands can reach.

6 Algothrims

6.1 Decision Tree

Decision tree [8] can be developed in general for quick contrasted with different techniques for characterization explanations can be built from tree that can be utilized to get to databases effectively [9]. Refer Fig. 1 which shows the decision tree classifiers acquire comparative or better precision when contrasted and other grouping strategies.

Various information mining methods have as of now been done on instructive information mining to enhance the execution of understudies like Regression, Genetic information mining methods can be utilized as a part of instructive field to improve our comprehension of learning procedure to concentrate on recognizing, extricating and assessing factors identified with the learning procedure of understudies [10]. Grouping is a standout amongst the most as often as possible. The C4.5, ID3, CART decision tree is connected on the information of understudies to foresee their execution.

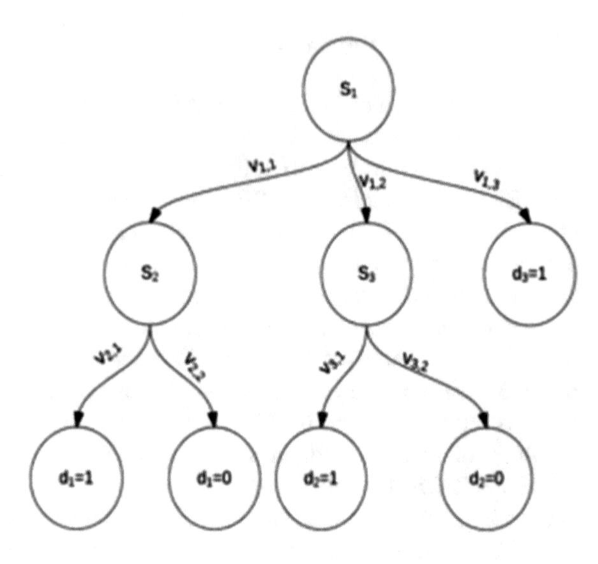

Fig. 1 Decision tree

6.2 CART Algorithm

CART is defined as Classification And Regression Trees [11]. The order tree development via CART depends on paired part of the traits. CART additionally in light of Hunt's calculation and can be actualized serially. Gini list is utilized as part measure as a part of selecting the part quality. CART is unique in relation to other Hunt's based calculation since it is additional use for relapse examination with the assistance of the relapse trees. The relapse investigation highlight is utilized as a part of estimating a needy variable given an arrangement of indicator factors over a given time frame. CART S bolsters constant and ostensible property information and have normal speed of handling.

6.3 K-Nearest Neighborhood Algorithm

Assume that a question is inspected with an arrangement of various characteristics, yet the gathering to which the protest has a place is obscure. Expecting its gathering can be resolved from its qualities; diverse calculations can be utilized to mechanize the grouping procedure [12]. A nearest neighbor classifier is a framework [7] for describing segments in perspective of the course of action of the segments in the arrangement set that are most similar to the experiment. From Fig. 2, which shows this kind of technique, we can get the closest nearing neighbors.

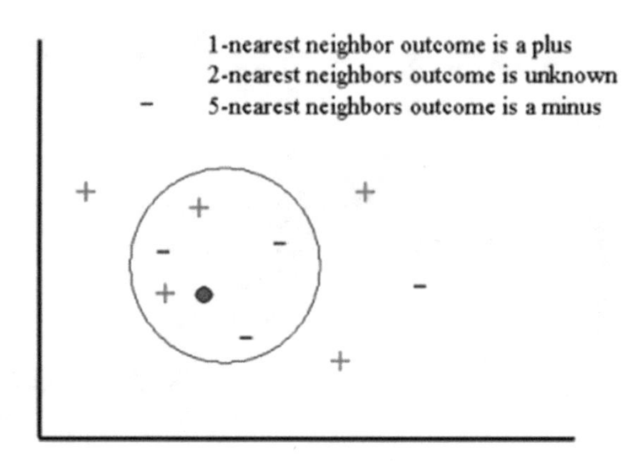

Fig. 2 K-Nearest neighbor

6.4 Naïve's Bayes Algorithm

If the inputs are independent, we will be using Naive Bayes technique [13] to solve the problem. Given a game plan of things, each of which has a place with a known class, and each of which has a known vector of components, us indicate is build up a lead which will allow us to dole out future articles to a class, given only the vectors of elements portraying the future things. Issues of this kind, called issues of directed request, are ubiquitous, and various techniques for building such standards have been delivered. One crucial one is the guileless Bayes procedure—in like manner called nitwit's Bayes which depicts Fig. 3, essential Bayes, and self-rule Bayes. This method is basic for a couple reasons. It is not hard to assemble, not requiring any convoluted iterative parameter estimation arranges. This infers it may be expeditiously associated with gigantic data sets. It is not hard to interpret, so customers clumsy in classifier development can grasp why it is making the portrayal it makes.

6.5 Support Vector Machines

In support vector machine the data in plotted in the n-dimensional space [14]. Figure 4 which shows after plotting the data in the n-dimensional space the data is to split separating the different classes which involves method of supervised learning classification in the n-dimensional space. Based on that we will be drawing a line which is called as hyper-plane since we are drawing the line in the n-dimensional space. After drawing the hyper-plane, we will be seeing the classes which is having the highest margin.

The classes which are best suited that is nearer to the hyper-plane that is having more distance between nearest data point and the hyper-plane [15].

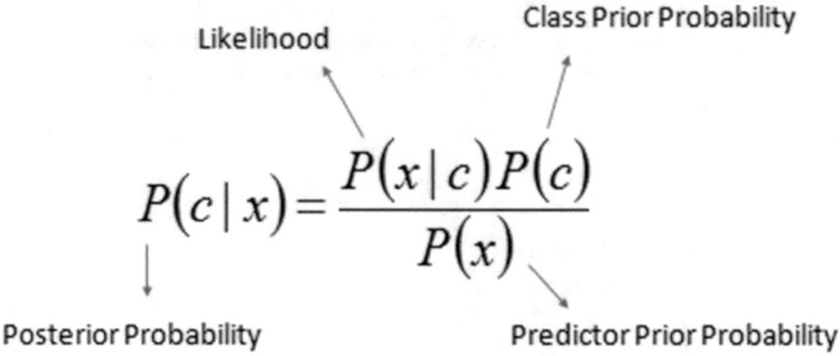

Likelihood
Class Prior Probability
Posterior Probability
Predictor Prior Probability

$$P(c \mid x) = \frac{P(x \mid c)P(c)}{P(x)}$$

$$P(c \mid X) = P(x_1 \mid c) \times P(x_2 \mid c) \times \cdots \times P(x_n \mid c) \times P(c)$$

Fig. 3 Naive Bayes

X2
Optimal hyperplane
Maximum
margin
X1

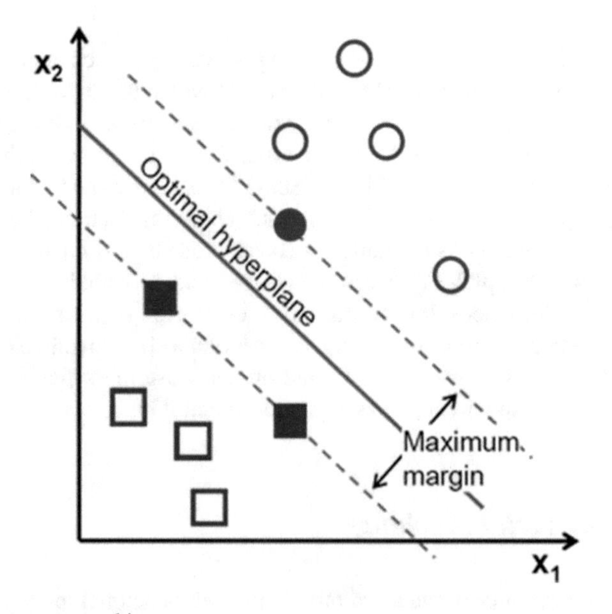

Fig. 4 Support vector machine

6.6 Apriori Algorithm

In Apriori algorithm, if we consider n itemset, then n set of rules are generated. Based on that among the n rules we need to find the rule which is having more support and confidence. For this we will using a best algorithm called Apirori algorithm which is referred to Fig. 5. Firstly we need to generate frequent itemset which is having more

Apriori(T, ϵ)
    $L_1 \leftarrow \{\text{large } 1 - \text{itemsets}\}$
    $k \leftarrow 2$
    while $L_{k-1} \neq \emptyset$
        $C_k \leftarrow \{a \cup \{b\} \mid a \in L_{k-1} \wedge b \notin a\} - \{c \mid \{s \mid s \subseteq c \wedge |s| = k-1\} \nsubseteq L_{k-1}\}$
        for transactions $t \in T$
            $C_t \leftarrow \{c \mid c \in C_k \wedge c \subseteq t\}$
            for candidates $c \in C_t$
                $count[c] \leftarrow count[c] + 1$
        $L_k \leftarrow \{c \mid c \in C_k \wedge \ count[c] \geq \epsilon\}$
        $k \leftarrow k + 1$
    return $\bigcup_k L_k$

Fig. 5 Apriori algorithm

support and we need to change the rules with having more confidence based on the splitting of items [16, 17].

7 Background

Machine Learning gives capacity for projects to learn without being expressly customized for a specific dataset. Edmondson's insight is that ML is part of a software engineering thread known as model-driven engineering. ML introduces a new category of model-building activities that can transform the software development life cycle. ML is coming to a mainframe near you, but it may be cloaked in predictive analytics. Last year Zementis, whose products leverage the Predictive Model Markup Language (PMML), announced availability for z/OS. Zementis models can be used to embed predictive models in z/OS CICS or Web Sphere settings. The models are "write once," meaning they can be deployed to z/OS SPSS, R, Python, or SAS. In a post on IBM Developer Works, Ravi Kumar outlines how z/OS users can now enable ML on OLTP applications, such as by embedding predictive models in DB2. One technique embeds the z/OS SPSS Scoring Adapter for DB2. Another approach combines a PMML model with business rules to make real-time decisions in DB2 or use Zementis-generated PMML to inject in-app scoring for CICS or Java apps.The IBM DB2 Analytics Accelerator for z/OS (IDAA) supports several major predictive analytics algorithms: K-Means, Naive Bayes, Decision Tree, Regression Tree, and Two-step.

8 Approach for Model Building

8.1 First Part: Data Extraction and Standardization

This phase consists of extracting relevant data from the log files and storing them into the database on which various analyses would be performed. To generalize the data for the whole day by linearly extending the data in 15 min time intervals. We track the entries in the log files which are of type MCP (Modify Current Plan) and JES2 (Job Entry Subsystem) commands. We standardize the entries into one format and store them as tables in a database. From the extracted data we are only interested in the following:

1. Date and time of occurrence 2. Application Number 3. Job-name 4. Job-number 5. Command-specific details

8.2 Second Part: Machine Learning Model

The objective of this phase is to relate each of the eights logical partitions taken at 15 min time interval across the six system properties to the time the changes have taken place in the logical partitions. To analyze this various machine learning classification algorithms can be used such as logistic regression, decision tree, support vector machine, etc.

9 Result Analysis

Figure 6 data analysis graph.

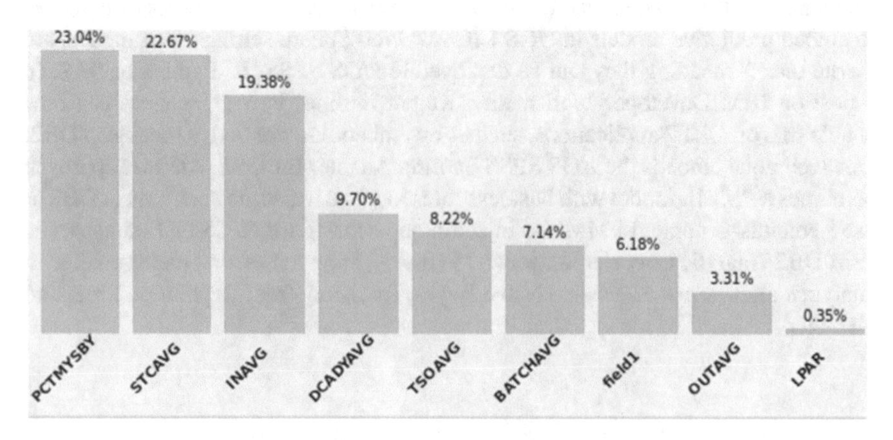

Fig. 6 Data visualization

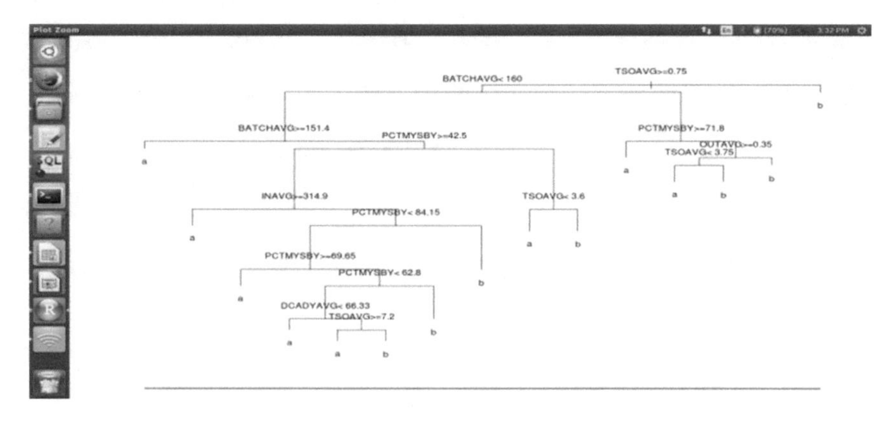

Fig. 7 Decision tree

Figure 7 shows the data is classified on the basis of decision tree for building a model.

Figure 8 which is residual analysis graph which is used for comparing how the data is skewed and how it is performing over the data.

It contains the snapshot of each of the 8 LARs (Logical Partitions) taken at 15 min intervals across 6 variables. This gives the health status of each of the LPAR (Logical partitions). The description of the six variables that gives the health status of each of the LPARs (logical partitions) is described below:

Principal component transform based CPU wait time measured which is helpful to pack all residual information to a small sets. In the Fig. 6 consider the LPAR system is hundred percent busy. Measure or total capacity being consumed as a percent of capacity available to that LPAR for executing work weight 15%.

Batchavg Average number of batch jobs in the system weight 20%.

Tsoavg Average number of TSO users (support associates that are on the system at the time) weight 5%.

Stcavg Average number of started tasks (usually system type jobs that are always up/active, compared to application batch jobs which executes based on time or other scheduling dependencies for a nite period) weight 5%. In avg Average number of address spaces in and executing (units of work, includes batch jobs and system started task) weight 20%.

Outavg Average number of address spaces out for some reason—waiting, not executing weight 20%.

Readyavg Average number of ready address spaces waiting to be dispatches to execute (not yet in the system) weight 15%.

The data captured at 15 min time intervals and the averages denote the average of these variables in that 15 min period. There is also relative importance to the time of the day by the complete analysis of the accuracy results in Fig. 9 and Fig. 10a have the weighting scale model 5-Highest weight i.e., most importance

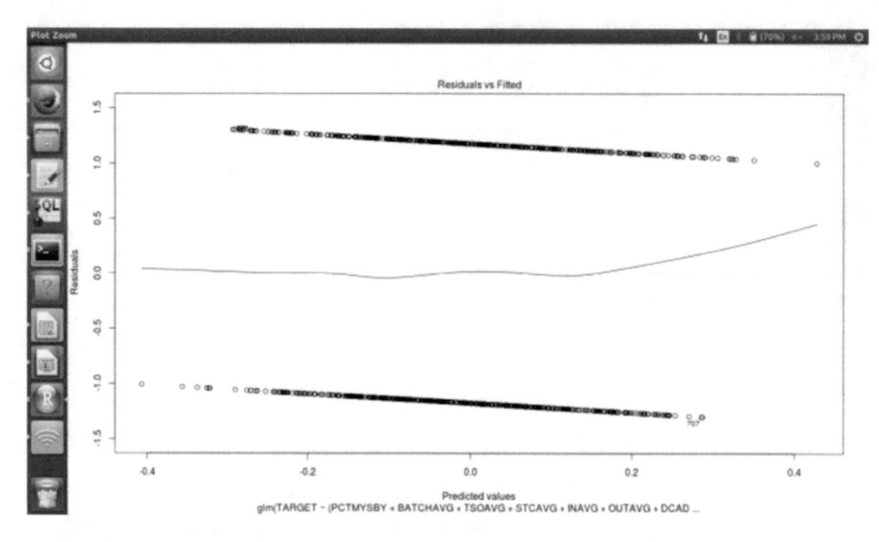

Fig. 8 Residual analysis

```
Confusion Matrix and Statistics

          Reference
Prediction   a   b
         a   0 287
         b   0 213

               Accuracy : 0.426
                 95% CI : (0.3822, 0.4707)
    No Information Rate : 1
    P-Value [Acc > NIR] : 1

                  Kappa : 0
 Mcnemar's Test P-Value : <2e-16

            Sensitivity :    NA
            Specificity : 0.426
         Pos Pred Value :    NA
         Neg Pred Value :    NA
             Prevalence : 0.000
         Detection Rate : 0.000
   Detection Prevalence : 0.574
      Balanced Accuracy :    NA

       'Positive' Class : a
```

Fig. 9 Linear discriminant analysis accuracy

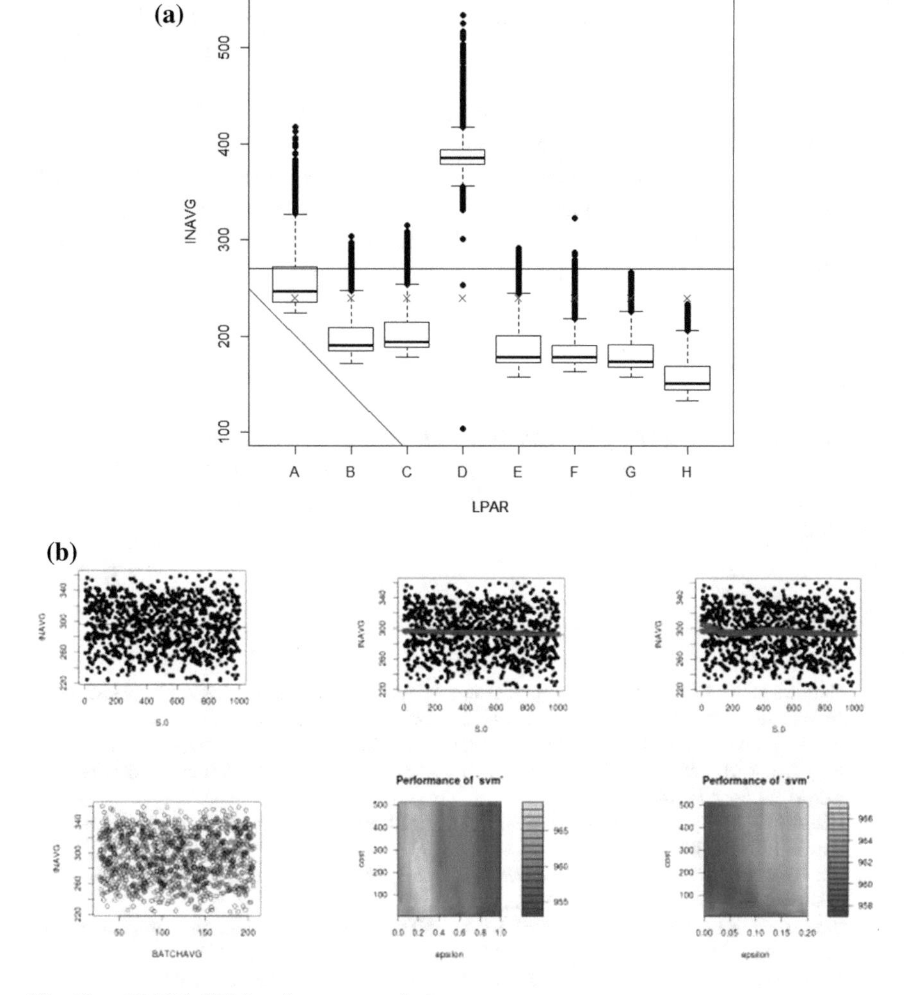

Fig. 10 **a** LPAR **b** SVM performance analysis

4–7 pm—2 (Market close 4 pm)
7–8 pm—3
8–10 pm—4
10 pm–2 am—5 (Peak batch processing period)
2–4 am—4
4–6 am—3
6–9 am—2 (market opens 9.30 am)

9 am–4 pm—1 SVM plot each data item in n-dimensional space with the value of each feature being the value of a coordinate. Then, we perform classification by ending the hyper-plane that differentiates the two classes very well. Support Vectors

are simply the co-ordinates of individual observation. Support Vector Machine is a frontier which best segregates the two classes (hyper-plane/line). In Fig. 10b shows the performance of SVM.

Linear Discriminant Analysis consists of statistical properties of your data calculated for each class. For a single input variable (x) this is the mean and the variance of the variable for each class. For multiple variables, this is the same properties calculated over the multivariate Gaussian, namely the means and the covariance matrix. These statistical properties are estimated from your data and plug into the LDA equation to make predictions M.

10 Conclusion

By identifying the high priority jobs which are having higher wait time, making them to allocate first so that the higher priority jobs will gets executed first. Therefore, we can characterize analyze and visualize the reasons for a manual change in the schedule.

References

1. Dumitru Diana (2009) Prediction of recurrent events in breast cancer using the Naive Bayesian classification, annals of university of craiova. Math Comput Sci Ser 36(2):92–96
2. Wu X, Kumar V, Quinlan JR, Ghosh J, Yang Q, Motoda H, McLachlan GJ, Ng A, Liu B, Yu PS, Zhou Z-H, Steinbach M, Hand DJ, Steinberg D (2008) Top 10 algorithms in data mining. Knowl Inf Syst 14(14):1–37
3. Ubeyli ED (2007) Comparison of different classification algorithms in clinical decision making. Expert syst 24(1):17–31
4. The IBM Archives, which contain a wealth of history of the mainframe, at www.ibm.com/ibm/history/exhibits/mainframe/mainframe_intro.html
5. Emerson WP, Lyle RJ, John HP (2002) IBM's 360 and Early 370 System, the definitive history of the development of the System/360. MIT Press. ISBN: 9780262517201
6. The IBM publications web site for z/OS. www.ibm.com/servers/eserver/zseries/zos/bkserv/
7. JES2 Commands-Version2 Release1 of z/OS (5650-ZOS), IBM Corporation (1997)
8. Delen D, Walker G, Kadam A (2005) Predicting breast cancer survivability: a comparison of three data mining methods. Artif Intell Med 34:113–127
9. Chen MS, Hans J, Yu PS (1996) Data mining: a overview from a data base perspective. IEEE Trans Knowl Data Eng 8(6): 866–883
10. Quinlan JR (1993) C4.5: programs for machine learning. Morgan Kaufmann, Amsterdam
11. Schwarzer G, Vach W, Schumacher M (2000) On the misuses of artificial neural networks for prognostic and diagnostic classification in oncology. Stat Med 19:541–561
12. Kaur H, Wasan SK (2006) Empirical study on applications of data mining techniques in healthcare. J Comput Sci 2(2):194–200
13. Jiawei H, Micheline K (1992) Data mining concepts and techniques. Elsevier
14. Hammerstrom D (1993) Neural networks at work. IEEE Spectr 26–32
15. Arun KP (2001) Data mining techniques. Universities Press (India) Ltd
16. Klosgen W, Zytkow JM (2002) Handbook of data mining and knowledge discovery. Oxford University Press
17. Nurnberger A, Pedrycz W, Kruse R (1990) Neural network approaches. In: Klosgen W, Zytkow JM (eds) Handbook of data mining and knowledge discovery. Oxford University Press

Multi-user Detection in Wireless Networks Using Decision Feedback Signal Cancellation

Monika Sharma and Balwant Prajapat

Abstract Wireless Networks are becoming increasingly ubiquitous in computer networks due to lesser cost and maintenance overhead. While some wireless networks may operate in regulated spectrum, the majority operate in the unregulated (ISM) band. It is highly challenging for a base station or control stations to successfully detect signals from multiple users in the same frequency range which may occur due to comparatively small frequency reuse distance. This paper proposes a technique based on decision feedback equalization (DFE) (Tu et al Proceedings of 44th Asilomar conference signals [1]) and strongest signal cancellation for multi-user detection (MUD) in wireless networks. It has can be seen that by employing the proposed system, the Bit Error Rate (BER) for strong (Stojanovic M Proceedings 137 of MTS/IEEE OCEANS conference, Boston, MA, 2006 [2]), average and weak users converge thereby indicating the fact that all the signals are detected with equal accuracy (Li et al IEEE J Ocean Eng 33(2):198–209, 2008 [3]).

Keywords Multi-user detection (MUD) · Decision feedback equalization (DFE) Signal to noise ratio (SNR) · Bit error rate (BER) · Industrial scientific and medical (ISM) band · Frequency reuse · Device to device (D2D) networks · Quality of service (QoS)

1 Introduction

With the sudden advent of Internet of Things (IOT) and extensive use of wireless networks, it has become a challenge to detect signals from multiple users with varying strengths in wireless networks. The problem becomes more challenging in the unregulated spectrum or ISM band as it may so happen that the frequency reuse factor

M. Sharma (✉) · B. Prajapat
VITM, Indore, India
e-mail: monika.sharma101989@gmail.com

B. Prajapat
e-mail: prajapat.balwant@gmail.com

D. K. Mishra et al. (eds.), *Data Science and Big Data Analytics*,
Lecture Notes on Data Engineering and Communications Technologies 16,
https://doi.org/10.1007/978-981-10-7641-1_33

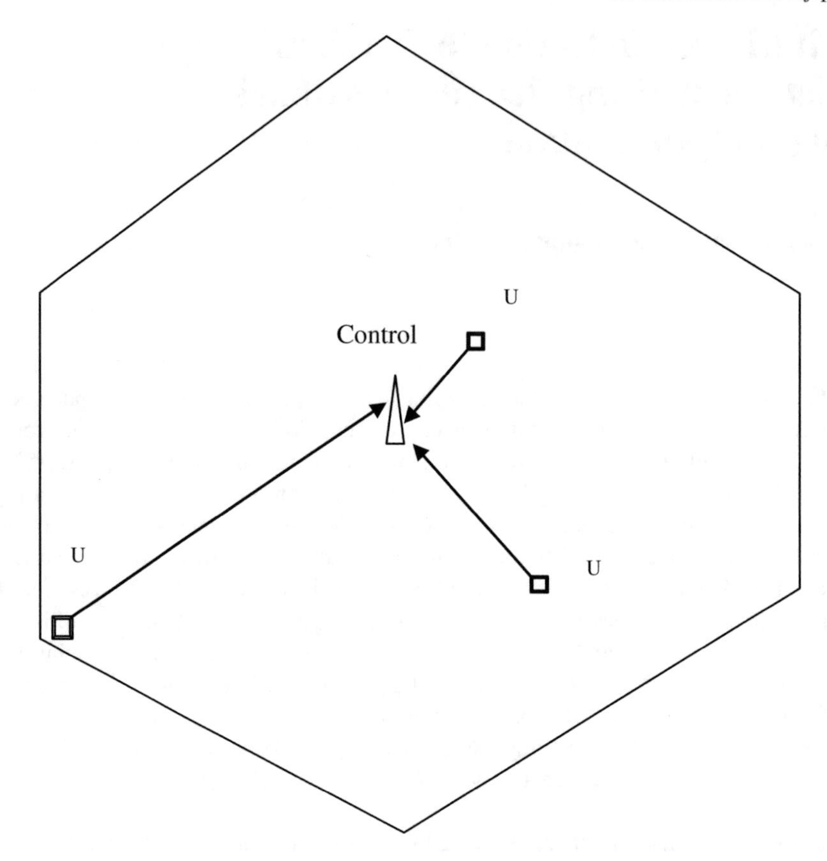

Fig. 1 A typical multi-user scenario in a wireless network

is kept relatively less [4, 5]. In such cases, signals may occupy overlapping bands. The scenario becomes more complex in case users are given same frequency bands for data transmission with orthogonal spreading sequences. In such a scenario, the strong signal make over ride the weaker ones and the BER of the weaker signals may encounter severe BER degradation due to low signal to noise ratio. Such problems can be mitigated using a decision feedback mechanism with successive cancellation of signals in a descending order since it is highly likely that the stronger signals will be detected with better BER performance. A typical MUD scenario is depicted in Fig. 1.

Figure 1 depicts a typical multi-user scenario in which the control station receives signals from three users at different distances and hence the amount of fading encountered by the signals is also different [6]. Considering spatial dependence of fading, the average level crossing rate of a signal can be defined as [7]:

$$N_R(r) = \int_0^{\infty} r^{\sim} \cdot pdf_{r,r^{\sim}}(r, r^{\sim}) dr^{\sim} \tag{1}$$

Here N_R represents the level crossing rate, i.e., the signal strength crossing a certain level in the positive direction.

$r^{\sim}$ Represents the temporal derivative of r given by $\frac{\partial r}{\partial t}$

$pdf_{r,r^{\sim}}$ Represents the joint probability density function of r and $r^{\sim}$

Considering the signals to undergo fading dips and considering an '**n-th**' order fading mechanism after a distance of $\mathbf{d_{break}}$, the received signal strength at the receiving end can be given by

$$P_{rx}(\mathrm{d}) = P_{rx} \cdot d_{break} \left(\frac{d}{\mathrm{dbreak}} \right)^{-\mathrm{n}} \tag{2}$$

It can be inferred that as the distance increases, the signal strength decays and as a result of which the Bit Error Rate (BER) degrades for weaker signals. The mathematical formulation for a matched filter mechanism at the receiver can be given by [7]

$$BER = Q\left[\sqrt{\frac{(b_1 - b_2)^2}{4\sigma_{n_0}^2}}\right] \tag{3}$$

If multiple signals are transferred through the matched filter, then the channel response h(t) can be formulated as

$$\mathrm{h(t)} = \mathrm{x}(\alpha - \mathrm{t}) \text{ where } \mathrm{x(t)} = \mathrm{x_1(t)} - \mathrm{x_2(t)}$$

and $\frac{(b_1-b_2)^2}{\sigma_{n_0}^2}$ represents the maxima of $\frac{\mathrm{E_D}}{\frac{N_0}{2}}$.

Correspondingly, the BER can be given by

$$\mathrm{P_e} = \mathrm{Q}\left[\sqrt{\frac{1}{4}\frac{\mathrm{E_b}}{\left(\frac{\mathrm{N_0}}{2}\right)}}\right] \tag{4}$$

Here, $\mathrm{E_b}$ represents the energy per bit.

$\mathrm{N_0}$ represents the noise power spectral density.

Q represents the Q-Function needed for the evaluation of integrals not intersecting the axis of the dependent variable.

Since the value of the Q function increases monotonically as the argument of the function decreases monotonically [8], hence the BER increases or rather degrades as the signal to noise ratio or SNR decays. This is atypical case of a multi-user detection mechanism in which some user is far away from the base station or control station. Since practical wireless channels typically tend to be frequency selective in nature, therefore, equalization needs to be provided for the receiving end. Practically, wireless channels do not satisfy the magnitude and phase relations for distortion less transmission given by

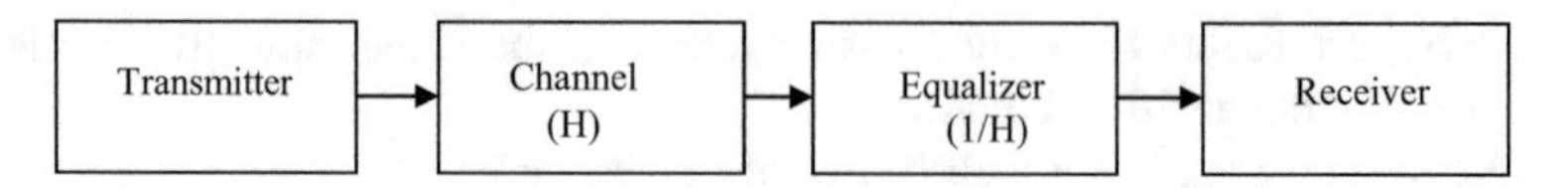

Fig. 2 Equalizer employed in a wireless channel

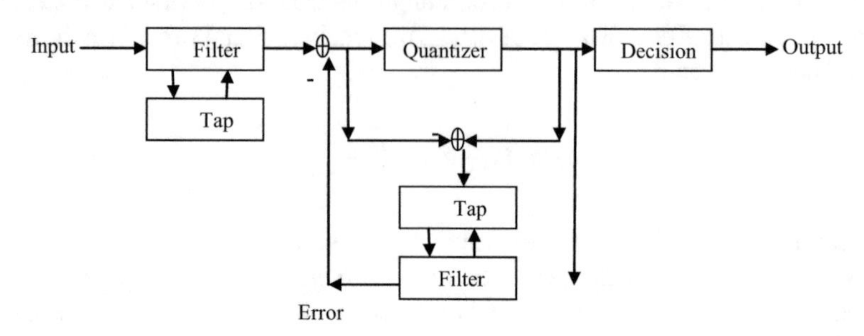

Fig. 3 Internal structure of decision feedback equalizer

$$\mathrm{Mod}(\mathrm{H}(\mathrm{f})) = \mathrm{k} \tag{5}$$

$$\Phi(\mathrm{H}(\mathrm{f})) = -\mathrm{k}(\mathrm{f}) \tag{6}$$

Here H(f) represents the frequency response of the channel.

Mod represents the magnitude response and ϕ represents the phase response. An equalizer tries to equalize the distortions inflicted by the practical channel. If the channel response is H, the equalizer response would be $\frac{1}{H}$ [9, 10].

2 Channel Equalization

Although different equalization techniques can be employed, one the most effective mechanisms of equalization is the decision feedback equalizer. The decision feedback equalizer designed can be represented in Figs. 2 and 3.

3 MUD System Design for Proposed System

Let the signal arriving at the receiver be designated by x(t). Considering x(t) as (1)

$$\mathrm{X}(\mathrm{t}) = \mathrm{x}_1(\mathrm{t}) + \mathrm{x}_2(\mathrm{t}) + \mathrm{x}_\mathrm{n}(\mathrm{t}) \tag{7}$$

(2) Detect the strongest among the arriving signals using a multi-level comparator. Let $x^*(t)$ be the strongest among the signals. Detect and store the information from $x^*(t)$ at the corresponding sampling time (t_s).
(3) Subtract the signal $x^*(t)$ from the composite signal X(t) and let the signal after cancellation be $X^{\sim}(t)$.
(4) Repeat steps 1–3 for $X^{\sim}(t)$ till $x_n(t)$ is reached where $x_n(t)$ is the weakest signal among all.
(5) Compute the BER for the following cases:

    (a) BER for all users individually for different path gains ($g_1, g_2 \ldots g_n$)
    (b) Compute the comparative BER for signals without proposed system.
    (c) Compute the comparative BER for signals with proposed system.

(6) Evaluate the performance of the proposed system by matching whether the BER curves for weak strong and average users converge.

4 Results

The proposed system is simulated on MATLAB 2017a. The results obtained are shown below. The simulations are carried out for a random binary data stream which is Quadrature Phase Shift Keying (QPSK) modulated. The channel chosen is Additive White Gaussian Noise (AWGN). The simulation considers 4 cases:

(1) BER performance of a user that is nearest to the control station and undergoes minimal fading. Hence it is the strongest.
(2) BER performance of a user that is at a larger distance compared to the near user and hence undergoes higher amount of fading.
(3) A user which is far away from the control station and undergoes the highest amount of fading.

Ideally the strongest user would show the quickest and steepest decrease in the BER performance, followed by the average and weak users. This is the case where the proposed system is not employed. The negative effect of such an approach is the low reliability and quality or service (QoS) for weak user. By employing the proposed technique, the BER curves of all the three users coincide indicating the fact that the BER performance or QoS for all conditions in a MUD scenario would remain almost identical (Figs. 4, 5, 6 and 7).

A comparative analysis of the results with and without the proposed system can be put forth in Table 1.

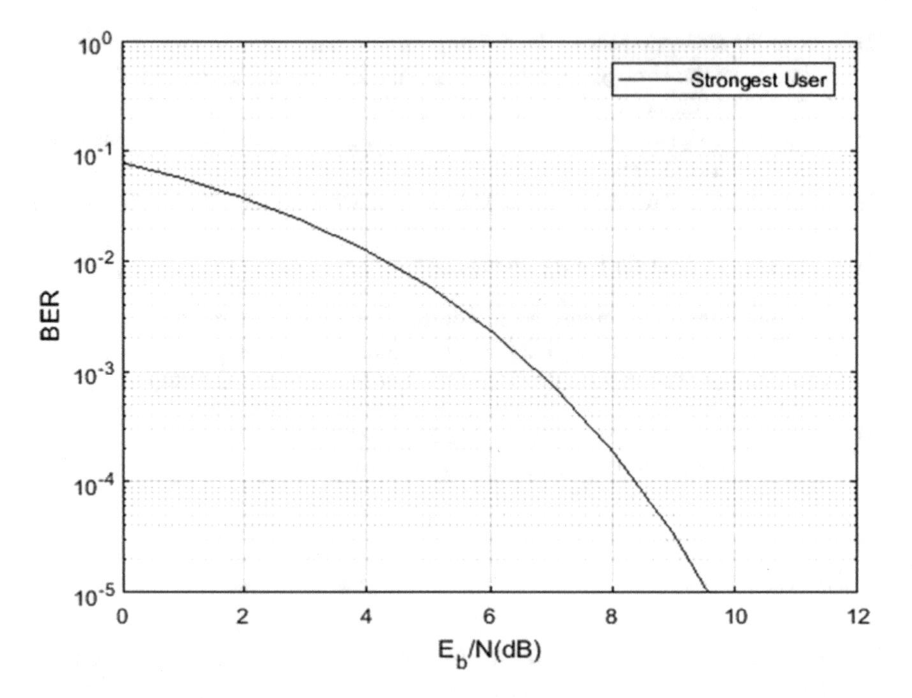

Fig. 4 BER performance of strongest user in the MUD scenario without proposed technique

Table 1 Comparative BER Performance with and without proposed system

User	SNR (dB)	BER (without proposed technique)	BER (with proposed technique)
Strong	10	10^{-5}	10^{-5}
Average	10	10^{-3}	10^{-5}
Weak	10	10^{-2}	10^{-5}

5 Conclusion

It can be concluded from the aforesaid discussions and obtained results that the proposed system achieves almost identical BER performance for different users (strong, average and weak) in a multi-user detection scenario in wireless networks. The results can be attributed to the fact that the proposed system uses a decision feedback equalization mechanism that effectively circumvents the distorting effects of a practical non-ideal channel. Moreover the signal cancelling mechanism that is employed iteratively detects the strongest signal at the outset and goes not detecting others with decaying signal strengths. The proposed system attains a BER performance of almost 10^{-5} for a SNR of around 10 dB. Identical BER performance indicates the fact that the all users are detected with almost equal accuracy even in a non-ideal fading scenario.

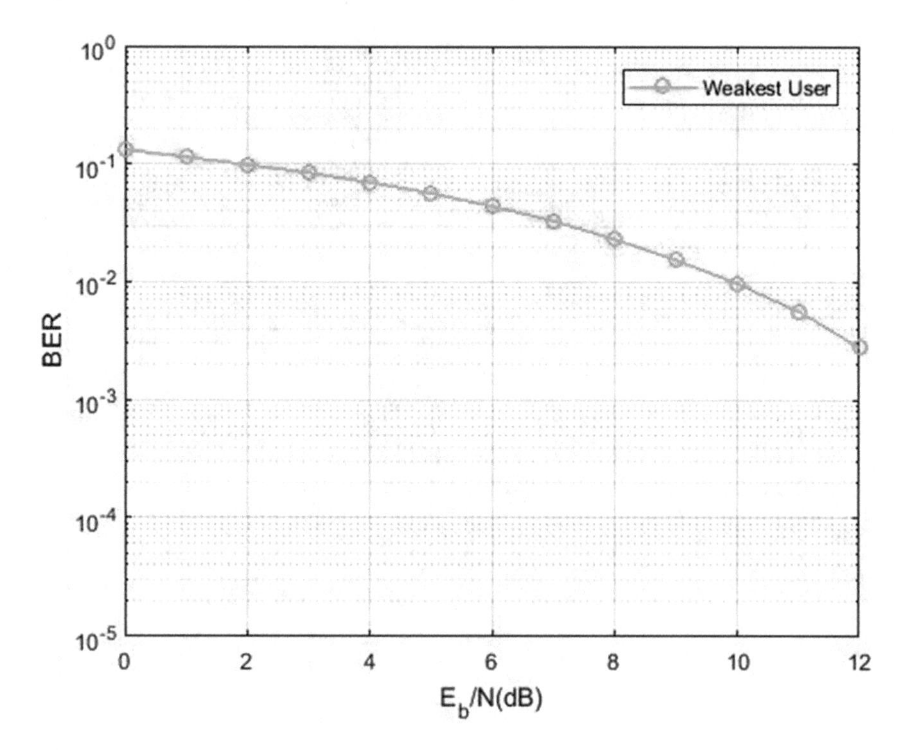

Fig. 5 BER performance of weakest user in the MUD scenario without proposed technique

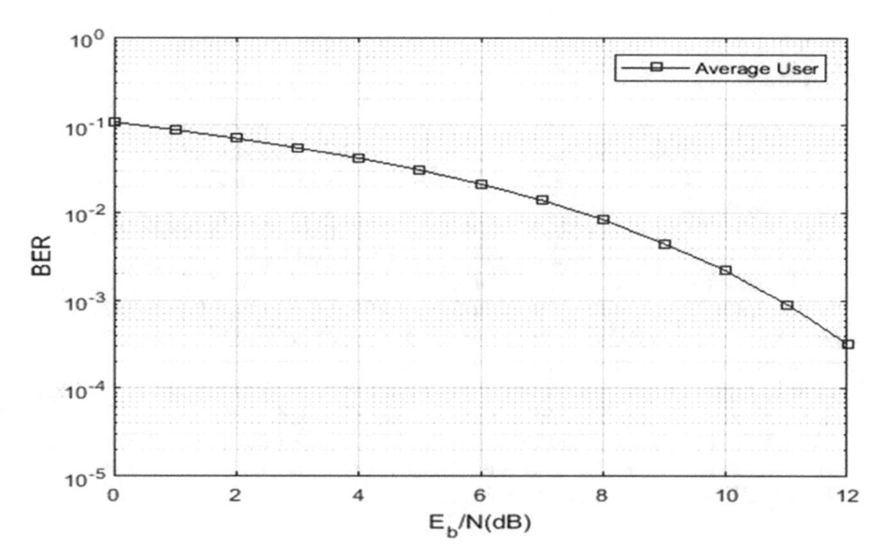

Fig. 6 BER performance of average user in the MUD scenario without proposed technique

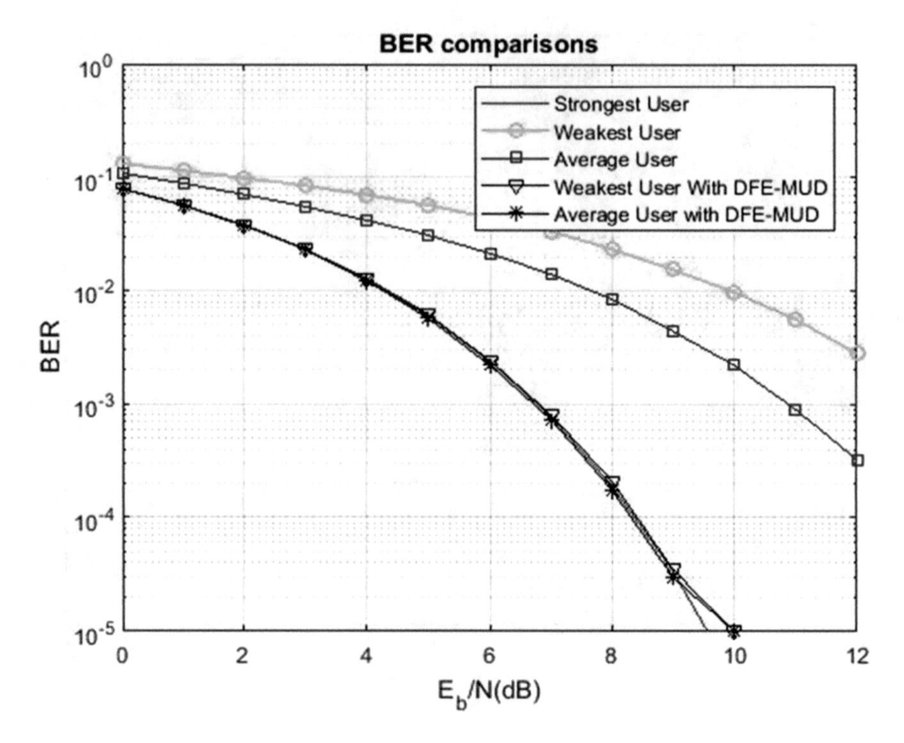

Fig. 7 Comparative BER performance of all users with proposed technique

References

1. Tu K, Duman T, Proakis J, Stojanovic M. Cooperative MIMOOFDM communications: receiver design for Doppler-distorted under wateracoustic channels. In: Proceedings of 44th Asilomar conference signals
2. Stojanovic M (2006) Low complexity OFDM detector for underwater channels. In: Proceedings of MTS/IEEE OCEANS conference, Boston, MA, 18–21 Sept 2006
3. Li B, Zhou S, Stojanovic M, Freitag L, Willett P (2008) Multicarrier communication over underwater acoustic channels with non uniform Doppler shifts. IEEE J Ocean Eng 33(2):198–209
4. Rahmatalla Y, Mohan S (2013) Peak to average power ratio in OFDM systems: a survey and taxonomy. IEEE
5. Otnes R, Eggen TH (2008) Underwater acoustic communications: long term test of Turbo equalization in shallow water. IEEE J Ocean Eng 33(3):321–334
6. Huang J, Zhou S, Willett P (2008) Non binary LDPC coding for multicarrier underwater acoustic communication. IEEE J Sel Areas Commun 26(9):1684–1696
7. Molisch AF. Wireless communication. Wiley, India

8. Leus G, van Walree P (2008) Multiband OFDM for covert acoustic communications. IEEE J Sel Areas Commun 26(9):1662–1673
9. Kang T, Iltis R (2008) Iterative carrier frequency offset and channel estimation for underwater acoustic OFDM systems. IEEE J Sel Areas Commun 26(9):1650–1661
10. Qu F, Yang L (2008) Basis expansion model for underwater acoustic channels? In: Proceedings of MTS/IEEE OCEANS conference, Quebec City, Canada, 15–18 Sept 2008

ANN-Based Predictive State Modeling of Finite State Machines

Nishat Anjum and Balwant Prajapat

Abstract Finite state machines have so many applications in the day-to-day life. Design of Finite State machines spread its role from the simple systems to complex systems. As Artificial Intelligence rule all over the technology world by its very effective applications, Finite state machines can also significantly use its essence in the process of next state prediction. The predictive analysis of Artificial intelligence helps to speed up the process of Finite state machines. This paper explores the design of anticipative state machines with the help of Artificial Neural Networks. To get the higher performance, less training time and low error prediction, Back propagation algorithm is used in ANN which helps to analyze the critical parameters in real time applications. Our proposed technique provides better results than the previously used technique and also provides less prediction and training time error with increasing number of inputs.

Keywords Artificial neural network (ANN) · Back propagation · Finite state machines (FSM) · Levengerg–Marquardt (LM) · Mean square error (MSE) Regression

1 Introduction

As we know the area of the digital application is raising at a very high speed day by day, so the need for the fast computation and prediction is necessary to grow the technology. Mostly the digital circuits have been used for the implementation of the digital systems. The digital systems can be designed only with the help of finite state machines. Now the technology is changing at a very rapid speed and the conventional finite state machines are not sufficient for that for the following reason:

(a) Finite state machines become very complex for the large number of data processing technology.

N. Anjum (✉) · B. Prajapat
VITM, Indore, India
e-mail: nishatanjum341@gmail.com

D. K. Mishra et al. (eds.), *Data Science and Big Data Analytics*,
Lecture Notes on Data Engineering and Communications Technologies 16,
https://doi.org/10.1007/978-981-10-7641-1_34

(b) For the applications of gaming, Human Machine Interfaces and prediction, the finite sate machines need an interactive and predictive nature.

The fields like statistical and stochastic computing use a predictive analysis with an efficient programming. The need for Artificial Intelligence is come to know in such approaches.

The steps which are incurred with the intelligence are given as

(a) To get input data
(b) To analyze the data
(c) To identify patterns or similarities in the data
(d) To take a decision on data
(e) To provide an output.

Humans are capable to analyze the data and to take a decision on it. Hence the living organism has a natural intelligence to do the work. If a person wants to design a system which uses a mechanism similar to the human intelligence, the above mentioned steps must be followed by its creation. Sometimes the intelligence evolve by the machine is called Artificial Intelligence. Artificial intelligence can be implemented with the help of Artificial Neural Networks (ANN). Artificial Neural Networks (ANN) is a technique which acquires the learning pattern of human brain to analyze the data and to take decision on them. Artificial Neural Networks (ANN) tries to establish a relation between the input data and target output. The structure of human brain contains so many neurons which are placed in a well formatted pattern in the brain. Neurons sense the input signal, analyze it and take a decision on it and then provide such information to the particular body part for a proper act with such a very high speed and accuracy. Hence we can say that the neurons are the processing unit of the brain which has the capacity to get and store the input signals and provide a response parallel to the body parts like a distributed processor. The human brains process at a very high speed than the modern computer; because the neurons do not pass the information from one unit to other else they encode such information in the neuron network and trained the brain. Therefore the second name of neural network is connectionism.

In this digital world, all the electronic systems are based on the digital circuits. Some of them are given below

(1) Vending Machine.
(2) ATM machine for credit and debit cash.
(3) Applications such as mobile phones, gaming consoles, etc.

Some of the digital circuits have a predefined input and predefined states, such digital circuits are called interactive finite state machines. The finite sate machines can be designed with the help of two types as mentioned below

1. With the help of internal circuit of the system.
2. With the help of truth table in which the mapping of input and output is mentioned.

Nowadays there are so many applications which need an approach of real time computing in which prediction of certain input and outputs are done with the help of forecasting of previous input and outputs. The systems which use such kind of techniques are mentioned below

1. Interactive gaming, for example playing cards with a machine.
2. Design of Human Machine Interface (HMI).
3. Digital machines designs PCB layouts.
4. Hardware level cryptography with the help of machine.

The above-mentioned systems required a prediction of the regularities for the machine response with the help of the previous data. All the above systems are digital systems and need a complex digital circuit for the implementation with a well defined technique with Artificial Intelligence (AI).

Development and implementation of Artificial Intelligence can be done with the help of Artificial Neural Network (ANN). The human brain consists of so many brain cells which are known as neurons. The network of such neurons is called neural network. To get the learning ability like human neural network a machine must have the characteristics similar to the human neural network and the network in a machine with such abilities is called artificial neural network.

2 Artificial Neural Networks (ANN)

Here,

X represents the inputs
W represents the Weights or Experiences
g represents the bias.

The applications of data mining majorly use Artificial Neural Network (ANN). There are two development periods of Neural Networks as the early 60s and the mid-80s. There were so many developments in the field of machine learning. As we know that the Artificial Neural Network finds a way related to the human neural network. In the human brain there are 10 billion neurons placed in a pattern in which they are connected to each other. Each neuron connected with 10,000 other neurons in network. The neurons gets the signal with the help of synapses which controls the signal. The synapses system is very useful for the working of the brain in the human body.

Figure 1 provides the Basic elements as a mathematical model of neurons in an Artificial Neural Network (ANN). The features of the figure are given below

(a) The data feeds parallel manner in the Artificial Neural Network as the human brain feeds the data.
(b) All the inputs are summed up for an instant.
(c) Some mathematical function is used for the analysis.

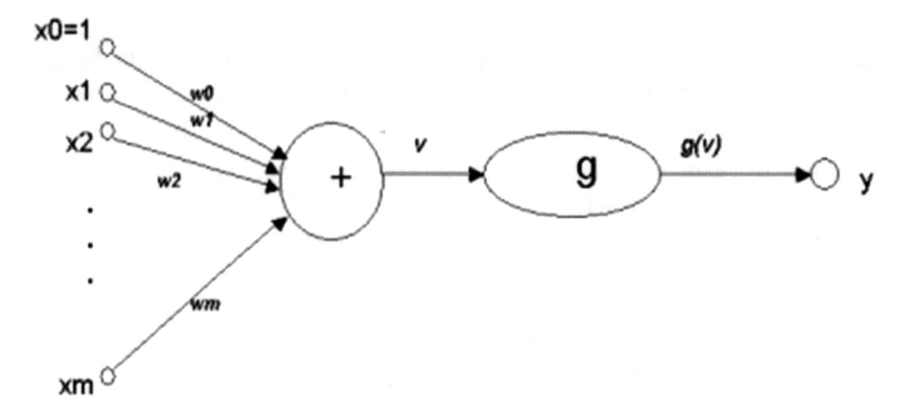

Fig. 1 Mathematical model of artificial neural network

(d) By the following above steps some output is achieved.
(e) Saves the experiences in the form of weights.

3 Back Propagation in Artificial Neural Networks

Although neural networks can be trained in several ways and there are several topologies of neural networks, yet one of the most effective techniques to train neural networks is by employing back propagation. In this mechanism, the errors in every iteration are fed back to the ANN architecture which affects the weight updating mechanism. The major advantage of back propagation is the training speed and reduction of error with respect to the number of epochs. The following Fig. 2 depicts the flow chart for back propagation.

4 Proposed Methodology

The proposed methodology uses the design of a finite state machine based on the Levenberg-Marquardt (LM) Back Propagation mechanism. The ANN designed is trained using the states of the finite state machine for a sequence of 1111. The LM algorithm is described below.

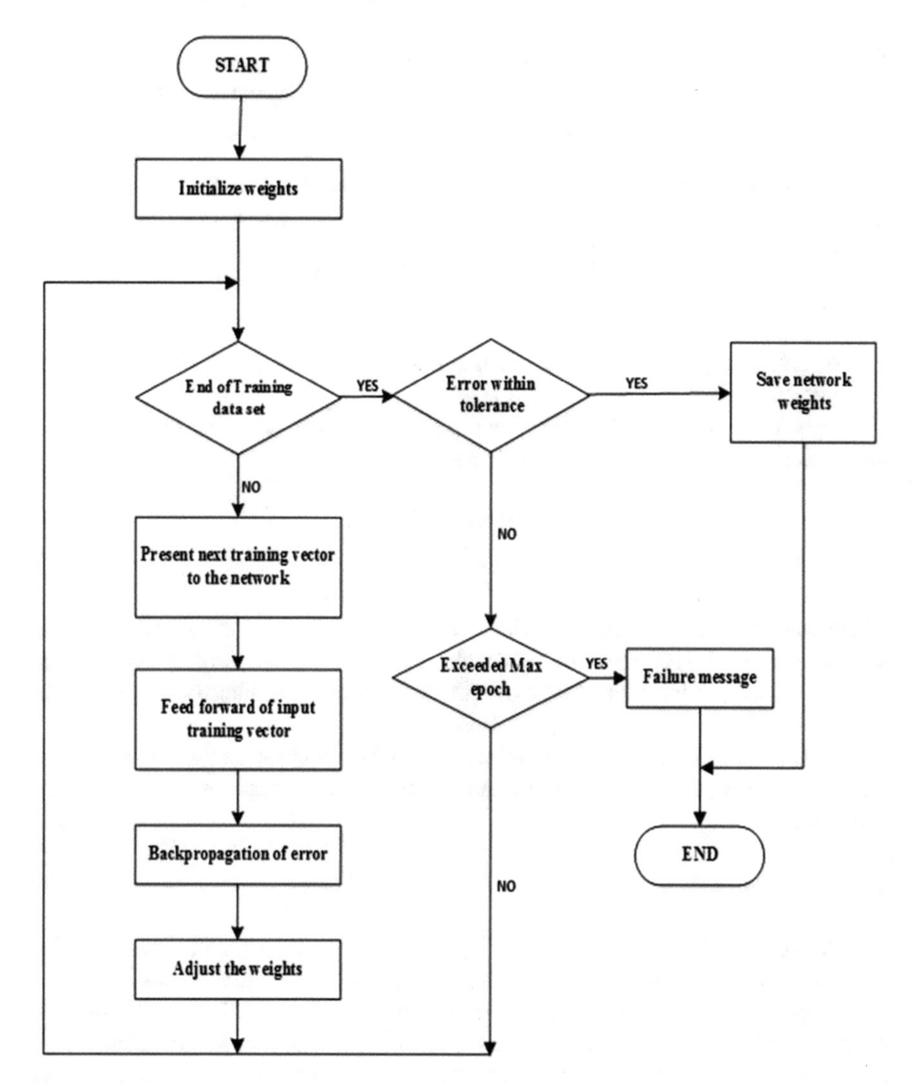

Fig. 2 Flowchart of back propagation in artificial neural networks

4.1 Levenberg–Marquardt (LM) Algorithm

In present study, we have utilized this algorithm because of its good ability to reduce error function. This algorithm is first proposed by Levenberg in 1944 and then later it is further modified by Marquardt in 1963 hence algorithm is named after both of them.

Main advantages of this technique are that first it is very fast which makes processing large data set very fast and secondly convergence is very stable which take

Table 1 The table illustrates the comparison of three algorithms

Algorithm	Weight updating rules	Convergence rate
Gradient Newton algorithm	$W_{k+1=W\_k} - \alpha g_x, \quad \alpha = \frac{1}{\mu}$	Stable, slow
Gauss–Newton algorithm	$W_{k+1=W\_k} - \left[J_K^T J_k\right]^{-1} J_K^T e_k$	Unstable, fast
Levenberg–Marquardt (LM) algorithms	$W_{k+1=W\_k} - \left[J_K^T J_k + \mu I\right]^{-1} J_k e_k$	Stable, fast

care of efficiency. This algorithm is basic a well-organized blend of two different methods one is hessian and other is a gradient.

When the sum of squares is the main performance function, then the two techniques are computed by the following relations:

$$H = J_k{}^T J_k \quad (1)$$

$$g = J_k{}^T e \quad (2)$$

where J_k is the Jacobian matrix, which comprises of first-order derivatives of the network errors with reference to the weights and biases is denoting network errors. Hence producing a Jacobian matrix with the help of a back propagation technique is far less complicated than forming the Hessian matrix.

The LM algorithm is a fine combination of the steepest descent method and the G–N algorithm. The following relation helps on understanding LM algorithm computation,

$$W_{k+1=W\_k} - \left[J_K{}^T J_k + \mu I\right]^{-1} J_K{}^T e_k \quad (3)$$

where, W_k represents current weight, W_{k+1} represents next weight, I represent the identity matrix and e_k represents last error, μ represents combination coefficient.

LM method attempts to combine the benefits of both the SD and GN methods hence it inherits the speed of the Gauss–Newton (GN) method and the stability of the Steepest Descent (SD) method. The factor μ is multiplied by some factor (β) whenever iteration would result in an increase in present error e_{k+1} and when epoch leads to reduction in present error e_{k+1}, μ is divided by β. In this study, we have used value of β as 10. When μ is large the algorithm converts to steepest descent while for small μ the algorithm converts to Gauss–Newton (Table 1).

Figure 3 shows the working of LM algorithm using block diagram. Initially, $M = 1$ is considered and random initial values of weights and bias values are taken in calculations. Now for this weights and bias value, the respective output is generated and error is calculated. According to this error matrix, Jacobian matrix (J_k) is computed and on the basis of this Jacobian matrix, next values or updated values of weights are calculated using Eq. 2. Now based on this updated weights and bias values updated or current error (e_{k+1}) is calculated.

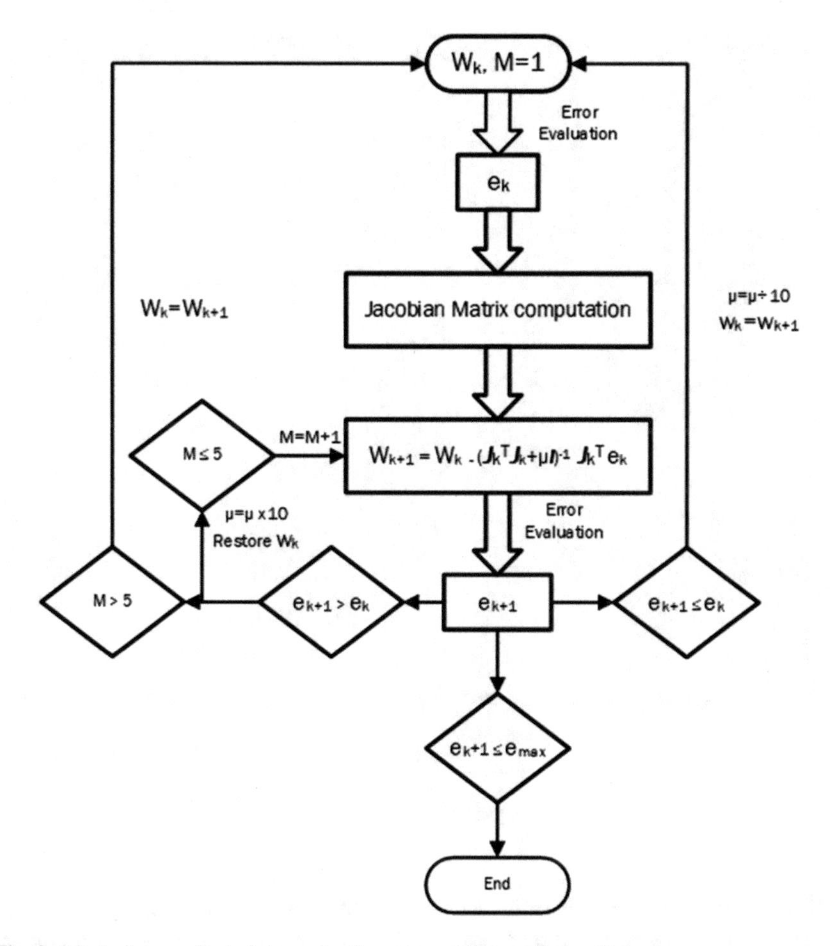

Fig. 3 Block diagram for training using Levenberg–Marquardt algorithm

Now in the next step comparison is made between present error and last error.

- If the present error is less than last error, then it indicates that the weights are updated in right direction. Hence combination coefficient (μ) will be divided by 10 and new weights now default initial weights, i.e., $W_k = W_{k+1}$ and computation will repeat for this value of weight from step 1 again.
- If the error calculated in this step are more than the previous values of weights are restored and combination coefficient (μ) will be multiplied by 10 and new weights are calculated using Eq. 3 with new combination coefficient with an increase in M by 1.
- Now if m>5 then the new weight will be made the default initial weight and computation again shifts to step 1 and the whole process is repeated again in search of required result.

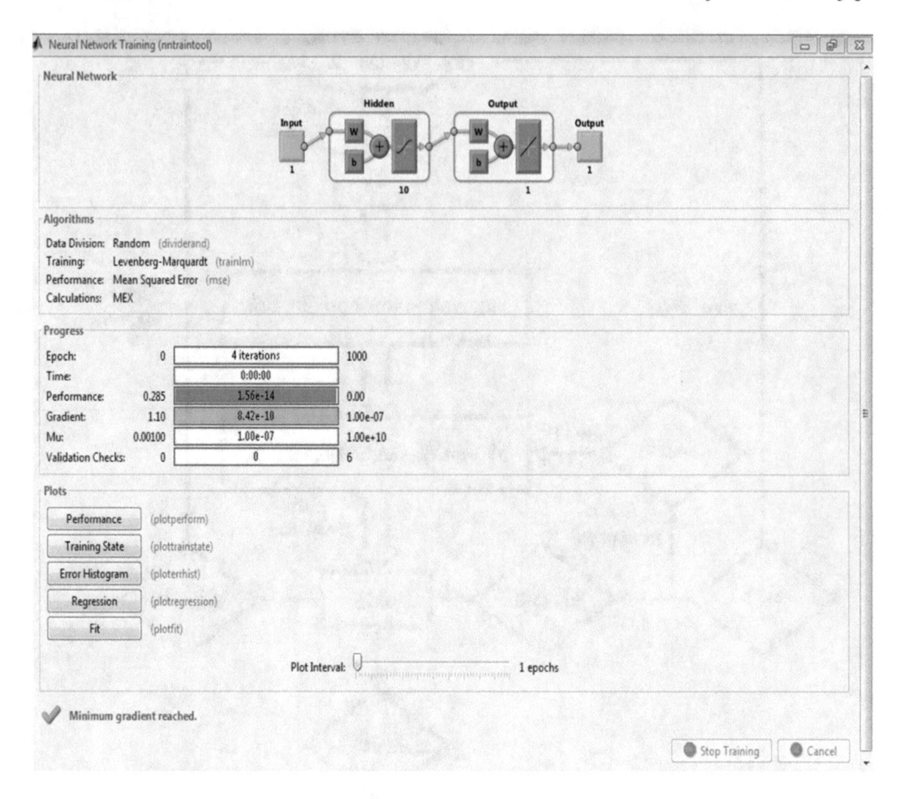

Fig. 4 Designed ANN with 10 hidden neurons

- If the value of the new error is less than the maximum allowed error value e_{max} than training is stopped at that moment and the present weights are saved as the chosen weights and the network will be finalized for further testing.

5 Results

The results obtained are based on the design of finite state machines and their predictive modeling. The number of inputs of the finite state machine is varied and the mean square error is observed. The regression analysis is also performed. The complexity of the finite state machine increases with increasing number of inputs. Hence the MSE increases with increasing number of inputs. The number of inputs is varied from 4 to 64. It is seen that as the number of inputs increases beyond 32, the MSE decreases which can be attributed to over-fitting (Table 2; Figs. 4, 5, 6 and 7).

By varying the number of inputs, the following MSE curve is obtained (Fig. 8).

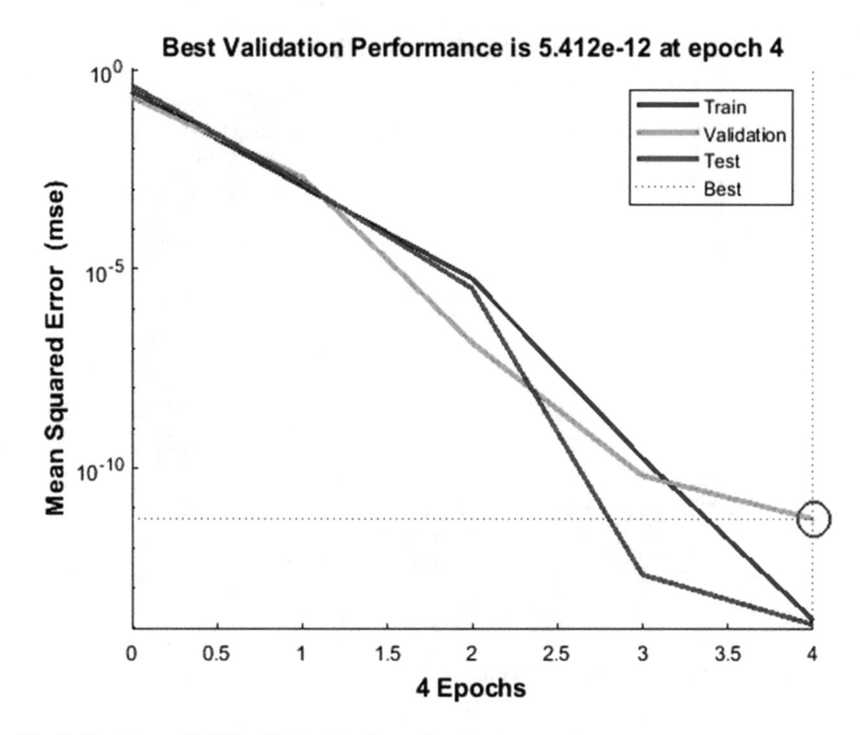

Fig. 5 Variation of MSE with number of epochs

Table 2 State logic of FSM with sequence detection of **1111**

Sequence	Present state	Produced output
1	0010	0
2	0011	0
3	0100	0
4	0101	0
5	0110	0
6	**1111**	**1**
7	0111	0
8	1000	0
9	1001	0
10	**1111**	**1**

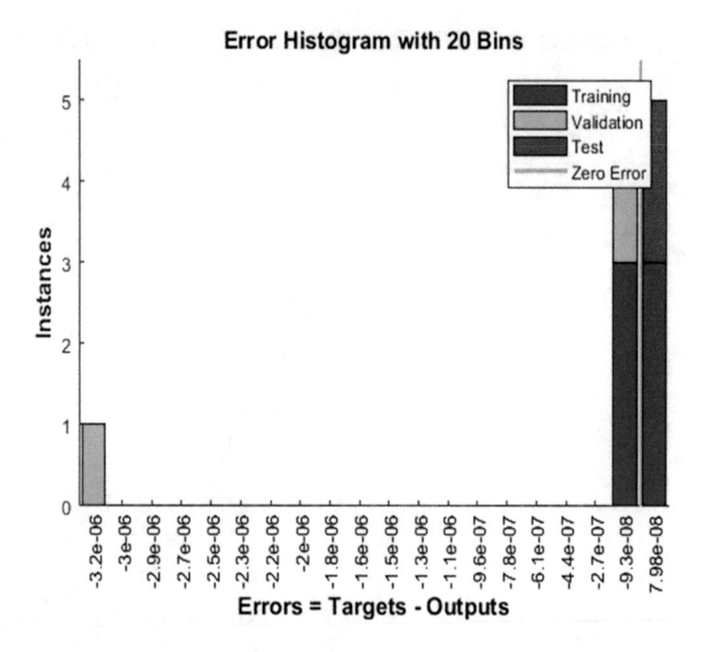

Fig. 6 Error histogram for proposed system

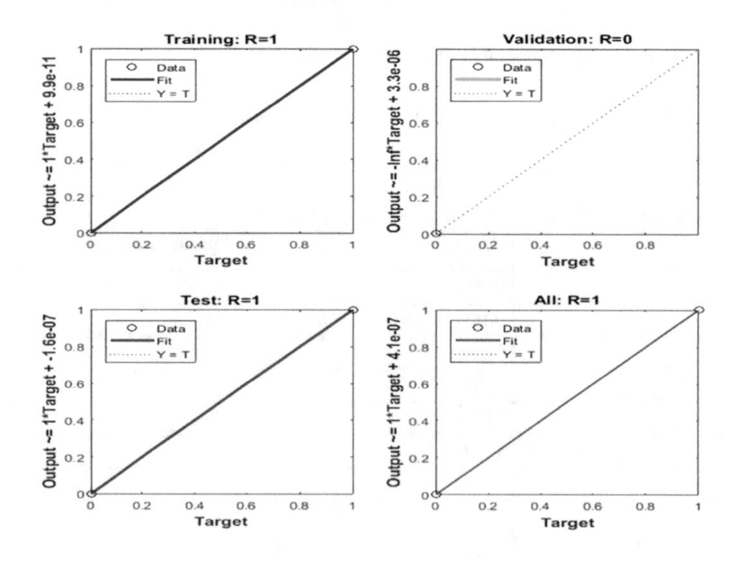

Fig. 7 Regression plot for proposed system

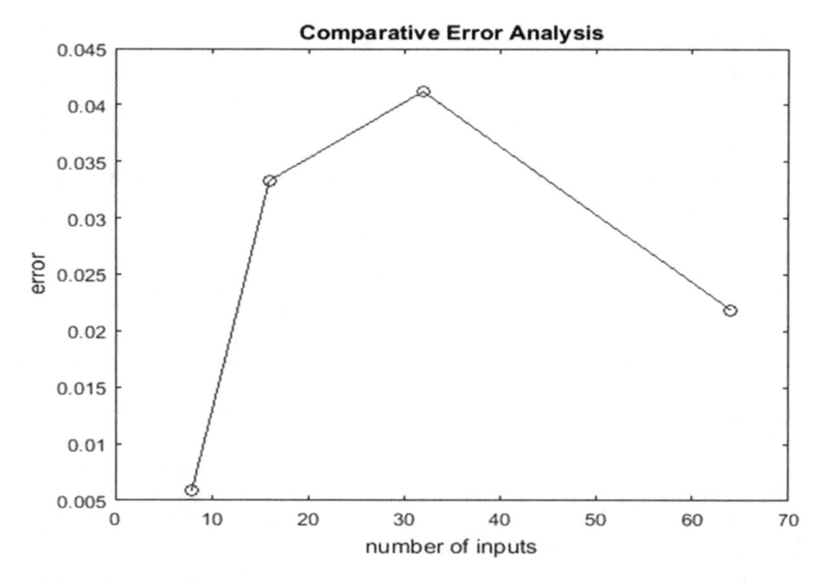

Fig. 8 Variation of MSE with number of inputs

6 Conclusion

The paper presents a technique to predict the state s of a finite state machine which can be excited by any sequence of binary input combinations. The major advantage of such a system is its ability to track a seemingly random input waveform that can be generated in several practical situations such as automatic vending machines, interactive gaming, cryptographic systems, etc. The proposed system attains an MSE of less than 1 even when the number of inputs is increased to 64. The substantially less value of MSE can be attributed to the fact that back propagation using LM algorithm is used. The MSE increases as the complexity of the FSM increases but as the number of inputs exceeds a certain level, the MSE decreases due to over-fitting in ANN.

References

1. Kushik N, El-Fakih K, Yevtushenko N, Cavalli AR (2014) On adaptive experiments for non-deterministic finite state machines. Springer
2. Schmidhuber J (2014) Deep learning in neural networks: an overview. Technical Report IDSIA-03-14/arXiv:1404.7828 v4 [cs.NE] (88 pages, 888 references). The Swiss AI Lab IDSIA Istituto Dalle Molle di Studisull' Intelligenza Artificiale. IEEE
3. El-Maleh AH, Sait SM, Bala A (2015) State assignment for area minimization of sequential circuits based on cuckoo search optimization. Elsevier

4. Wysocki A, Ławry´nczuk M (2015) Jordan neural network for modelling and predictive control of dynamic systems. IEEE
5. Ardakani A (Student Member, IEEE, François), Leduc-Primeau F, Onizawa N (Member, IEEE), Hanyu T (Senior Member, IEEE), Gross WJ (Senior Member, IEEE) (2016) VLSI implementation of deep neural network using integral stochastic computing
6. Kayri M (2016) Predictive abilities of bayesian regularization and Levenberg–Marquardt algorithms in artificial neural networks: a comparative empirical study on social data. MDPI
7. Rastogi P, Cotterell R, Eisner J (2016) Weighting finite-state transductions with neural context. In: NAACL-HLT Proceedings
8. Song T, Zhen P, Wong MLD, Wang X (2016) Design of logic gates using spiking neural P systems with homogeneous neurons and astrocytes-like control. Elsevier
9. Duan S (Member, IEEE), Hu X (Student Member, IEEE), Dong Z (Student member, IEEE) (2015) Memristor-based cellular nonlinear/neural network: design, analysis, and applications. Transaction
10. Reddy PR, Prasad D (2015) Low-power analysis of VLSI circuit using efficient techniques. IJNTSE
11. Giles CL, Ororbia II A Recurrent neural networks: state machines and pushdown automata. The Pennsylvania State University, University Park, PA, USA
12. Goyal R, Vereme V (2000) Application of neural networks to efficient design of wireless and RF circuits and systems. AMSACTA
13. Reynaldi A, Lukas S, Margaretha H (2012) Back propagation and Levenberg-Marquardt algorithm for training finite element neural network. IEEE
14. Soeken M, Wille R, Otterstedt C, Drechsler R (2014) A synthesis flow for sequential reversible circuits. IEEE

Deep Dive Exploration of Mixed Reality in the World of Big Data

Prajal Mishra

Abstract With the exponential growth of data volumes in current scenario, it's becoming more difficult to incorporate the increasing need of data storage and analysis with the existing available systems. Methods to deal with Big Data and analyzing it, comes in play here. In the real world, higher revenues are generated from processing of big data in comparison to the costs involved in processing it which attracts all big organizations in the world. Visualization techniques and methods are improving regularly to cope with the increasing complexity of Big Data. A new perspective solution can be seen here which involves the use of Virtual Reality, Augmented Reality or Mixed Reality to make use of human perception and cognition for more effective and useful ways to utilize the information gathered from Big Data.

1 Introduction

From a long period of time, machines have been considered an important extension of human resources and are increasing continuously till date at a rapid growing rate. Many organization or companies came into existence which deals with data stored digitally on these machines. In the current scenario, data is growing tremendously every second and this flow of data is Big Data coming from different corners and dimensions of the world. With the presence of multiple definitions of Big Data present all over the web, the paper focuses on modified Gartner Inc. definition [1, 2]: *Big data is a technology to process high-volume, high-velocity, high-variety data or data-sets to extract intended data value and ensure high veracity of original data and obtained information that demand cost-effective, innovative forms of data*

P. Mishra (✉)
Computer Science & Engineering, University of Texas at Arlington, Arlington, USA
e-mail: prajalmishra24@gmail.com

D. K. Mishra et al. (eds.), *Data Science and Big Data Analytics*,
Lecture Notes on Data Engineering and Communications Technologies 16,
https://doi.org/10.1007/978-981-10-7641-1_35

and information processing (analytics) for enhanced insight, decision making, and processes control [3].

The biggest challenge faced at this moment is to cope with the complexity of big data and to provide meaningful results to the end users or organizations which can be used for critical decision-making process. Visualization plays an important role for better and efficient understanding of the analysis presented by the Big Data spectrum. Visualization allows end users to provide correlations between different various entities and dimensions of the data present. Multiple visualization techniques and methods exists at this moment which will soon be outdated with the increasing dimensions of data coming into the scenario. Therefore, new approach is required to deal with this situation and a new method is proposed using Augmented Reality (AR), Virtual Reality (VR) or Mixed Reality (MR) [4] concepts which can effectively overcome the barrier in the current system and can utilize the human perception and cognition abilities to deal with data of higher dimensions.

2 Current Scenario

After detailed study from a focus group of appointed professionals of different backgrounds, the major limitation for effective visualization results were the display sizes of end user's smartphones, laptops, desktops, etc. which are considered to limit the human perception and cognitive skills and it was difficult to incorporate multiple screens for the showcase of analysis result as it increases cost significantly and is not considered a suitable working environment for human health. Looking forward in the current scenario, end users can change and customize their data view presented in 2D interfaces while allow going through different level of detail.

Tools like MS Excel are currently used for visualization of data analysis which hinders synchronous collaboration for versatile tasks to explore and identify behavioral patterns and signatures [5]. It also restricts the allowable capability of maximum dimensions which can be displayed to end users while making sense out of it which can be used for decision making process. The overall time utilized to process the analysis varies with the volume of data sets present and is a major concern to deal with while looking for upgrades in the visualization methods.

The main drawbacks with the current scenario comes forward with the use of complex data structures which results in more complex visualizations coming from 2D/3D standard figures like bar graphs, line graphs, etc. More methods like *Hierarchical images, Methods focused on the pixels, Display Icons and Geometric transformations* [6] are used for visualizing the data in different formats which will make more sense to the end users or organizations. Everything was working desirably till the discovery of new dimensions with the excessive flow of input data with unbounded valuable information. There was a need to overcome this scenario for better functioning of the system.

3 Proposed Scenario

Integrating augmented and virtual reality with big data analytics to deal with the problems in the existing scenario. Augmentation can into play when researchers realized the potential of superimposing virtual data with real life as it allowed end users or organizations to view the virtual data from different angles giving a new perspective to the problems as well as their solutions. With the mixed reality concept, physical environment of the users can play an important role to overcome the issues with the small and fixed sizes of their screens and can potentially display huge chunks of data in a meaningful order. With the advent of egocentric navigation provided by MR interface, users will be able to view their data from different angles and perspectives and can understand it completely new scenarios and infer meanings which were non-existent previously.

The major concern with this advancement are the functionalities present in the 2D/3D existing system since work is still pending to realize the workability of changing level of detail and to customize the data view with ease as it was available previously. Although, with the utilization of physical environment, users can process more data concurrently and can be more effective in time, cost and efforts. Clustering of data can be done based on the common behavioral patterns, differentiated by distinct places in the physical world. To overcome with all the barriers in the current scenario, the potential of AR/VR has been seen significantly but the important question here is how to manipulate the data to visualize it in a way which will use it to maximum potential and can result in effective analysis outcomes in minimum time duration [7].

The use of augmented and virtual reality can be effective in various fields such as construction, health-care, mechanical engineering, gaming, military, education, etc. Looking deeply into construction industry, it has improved the working efficiency at a great level as using MR allows to deal with real-world projects and to analyze and fix the defects present in the system even before they occur in real world saving lots of time, money, efforts and lives of human beings [8]. Communication is another factor which can be benefitted from the use of MR as information regarding the project is readily available to everyone with precision and there are minimum possibilities of miscommunication that can take place in this environment. Increasing the safety of the project site as well by training the individuals in an environment with real objects but virtual scenarios and hence helping to improve the overall performance throughput of the entire scenario.

To better understand the objects in the world and to bring more independency in the MR world, use of Simultaneous Localization and Mapping (SLAM) [9] based systems are preferred as this technology makes it possible for AR applications to recognize objects to overlay digital interactive augmentations which will allow an end user or organization to better interact with Big Data analytics and to tweak the results based on various customizable parameters for better understanding and effective decision making outcomes. Using various HMD Type or Glasses-Based devices which are still in development like Oculus rift and Google Glass, etc. can be upgraded with new innovative technologies keeping in consideration the importance of Big Data and the necessity of visualizing Big Data analytics for efficient work flow of the system.

4 Conclusions

Understanding the potential opportunities and risks in bringing virtual reality, augmented reality or mixed reality for solving issues with the visualization of Big Data analysis with increasing dimensions and complexity and using SLAM based systems for more efficient mapping of virtual objects in real world helps in understanding the Big Data analytics efficiently. Still there are multiple concerns which were discussed in integrating the technologies for better visualization and understanding using human perception and cognition skills to overcome the problems in current scenario of data bombing with increasing volume, velocity, and veracity. More up gradations will be suggested in future work which will help to overcome the problems faced by data scientists and industry experts in the field of visualizations of Big Data analytics.

In this paper we have obtained relevant Big Data Visualization methods classification and have discussed the recent work and practices in visualization-based tools adoption for different applications and business support in varied significant fields. Current and futuristic states of data visualization were described and supported by analysis of benefits and challenges. The approach of utilizing VR, AR and MR for Big Data Visualization is presented and the advantages, disadvantages and possible optimization strategies of those are discussed.

References

1. Gartner—IT Glossary. Big Data defintion. http://www.gartner.com/it-glossary/big-data/
2. Gartner SS (2013) Big Data definition consists of three parts, not to be confused with three "V"s. Gartner, Inc., Forbes
3. Demchenko Y, De Laat C, Membrey P (2014) Defining architecture components of the Big Data Ecosystem. In: Proceedings of international conference on collaboration technologies and systems (CTS). IEEE, pp 104–112
4. Müller J, Butscher S, Reiterer H (2016) Immersive analysis of health-related data with mixed reality interfaces: potentials and open questions
5. Ştefan L, Moldoveanu F, Gheorghiu D (2016) Evaluating a mixed-reality 3D virtual campus with Big Data and learning analytics: a transversal study
6. Olshannikova E, Ometov A, Koucheryavy Y, Olsson T (2015) Visualizing Big Data with augmented and virtual reality: challenges and research agenda. J Big Data (SpringerOpen)
7. West R, Parola MJ, Jaycen AR, Lueg CP (2015) Embodied information behavior, mixed reality and Big Data
8. Behzadi A (2016) Using augmented and virtual reality technology in the construction industry. Am J Eng Res
9. Ritsos PD, Mearman JW, Jackson JR, Roberts JC (2017) Synthetic visualizations in web-based mixed reality

Author Index

D. K. Mishra et al. (eds.), *Data Science and Big Data Analytics*,
Lecture Notes on Data Engineering and Communications Technologies 16,
https://doi.org/10.1007/978-981-10-7641-1

Printed in the United States
By Bookmasters